*High-Pressure Shock Compression of
Condensed Matter*

Editors-in-Chief
Lee Davison
Yasuyuki Horie

Founding Editor
Robert A. Graham

Advisory Board
Roger Chéret, France
Vladimir E. Fortov, Russia
Jing Fuqian, China
Yogendra M. Gupta, USA
James N. Johnson, USA
Akira B. Sawaoka, Japan

Springer
*New York
Berlin
Heidelberg
Hong Kong
London
Milan
Paris
Tokyo*

High-Pressure Shock Compression of Condensed Matter

L.L. Altgilbers, M.D.J. Brown, I. Grishnaev, B.M. Novac, I.R. Smith, I. Tkach, and *Y. Tkach:* Magnetocumulative Generators

T. Antoun, L. Seaman, D.R. Curran, G.I. Kanel, S.V. Razorenov, and *A.V. Utkin:* Spall Fracture

J. Asay and *M. Shahinpoor* (Eds.): High-Pressure Shock Compression of Solids

S.S. Batsanov: Effects of Explosion on Materials: Modification and Synthesis Under High-Pressure Shock Compression

R. Cherét: Detonation of Condensed Explosives

L. Davison, D Grady, and *M. Shahinpoor* (Eds.): High-Pressure Shock Compression of Solids II

L. Davison, Y. Horie, and *T. Sekine* (Eds.): High-Pressure Shock Compression of Solids V

L. Davison, Y. Horie, and *M. Shahinpoor* (Eds.): High-Pressure Shock Compression of Solids IV

L. Davison and *M. Shahinpoor* (Eds.): High-Pressure Shock Compression of Solids III

A.N. Dremin: Toward Detonation Theory

V.E. Fortov, L.V. Al'tshuler, R.F. Trunin, and *A.I. Funtikov* (Eds.): High-Pressure Shock Compression of Solids VII

R. Graham: Solids Under High-Pressure Shock Compression

Y. Horie, L. Davison, and *N.N. Thadhani* (Eds.): High-Pressure Shock Compression of Solids VI

J.N. Johnson and *R. Cherét* (Eds.): Classic Papers in Shock Compression Science

G.I. Kanel, S.V. Razorenov, and *V.E. Fortov:* Shock-Wave Phenomena and the Properties of Condensed Matter

V.F. Nesterenko: Dynamics of Heterogeneous Materials

M. Sućeska: Test Methods for Explosives

J.A. Zukas and *W.P. Walters* (Eds.): Explosive Effects and Applications

G.I. Kanel S.V. Razorenov
V.E. Fortov

Shock-Wave Phenomena and the Properties of Condensed Matter

With 209 Illustrations

Springer

G.I. Kanel
Institute for High Energy Densities
Russian Academy of Sciences
IVTAN
Izhorskaya 13/19
Moscow 125412
Russia
kanel@ficp.ac.ru

S.V. Razorenov
Institute of Problems of Chemical Physics
Russian Academy of Sciences
Chernoglovka
Moscow region 142432
Russia
razsv@ficp.ac.ru

V.E. Fortov
Russian Academy of Sciences
32a Leninsky Prosp.
Moscow 117993
Russia
fortov@ras.ru

Editors-in-Chief:

Lee Davison
39 Cañoncito Vista Road
Tijeras, NM 87059
USA
leedavison@aol.com

Yasuyuki Horie
MS F699
Los Alamos National Laboratory
Los Alamos, NM 87545
USA
horie@lanl.gov

Library of Congress Cataloging-in-Publication Data
Kanel, G.I. (Gennadii Isaakovich)
    Shock-wave phenomena and the properties of condensed matter / Gennady I. Kanel,
Sergey V. Razorenov, Vladimir E. Fortov.
      p. cm. — (High pressure shock compression of condensed matter)
    Includes bibliographical references and index.
    ISBN 0-387-20572-1 (alk. paper)
      1. Condensed matter. 2. Shock waves. I. Razorenov, Sergey V. II. Fortov, V.E.
III. Title. IV. Series.
    QC173.454.K36 2004
    530.4′12—dc22
                                                        2003063816

ISBN 0-387-20572-1   Printed on acid-free paper.

© 2004 Springer-Verlag New York, Inc.
All rights reserved. This work may not be translated or copied in whole or in part without the written permission of the publisher (Springer-Verlag New York, Inc., 175 Fifth Avenue, New York, NY 10010, USA), except for brief excerpts in connection with reviews or scholarly analysis. Use in connection with any form of information storage and retrieval, electronic adaptation, computer software, or by similar or dissimilar methodology now known or hereafter developed is forbidden. The use in this publication of trade names, trademarks, service marks, and similar terms, even if they are not identified as such, is not to be taken as an expression of opinion as to whether or not they are subject to proprietary rights.

Printed in the United States of America.   (SBA)

9 8 7 6 5 4 3 2 1   SPIN 10951279

Springer-Verlag is a part of *Springer Science+Business Media*

springeronline.com

Preface

One of the main goals of investigations of shock-wave phenomena in condensed matter is to develop methods for predicting effects of explosions, high-velocity collisions, and other kinds of intense dynamic loading of materials and structures. In the modern view, complete predictability is achieved when computers can accurately simulate processes of interest. The simulations require thermophysical equations of state (relationships among pressure, density, phase composition, and internal energy or temperature of the material) and constitutive relationships to describe processes of elastic–plastic deformation and fracture, as well as chemical, phase, and polymorphic transformations under these conditions. The main research directions of shock-wave physics concern the study of these phenomena. Other categories of general tasks are motivated by the high pressures and temperatures associated with compression of materials by strong shock waves and by the extremely high rates of variation of these quantities in shock waves. These circumstances open unique opportunities for investigations in the fields of thermophysics, physics of condensed matter, chemical physics, and physics of strength and plasticity. In this regard, it is important to mention the high precision and validity of the information obtained from shock-wave tests because the measurements and their interpretation are directly based on fundamental physical laws.

The physics of shock waves in condensed matter deals with pressures from ~100 MPa to several hundreds of GPa (and up to ten TPa in some unique experiments), temperatures up to tens of thousands of K, and durations of load application of 10^{-9} to 10^{-5} s. The beginning of this science dates from the 1950s and is associated with the names of Yakov B. Zel'dovich, Lev V. Altshuler, and Samuel B. Kormer in Soviet Union and Melvin H. Rice, John M. Walsh, Robert G. McQueen, and George E. Duvall in the United States. At that time, the investigations were concentrated on the problems of determining wide-range equations of state of condensed matter and on the parameters of detonation of high explosives. The early development of experimental techniques stimulated new research tasks related to phase transitions, dynamic strength, electrical and optical properties of shock-compressed matter, and technological shock-wave treatment of materials. Computational methods of simulating impact and shock-wave phenomena exert a strong influence on this science. Computational methods have many practical applications and promise more as the capability to apply them increases. The needs of computer simulations have made the research goals more explicit. On the other hand, computer simulations have made it possible to investigate the kinetics of time-dependent physical and chemical transformations, inelastic deformation, and fracture.

Figure 1 illustrates the current distribution of effort devoted to the investigation of shock-wave physics and detonation phenomena in condensed matter. The diagram is based on the relative number of reports presented by over 400 scientists in the largest recent meetings in this area. It is almost impossible to discuss all achievements in this science, so we have limited ourselves to those problems that represent our main areas of interest.

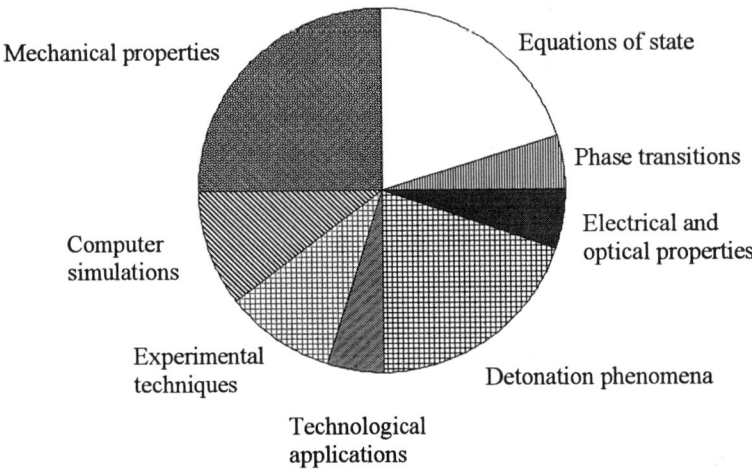

Figure 1. Research directions in the physics of shock waves in condensed matter.

Determining equations of state for materials at high pressures and temperatures is a problem of great importance for tasks of geophysics, astrophysics, and thermonuclear fusion, for prediction of effects of hypervelocity collisions of space bodies with planets and spacecraft, etc. To the present time, the greatest number of measurements has been carried out for metals and frequently encountered chemical compounds. These measurements have covered a wide range of the phase diagrams of these materials. Measurements of shock compressibility are augmented by optical and electro-physical data. The existing semi-empirical equations of state accurately describe the thermodynamic properties, including melting and vaporization, of most important substances and have correct asymptotics to super-high pressures and temperatures. Now, the main emphasis of investigations is shifting from measurement of compressibility to the deeper study of the properties of matter in states that can be reached by shock compression. Plasma models developed for dense matter have allowed correct estimation of the relative contributions to the equation of state of different kinds of inter-particle interaction. Several excellent monographs and review articles have been published in this voluminous field of science.

Experiments with shock waves permit measurement of the fundamental strength properties of matter. The high rate of load application causes overstressing of the material and, as a result, may lead to activation of new mechanisms of deformation and fracture. In contrast to loads applied at moderate and low rates, shock-wave loads are accompanied by the activation of secondary slip planes and a greater contribution of stacking faults and twins to deformation, even in those materials that are normally deformed without them. Reflection of a compression pulse from a stress-free surface of a body causes spall fracture at small values of one-dimensional strain and in stress states close to three-dimensional tension. During the dynamic fracture process, many microvoids or microcracks undergo more or less simultaneous nucleation, growth, and coalescence within the material. One can certainly say that the value of investigations of elastic–plastic and strength properties of solids under shock-wave loading is not limited by practical tasks involving impacts and explosions. Unique data can be obtained from such tests at the highest and most reliably measured strain rates and are undoubtedly valuable for studying the physics of strength and plasticity.

Since the development of high-resolution methods for monitoring stress and particle velocity histories in shock waves, numerous investigations of mechanical properties of different classes materials have been conducted. This has motivated the development of numerous phenomenological and micromechanical models for describing material behavior under shock-wave loading conditions. New opportunities for studying mechanisms of high-rate plastic deformation and fracture are opened by introducing the temperature as a variable parameter in shock-wave experiments. Experiments conducted at elevated temperatures have revealed a transition in the rate-controlling mechanisms of plastic deformation and have allowed study of melting under tension. Exotic behaviors of metal single crystals, such as increase of the yield stress with increasing temperature and generation of superheated solid states, were observed. During the last decade, considerable attention has been accorded to the behavior of brittle materials such as ceramics, glasses, and single crystals. A new phenomenon, the failure wave, was discovered during investigation of the response of glasses to shock compression. Failure waves present a mode of catastrophic fracture in an elastically stressed medium that is not limited to impact events.

Systematic discussions of various aspects of mechanical behavior of solid materials of different classes under shock-wave loading comprise the main part of this book. The discussions include a comprehensive analysis of the structure of compression and rarefaction waves in solids, some summary of experimental data on the dynamic yield and tensile strengths of metals and alloys at normal and elevated temperatures, constitutive relationships and models of high-rate yielding, static and dynamic compressive fracture of brittle materials, and failure-wave phenomena. One should say that, despite the development of quite sufficient general understanding of these phenomena, experiments, theory, and material models still do not agree in many of the details. Mechanical yielding and strength behavior in shock waves show complexities that are not yet fully understood.

The high pressures and temperatures of shock-compressed solids may cause transformations of their crystal structure, melting, and vaporization. The shock-wave investigations in this field are not usually able to compete with modern quasistatic researches. On the other hand, any observation of structural rearrangement that occurs on a sub-microsecond time scale is very interesting and intriguing by itself. There are practical reasons to study shock polymorphism which are related to explosive or impact production of super-hard materials such as diamond and diamond-like boron nitride. The essential influence of fast polymorphic transformations on the response of materials to high-velocity impact or explosion also stimulates investigation of these phenomena. In this book, polymorphic transformations are discussed for several of the most practically important materials: iron and steels, titanium, carbon, and boron carbide. The discussions of phase transitions of shock-compressed solids are completed by some non-traditional observations of melting and vaporization during stress release.

The natural goals of studying detonation phenomena are to permit their predictability, to uncover microscopic mechanisms of the initiation and development of the chemical energy release, to find methods of controlling the sensitivity of explosive materials, and to provide a theoretical basis for solving the safety problems encountered when working with explosive materials.

Quantitative analysis of detonation phenomena requires knowledge of the equations of state of the unreacted explosive material and the detonation products, and the kinetics of transformation of the initial explosive materials to the detonation products. Usually, empirical or semi-empirical relationships are used for this purpose. The relationships summarize known experimental data in a formalized way.

Methods have now been developed for obtaining information about the macroscopic kinetics of energy release in detonation waves. Good understanding has been achieved of the mechanisms of localization of the shock-wave energy in hot spots where ignition of the explosive occurs. Ways of varying the sensitivity of explosive materials are more or less clear and the reasons that these methods are effective are understood. Nevertheless, the mechanisms and macroscopic kinetics of the energy release in multi-component explosive materials, in particular in explosives with metal additives which add to the heat of explosion, are still under discussion. An interesting and sophisticated task is determining the influence of endothermal processes upon the structure of shock and detonation waves in multi-component explosives. These issues have not yet been reviewed in the literature, so we have found it reasonable to discuss them systematically in this book.

This book is addressed not only to experts in shock-wave physics but also to interested representatives from adjacent fields of activity and to students. With this goal in mind, we begin our discussion with a brief account of the theoretical background and a short description of experimental techniques.

Acknowledgments

This book is based on results of investigations conducted over the past 30 years which were funded by Russian Academy of Sciences, Russian Foundation for Basic Research, Russian–German Scientific and Technological Cooperation program WTZ, US Army Research Office, and European Office of Aerospace Research and Development. We appreciate very much the contributions of Kurt Baumung (Forschungszentrum Karlsruhe, Germany), Stephan J. Bless (the University of Texas at Austin, Institute for Advanced Technology), Andrei A. Bogatch and Alexander V. Utkin (Institute of Problems of Chemical Physics, Russia), Zhen Chen (University of Missouri–Columbia), Eugene B. Zaretsky (Ben-Gurion University of the Negev, Israel) and other numerous colleagues from Russia, the United States, Germany, Israel, and China, who participated in the experiments and discussions described in this book. To them we express our sincere gratitude. We especially thank Lee Davison, who helped edit our text.

Contents

Preface v

CHAPTER 1
Introduction to the Theoretical Background and
Experimental Methods of Shock Physics 1
    1.1. Uniaxial Isentropic Flows 1
    1.2. Shock Waves 3
    1.3. Decomposition of Discontinuities and Wave Interactions 6
    1.4. Shock-Wave Processes in Layered Plates 15
    1.5. Generation of Negative Pressures During Reflection
         of a Compressive Pulse From a Plate Surface 17
    1.6. Generation and Measurement of Shock Waves in
         Condensed Matter 20
    1.7. Analysis and Interpretation of Measurements 23
    References 25

CHAPTER 2
Elastic–Plastic Response of Solids Under Shock-Wave Loading 29
    2.1. The Main Relationships 30
    2.2. Rate-Independent Strain Hardening Effects 34
    2.3. Development of Elastic Precursor Waves in Relaxing Materials ... 39
    2.4. Interpreting Free Surface Velocity Histories 45
    2.5. Hugoniot Elastic Limits and Dynamic Yield Stresses of some
         Solid Materials 50
    2.6. Structure of Plastic Waves 52
    2.7. The Stressed States of Shock-Compressed Solids. 55
    2.8. Unloading of Shock-Compressed Solids 58
    2.9. Sound Speeds in Shock-Compressed Solids 64
    2.10. Behavior of Rubber Under Shock Compression 67
    2.11. Behavior of Metals Under Stepwise and
         Repetitive Shock Compression 68
    2.12. On Constitutive Relationships and Models of High-Rate Yielding . 71
    References 78

CHAPTER 3
Yield and Strength Properties of Metals and Alloys at
Elevated Temperatures 83
    3.1. Spall Strength at Melting 84
    3.2. High-Temperature Yielding 94

- 3.3. Shock Yielding and Fracture of Some Alloys 97
- 3.4. Discussion 105
- References 107

CHAPTER 4
Behavior of Brittle Materials under Shock-Wave Loading 111
- 4.1. Introduction 111
- 4.2. General Behavior of Brittle Materials Under Compression 112
- 4.3. Possible Mechanisms of Microcracking Under Compression 115
- 4.4. Dynamic Strength Properties of Brittle Single Crystals 119
- 4.5. Shock Wave Properties of Silicate Glasses 125
- 4.6. Failure Waves in Glasses 132
- 4.7. Dynamic Strength Properties of Polycrystalline Ceramics 145
- 4.8. Brittle Failure Criteria and Models 164
- 4.9. Comparison of 1D Stress and 1D Strain Data for Ceramics at Various Strain Rates 165
- 4.10. Evidence of Ductile Response of Alumina Ceramic Under Shock-Wave Compression 167
- 4.11. Discussion 170
- References 171

CHAPTER 5
Two Examples of Spatially Resolved Shock-Wave Tests 179
- 5.1. Dynamic Strength Variations in Metals 179
- 5.2. Measurements of Adhesion Strength Using the Spall Technique .. 182
- 5.3. Conclusion 187
- References 187

CHAPTER 6
Polymorphic Transformations and Phase Transitions in Shock-Compressed Solids 189
- 6.1. Introduction 189
- 6.2. The $\alpha \leftrightarrow \varepsilon$ Polymorphic Transformation in Shock-Compressed Iron and Steels 191
- 6.3. The $\alpha \to \omega$ Transformation in Shocked Titanium 196
- 6.4. The Graphite to Diamond Transition under Shock Compression .. 200
- 6.5. On the Possibility of Polymorphic Transformations Occurring in the Negative-Pressure Region 202
- 6.6. Melting of Shock-Compressed Metals During Decompression 204
- 6.7. Measuring Unloading Isentropes and Vaporization of Shock-Compressed Polymers 209
- References 213

CHAPTER 7
Equations of State and Macrokinetics of Decomposition of
Solid Explosives in Shock and Detonation Waves 217
- 7.1. Introduction 217
- 7.2. General Structure of Plane Steady Detonation Waves 217
- 7.3. Detonation Failure Diameter 227
- 7.4. Initiation of Detonation by Shock Waves 229
- 7.5. Sensitivity of Solid Explosives to Shock-Wave Effects 236
- 7.6. Detonation Properties of HE Single Crystals 241
- 7.7. Evolution of Shock Waves During Initiation of Detonation of Solid Explosives 244
- 7.8. Ignition and Growth of Reaction Nuclei During Shock-Wave Initiation of Detonation 249
- 7.9. Macrokinetics of Decomposition of Solid Explosives in Shock Waves 252
- 7.10. Equations of State for Explosives 263
- 7.11. Equations of State for Detonation Products 270
- 7.12. Calculation of States of Mixtures of Explosives and Detonation Products 276
- 7.13. Detonation Properties of High Explosives Containing Metal Particles 277
- 7.14. Conclusion 288
- References 289

CHAPTER 8
Shock Waves and Extreme States of Matter 301
- 8.1. On Wide-Range Equations of State 301
- 8.2. Shock Waves and Non-Ideal Plasmas 305
- 8.3. Generation and Diagnosis of Dense Plasma States 307
- 8.4. On the Metal−Insulator Transition in Shock-Compressed Lithium 315
- 8.5. Conclusion 316
- References 317

Index 320

CHAPTER 1

Introduction to the Theoretical Background and Experimental Methods of Shock Physics

This chapter deals with the laws of one-dimensional motion of compressible continuous media to the extent necessary for subsequent discussion of dynamic experiments. A comprehensive account of the fundamentals of the mechanics of continuous media can be found, for example, in textbooks by Courant and Friedrichs, 1948, and Zel'dovich and Raizer, 1967.

Strong shock waves in solids are usually considered in the hydrodynamic approximation, which does not account for the material yield strength. If the pressure is much higher than the yield strength of the material, the hydrodynamic approximation provides a sufficiently accurate description of shock-wave parameters and states of shock-compressed matter. Since, in the case of condensed matter, almost all measurements of flow parameters are connected with material particles, the analysis of wave dynamics is preferably effected in terms of Lagrange coordinates. We shall take the Lagrange coordinate h to be the spatial coordinate, x, of a particle at the initial time:

$$h = \frac{1}{\rho_0} \int_0^x \rho \, dx, \quad (1.1)$$

where ρ and ρ_0 are, respectively, the current and initial densities of the material. The partial derivative with respect to time, t, and coordinate, h, will be denoted by $\partial(\cdots)/\partial t$ and $\partial(\cdots)/\partial h$, respectively, and the directional derivative of a function $f(h, t)$ is given by

$$\frac{df}{dt} = \frac{\partial f}{\partial t} + \frac{\partial f}{\partial h}\frac{dh}{dt}$$

$$\frac{df}{dh} = \frac{\partial f}{\partial h} + \frac{\partial f}{\partial t}\left[\frac{dh}{dt}\right]^{-1}. \quad (1.2)$$

1.1. Uniaxial Isentropic Flows

If we neglect the contributions of the elastic–plastic properties and relaxation processes, one-dimensional motion of a compressible medium is described by a set of partial differential equations expressing the laws of conservation of mass, momentum, and energy, supplemented by an equation of state:

$$\rho_0 \frac{\partial V}{\partial t} - \frac{\partial u_p}{\partial h} = 0$$

$$\rho_0 \frac{\partial u_p}{\partial t} + \frac{\partial p}{\partial h} = 0 \qquad (1.3)$$

$$\frac{\partial \mathscr{E}}{\partial t} + p \frac{\partial V}{\partial t} = 0$$

$$\mathscr{E} = \mathscr{E}(p, V),$$

where p is the pressure, u_p is the particle velocity, V is the specific volume, and $\mathscr{E}$ is the specific internal energy.

The equations of one-dimensional motion of an elastic–plastic medium are obtained from Eqs. (1.3) by replacing the pressure with the normal stress, σ_x, acting in the axial direction (taken positive in compression):

$$\rho_0 \frac{\partial u}{\partial t} + \frac{\partial \sigma_x}{\partial h} = 0 \qquad (1.4)$$

$$\frac{\partial \mathscr{E}}{\partial t} + \sigma_x \frac{\partial V}{\partial t} = 0.$$

If the change of state is accompanied by a process of relaxation of the deviatoric stresses, the density, or the internal energy, then Eqs. (1.3) or (1.4) must be supplemented with a relation describing the kinetics of the relaxation process.

For one-dimensional motion, we introduce the Lagrangian sound velocity, a, related to the sound velocity in the laboratory coordinate system, c, by the simple equation

$$a = \frac{\rho}{\rho_0} c = \frac{\rho}{\rho_0} \left[\left(\frac{\partial p}{\partial \rho} \right)_S \right]^{1/2}, \qquad (1.5)$$

where the subscript S designates a derivative taken at constant entropy.

For isentropic flow there are two sets of characteristics, which are the trajectories of perturbations to the motion that propagate in the positive and negative directions. They are given by the equations

$$\frac{\partial h}{\partial t} = a \quad \text{and} \quad \frac{\partial h}{\partial t} = -a, \qquad (1.6)$$

and are called the C_+ and C_- characteristics, respectively. The variation of the state of matter along these characteristics is described by the Riemann integrals

1. Theoretical Background and Experimental Methods of Shock Physics

$$u_p = u_0 - \int_{p_0}^{p} \frac{dp}{\rho_0 a} \quad \text{along } C_+$$

$$u_p = u_0 + \int_{p_0}^{p} \frac{dp}{\rho_0 a} \quad \text{along } C_-, \tag{1.7}$$

where u_0 and p_0 are the integration constants. The characteristic directions in relaxing media are determined by the frozen velocity of sound; however, the change of state along the characteristics deviates from the Riemann integrals in this case.

A flow in which all disturbances propagate in the same direction takes the form of a simple, or progressive, wave. In this simple wave, the states along the characteristics pointing in the direction of wave propagation remain constant, whereas all states along any other path in the x–t plane are described by the single function $p(u_p)$ that corresponds to the Riemann invariant. We shall call the trajectory in p–u_p coordinates which describes states in a simple wave as the rarefaction isentrope or Riemann's isentrope. Generally speaking, this is not quite correct, but it has been accepted in the literature on shock compression of condensed matter. When all characteristics originate at a single point in the x–t plane, the wave is referred to as a centered wave. The slope of Riemann's isentrope,

$$\frac{dp}{du_p} = \pm \rho_0 a, \tag{1.8}$$

is the dynamic impedance of the material.

In normal media, the sound velocity increases with pressure so rarefaction waves diverge during their propagation. On the contrary, compression waves become more and more steep and are transformed into discontinuities or shock waves at a sufficiently great propagation distance.

1.2. Shock Waves

The Rankine–Hugoniot equations express the conservation laws for shock waves in the form

$$V = V_0 \frac{U_s - (u_p - u_0)}{U_s}$$

$$p = p_0 + \rho_0 U_s (u_p - u_0) \tag{1.9}$$

$$\mathscr{E} = \mathscr{E}_0 + \frac{1}{2}(p + p_0)(V_0 - V),$$

where U_s is the shock front velocity relative to the undisturbed medium of specific volume V_0. The set of conservation equations for a shock discontinuity, together with an equation of state, define the shock adiabat or Hugoniot of the material. Figure 1.1 shows the relative positions of the Hugoniot, isentrope, and isotherm in the pressure–volume plane. The shock compression is accompanied by additional irreversible heating of the material that increases the pressure.

The line

$$p = \rho_0^2 U_s^2 (V_0 - V) \tag{1.10}$$

connecting the initial state to the shocked state is termed the Rayleigh line. Because the Rankine–Hugoniot conservation equations can be applied to any part

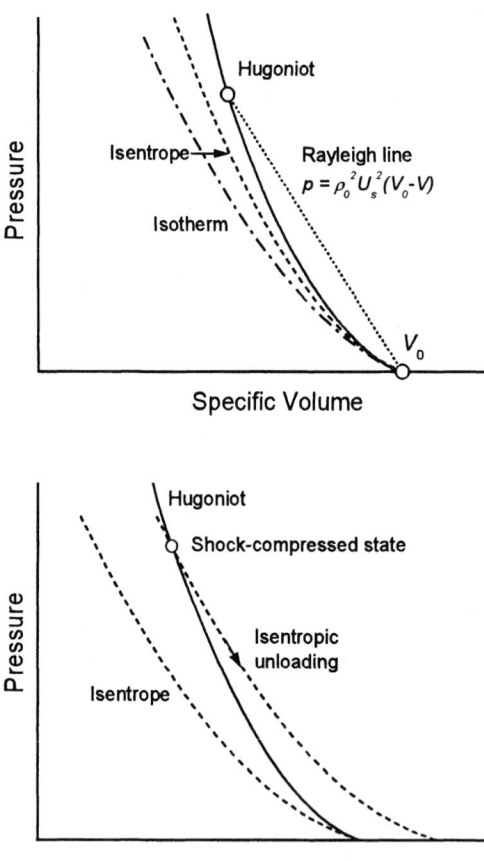

Figure 1.1. Relative positions of the Hugoniot, isentropes, and isotherm of a normal material in pressure–volume coordinates.

of a steady wave, all intermediate states within a steady shock wave correspond to $p-V$ states on the same Rayleigh line. The shock waves are supersonic relative to the matter ahead of them and subsonic relative to the shock-compressed matter behind them. Shock waves of a finite thickness can occur in relaxing media.

The Hugoniot, which is a relationship between the parameters of states reached by shock compression, is used to construct the equation of state (EOS) of the material. Determining this equation of state, which is the basis for the analysis of any high-energy-density process, is one of main goals of shock-wave physics investigations. At moderate pressures, Hugoniots of many condensed materials are well represented by the linear relationship,

$$U_s = c_0 + s u_p, \qquad (1.11)$$

between the propagation velocity of a shock wave, U_s, and the particle velocity, u_p, behind the shock discontinuity. In this equation, the first term, c_0, is close to the sound speed at zero pressure and the constant factor s has a value between 1 and 1.7 for most materials. Using the conservation equations, this expression for the Hugoniot can be transformed into relationships between the pressure and specific volume, the pressure and particle velocity, etc.

Experiments show that, in the pressure vs. particle velocity plane, the release isentrope of compressed matter deviates from the Hugoniot by not more than 3% for pressures of up to at least 50 GPa. Because the Hugoniot and isentrope are almost coincident, the quasi-acoustic approximation for the shock-wave velocity (Landau and Lifshitz, 1959) is quite accurate over a wide pressure range in the case of condensed matter. According to this approximation, the velocity of a shock wave is the average of the velocities of weak perturbations ahead of the shock discontinuity, c_0, and behind it, a: $U_s = (c_0 + a)/2$.

The fact that shock waves are subsonic relative to the shock-compressed matter is a reason for their stability. Any compression wave transforms into a shock discontinuity as a result of the increase of the sound speed with compression that occurs for most materials. However, some materials have non-monotonic compressibility. Figures 1.2 and 1.3 illustrate peculiarities of evolution of a compression pulse in such a material. The part BCDE of the Hugoniot is a forbidden region that is not accessible to single shock-wave transitions from the initial state at point A, and shock waves of peak pressure $p_B < p < p_E$ become unstable and split into two shock waves AB and CD with a ramped part BC between them. When the compressed matter is released, the change of state down to the point m follows the compressibility curve. The pressure region $p_m < p < p_n$ is anomalous for release waves in the sense that the velocity of sound varies non-monotonically with decreasing pressure, and is greater along the segment Bn than along the segment mC. A rarefaction shock wave mn is therefore formed during the release.

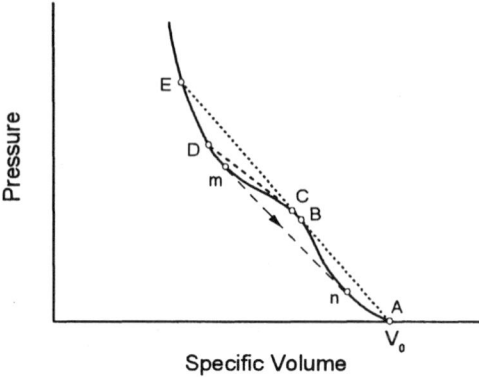

Figure 1.2. Example of an anomalous Hugoniot.

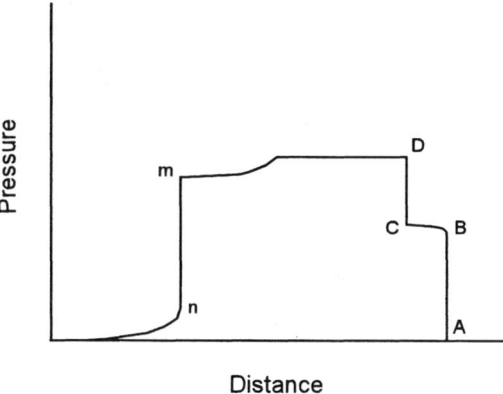

Figure 1.3. Stress pulse in a material with the anomalous Hugoniot shown in Fig. 1.2.

1.3. Decomposition of Discontinuities and Wave Interactions

Shock waves and simple Riemann waves constitute an important class of self-similar flows that forms the basis of dynamic studies. Measurement of the states being studied is based on the solution for the decomposition problem. This solution is a combination of shock and centered rarefaction waves emitted from the original discontinuity and separated by a region of constant parameters of state. Figures 1.4 and 1.5 illustrate decompositions of discontinuities in particle velocity and in pressure.

After decomposition of the discontinuity, the state of the medium is determined by the condition that pressures and particle velocities must be equal on the two sides of the initial position of the discontinuity. In other words, the compression and rarefaction waves formed as a result of the decomposition of the discontinuity should transform the medium to states with the same p and u_p in the pressure–particle velocity plane. We must remember that, for waves propagating in the positive direction, the derivative $\partial p / \partial u_p > 0$, whereas for waves traveling in the negative direction $\partial p / \partial u_p < 0$.

As a result of decomposition of the discontinuity in particle velocity from 0 to u_0, two shock waves are formed; one propagates forward in the positive direction of the x axis while the other propagates backward in the negative direction. The states behind these shock waves must lie on the corresponding Hugoniots. For the forward propagating shock wave, the Hugoniot has positive slope in the p–u_p coordinates and passes through the point $p = p_0$, $u_p = u_0$ of the initial state on the right side of the discontinuity. For the shock wave moving backward, the Hugoniot has negative slope and passes through the point $p = p_0$, $u_p = 0$ of the initial state on the left side of the discontinuity. The condition of equal pressures and particle velocities is satisfied at the point $p = p_1$, $u_p = u_1$ of intersection of these two Hugoniots.

As a result of decomposition of the discontinuity in pressure from 0 to p_0, a shock and a rarefaction wave are formed. In the case illustrated, the shock wave propagates forward into uncompressed matter whereas the rarefaction wave propagates backward into compressed matter. This rarefaction wave is simple, so the unloading process is described by the Riemann isentrope of negative slope which passes through the point $p = p_0$, $u_p = 0$. The result of decomposition of the pressure discontinuity corresponds to the point of intersection of the Hugoniot and the Riemann isentrope in the p–u_p coordinates.

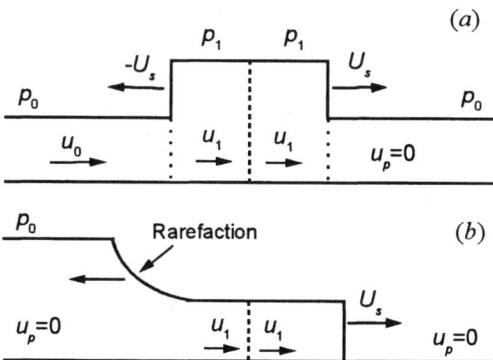

Figure 1.4. Pressure profiles formed as a result of decomposition of a discontinuity in particle velocity (*a*) and in pressure (*b*). Dashed lines show the initial positions of the discontinuities, arrows at the bottom show the directions of motion of the medium, and arrows at the top show the directions of wave propagation.

8   Shock-Wave Phenomena and the Properties of Condensed Matter

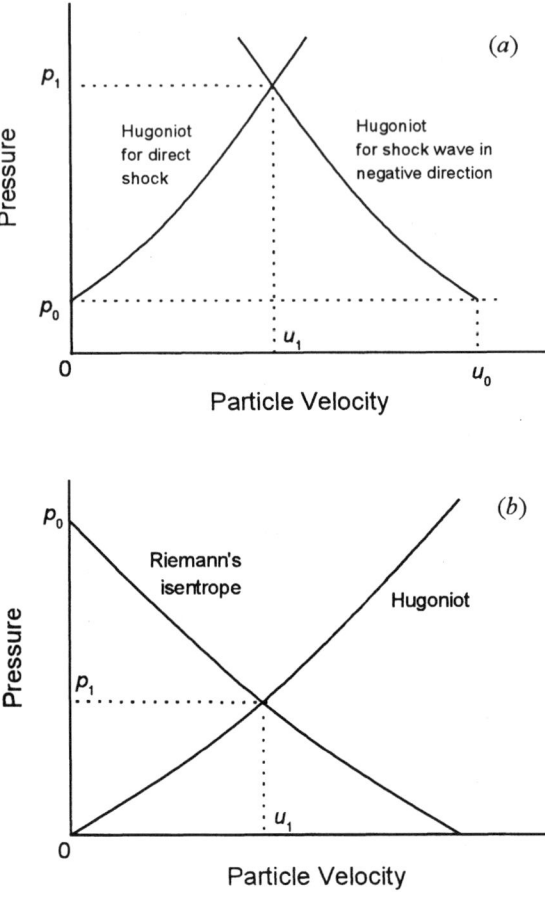

Figure 1.5. Pressure versus particle velocity diagram for the two cases of decomposition of discontinuity shown in Fig. 1.4.

Now, let us consider several practical examples of decomposition of discontinuities and wave interactions. Figure 1.6 illustrates the way to calculate the parameters of a shock wave generated in a target plate by impact of a flyer plate having a velocity u_i. At the moment of collision, the impact surface constitutes a discontinuity in particle velocity: The particle velocity is zero inside the target and is u_i inside the flyer plate and the pressure is zero everywhere. Decomposition of this discontinuity creates shock waves that propagate from the impact surface into the target and the flyer plate. As above, the shock pressure and particle velocity correspond to the intersection point of the Hugoniot of the target material and the Hugoniot of flyer plate material. Because the shock wave in the flyer plate moves backward ($U_s < 0$), the Hugoniot of the flyer plate is taken with negative slope. The diagram shows that the shock pressure is higher if the flyer plate is made of material with higher dynamic impedance, ρc.

The shock pressure, p, and particle velocity, u_p, in the target must satisfy the equation of conservation of momentum, $p = \rho_0 U_s u_p$, where ρ_0 is the initial density of the target material and U_s is the propagation velocity of the shock wave in the target. Thus, the pressure and particle velocity of the shock-compressed matter corresponds to the intersection point of the Hugoniot of the flyer plate and the Rayleigh line $p = \rho_0 U_s u_p$. In other words, measurements of the impact velocity and the velocity of the shock wave in the target and the known Hugoniot of the flyer plate give us p and u_p on the Hugoniot of the target material. If the target and the flyer plate are made of the same material, the Hugoniots are symmetrical to each other and $u_p = u_i/2$. With the values of p and u_p thus determined, the Rankine–Hugoniot equations (1.9) give us the remaining parameters of the shock wave: the specific volume, V, and the specific internal energy, $\mathscr{E}$. By varying the impact velocity we may obtain, point by point, the Hugoniot of the target material.

Besides the Hugoniots, the experimental basis for determining an EOS includes isentropes corresponding to unloading from shock-compressed states. Shock compression is accompanied by heating of the medium. At a certain peak pressure, the entropy increase in the shock wave results in melting and, at even higher peak pressure, vaporization of solids during isentropic unloading. Experimental isentrope data are especially important in the region of vaporization and in the vicinity of the critical point.

The unloading isentrope is recovered from a series of experiments in which the shock-wave parameters in plates of standard low-impedance materials placed

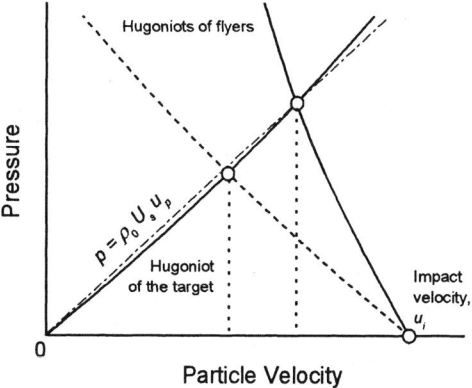

Figure 1.6. Parameters of shock waves generated by impacts of flyer plates of different dynamic impedances at the same impact velocity.

behind the sample are measured (Skidmore and Morris, 1962, Bushman and Fortov, 1983). The idea is illustrated in Fig. 1.7. Low-density foams and gases at different initial pressures are used in these experiments as standards to provide measurements of unloading down to the low-pressure region of vaporization. The experimentally determined $p(u)$ relationship corresponds to the Riemann integral. The specific internal energy and specific volume are calculated from the measured release curve, $p(u)$.

Figure 1.8 illustrates the process of shock-wave acceleration of a foil of high-impedance material that covers a low-impedance plate. As the initial shock wave reaches the interface with the covering foil, it is reflected as a shock-compression wave. The pressure and particle velocity in the reflected shock and in the shock wave passed into the foil must be the same and correspond to the intersection point B of the Hugoniot 0B of the foil material and the Hugoniot AB for reflected shock waves in the plate. Interaction of the shock wave in the foil with its free surface results in the appearance of a release fan which unloads the foil to a state of zero pressure and the particle velocity u_1. However, at this particle velocity the pressure in plate is not zero if unloading follows the Riemann isentrope BCD. The disagreement of dynamic impedances is resolved by production of a new equilibrium state, point C, in both the foil and the plate. In the plate, the new equilibrium state is reached through an unloading wave whereas in the foil the new equilibrium state is reached through a compression wave. In other words, the rarefaction wave is reflected from the plate–foil interface as a compression wave. Thus, the wave reflections from the interface with a softer material occur with change in the load sign. As a result of multiple wave reflections, the release of pressure in the plate occurs by steps of duration corresponding to the time for the shock wave to go back and forth in the foil. The foil surface approaches the final velocity u_f of the unloaded plate after several steps.

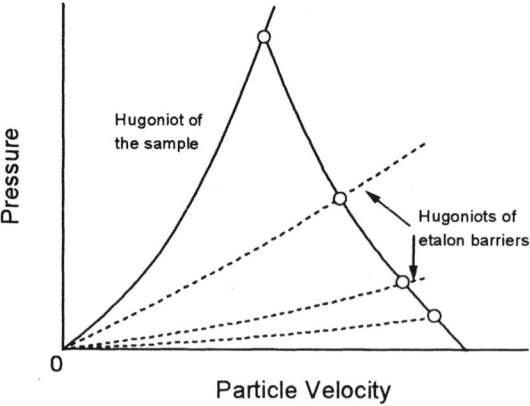

Figure 1.7. Measurements of the unloading isentrope by the method of etalon barriers.

1. Theoretical Background and Experimental Methods of Shock Physics 11

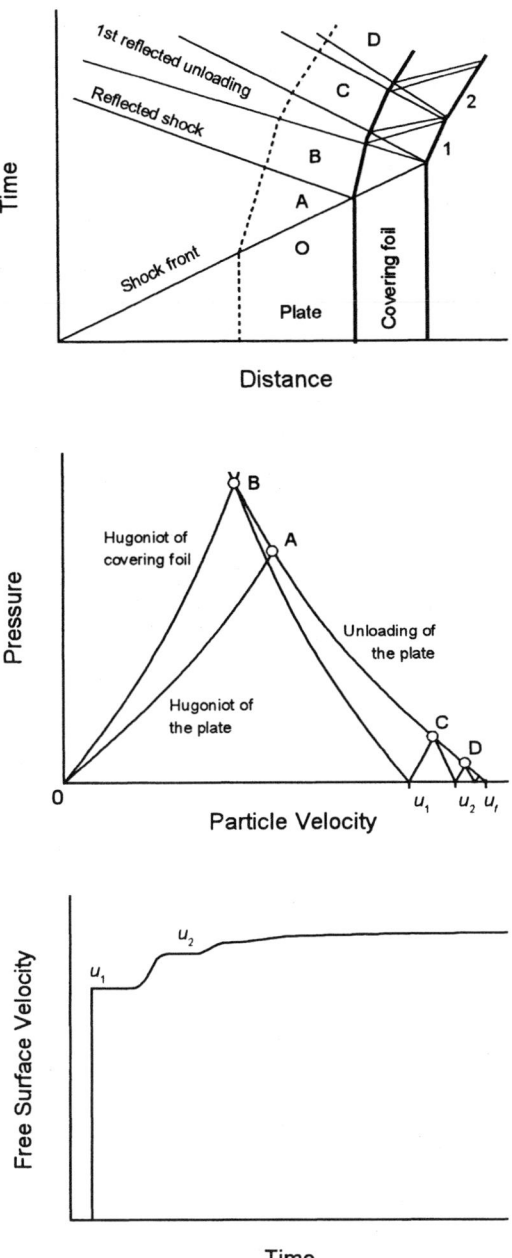

Figure 1.8. Shock-wave acceleration of a foil of high-impedance material which covers a low-impedance plate.

12   Shock-Wave Phenomena and the Properties of Condensed Matter

Joint consideration of time–distance and pressure–particle velocity diagrams is a common way of analyzing shock-wave interactions. Figure 1.9 presents such a diagram for the case of interaction of a shock wave with a foil of low-impedance material that covers a high-impedance plate.

Reflection of a shock wave from the plate–foil interface causes an unloading wave to propagate back into the plate and a compressive shock wave with peak pressure p_B to propagate forward into the foil. Interaction of the shock wave with the free rear surface of the foil causes a release fan to emerge. Behind this release fan, the foil material is in a state of zero pressure and the foil particles acquire the velocity u_D equal to twice the jump in the particle velocity across the shock wave ($u_D = 2u_B$). Interaction of the release fan with the foil–plate interface should lead to a new equilibrium state. The condition of continuity across

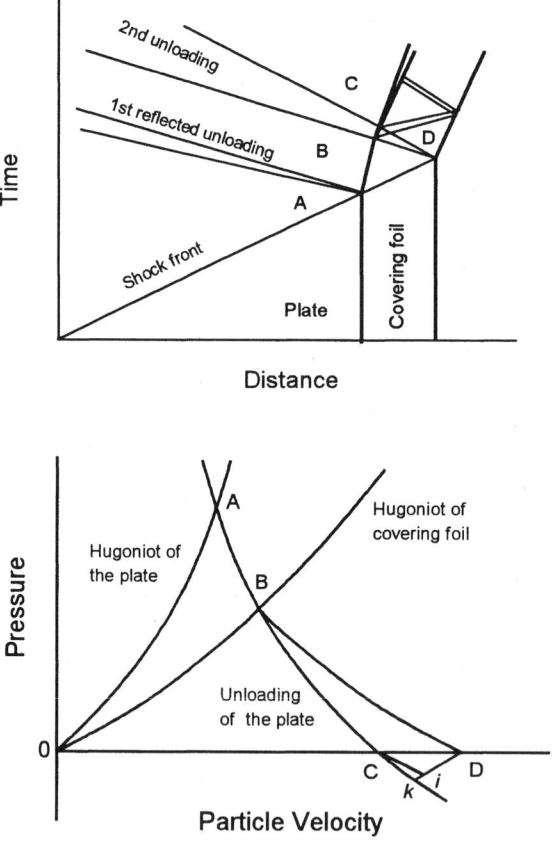

Figure 1.9. Time–distance and pressure–particle velocity diagrams for the case of interaction of a shock wave with a foil of low-impedance material that covers a high-impedance plate.

the interface requires that both the plate and the foil reach the same pressure and particle velocity which corresponds to the intersection point k of Riemann's isentropes Dk and BCk for simple rarefaction waves propagating forward into the foil and back into the plate. However, this intersection occurs in the negative pressure region and is physically impossible if the interface cannot support tensile stresses. For this reason, the foil separates from the plate and the latter unloads to a stress-free state. Particles of the plate in the reflected rarefaction wave acquire a velocity $u_C = 2u_A < u_D$. In other words, the covering foil jumps forward from a plate of higher dynamic impedance. After the separation, the stress-free surface appears instead of the foil–plate interface and the remaining part of the unloading wave reflects from this interface as a compression wave.

We saw that the wave reflections from the interface with material of higher impedance occur with conservation in the load sign, so reflection of a rarefaction wave produces a new rarefaction wave. As a result, a short tensile pulse is generated in the unloaded foil, which, in turn, should produce a negative pullback in the velocity history of the rear surface of the foil. The pressure and particle velocity at maximum tension correspond to the intersection point i of Riemann's isentropes Dk and Ci. Figure 1.10 presents an experimental observation (by Kanel et al., 1988) of the tensile pulse created by interaction of a rarefaction wave with a higher impedance material. It is interesting that even a thin film of a low-impedance material reverses the sign of this pulse.

The relationship between the velocities of the covering foil and the plate surface occurs only when the shock reflection at the contact surface can be regarded as the decomposition of a discontinuity. If, on the other hand, the thickness of the foil is less than the shock thickness, its free surface will influence the process of reflection of the incident wave by the contact surface. The foil thus gains the velocity u, $u_C < u < u_D$ (Fig. 1.9). As a result, the foil velocity depends on its thickness. Figure 1.11 presents results of measurements by Razorenov et al., 1985, 1987, which were used for determining the thickness of the shock waves in copper. The dependencies consist of two parts: the velocities of relatively thick foils are independent of their thickness whereas the velocities of thinner foils increase from the velocity u_C of the copper plate surface up to the ultimate value measured for thick foils.

The velocity of the foil is determined by the pressure pulse communicated to it through the separation boundary:

$$u = \int_0^{t'} \frac{p\,dt}{\rho_{Al}\,\delta_f},$$

where ρ_f and δ_f are the foil density and thickness and t' is the total duration of the pressure pulse on the interface. Analysis based on the acoustic approximation shows that, for a linear increase of parameters within the shock wave, the

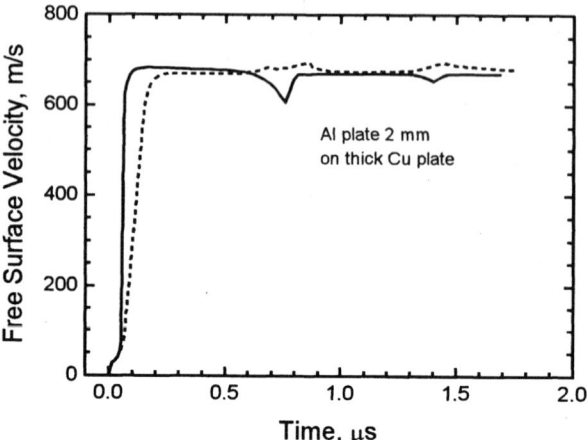

Figure 1.10. The free surface velocity history of a 2-mm-thick aluminum plate impacted through a thick copper base plate. The dashed line presents a shot with a thin polymer film between the aluminum and copper plates. Measurements by Kanel et al., 1988.

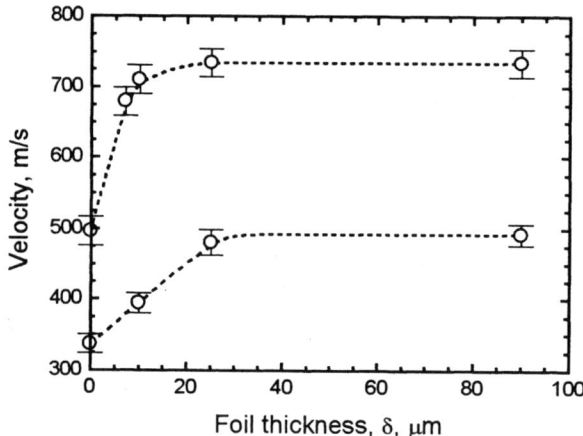

Figure 1.11. Velocity of aluminum foils as functions of foil thickness. The foils were in contact with copper plates in which shock waves with peak stresses of 6.6 GPa and 9.6 GPa were generated. Measurements by Razorenov et al., 1987.

average velocity value $u = 0.5(u_C + u_D)$ corresponds to the foil thickness for which the wave reverberation time is equal to the rise time t_f of the shock front. For the experimental data shown in Fig. 1.11, the average foil velocity is reached at $\delta_f \cong 15$ μm for a peak pressure of 6.6 GPa and $\delta_f \cong 5$ μm for 9.6 GPa that corresponds to the rise times ~6 ns and 2 ns, respectively.

1.4. Shock-Wave Processes in Layered Plates

Shock-wave experiments often involve targets composed of several layers. Figure 1.12 illustrates the wave process in a thin layer sandwiched between two thick plates of another material. During the process of multiple reverberations of waves, the pressure and particle velocity inside the layer evolve toward equality with the values of these quantities in the initial shock wave in the thick plate. The compression process depends on the relationship between the dynamic impedances of the layer material and the material of the thick plates. A layer of relatively soft material is compressed monotonically in a step-like mode, whereas

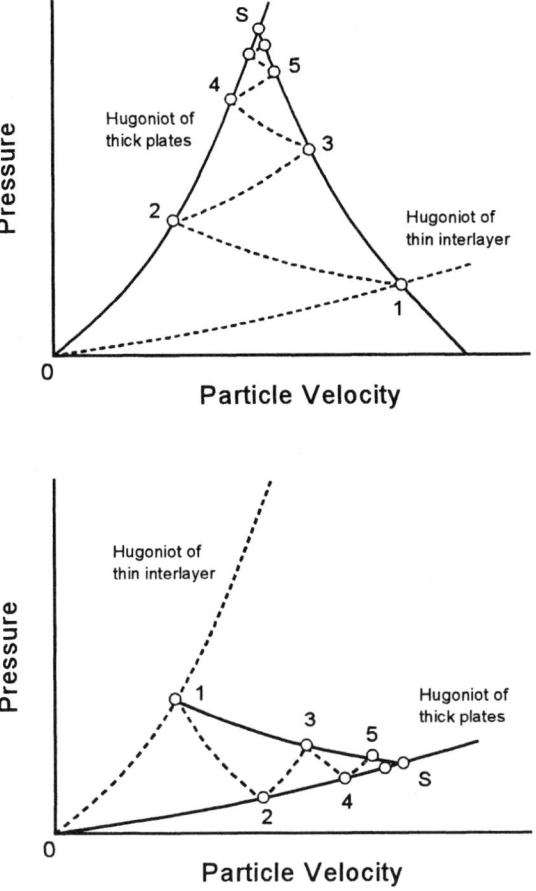

Figure 1.12. The wave processes in a thin layer sandwiched between two thick plates of another material. The point S corresponds to the parameters of the initial shock wave in the first plate and the numbers correspond to the order of wave reflections within the thin interlayer.

the pressure within a layer of relatively stiff material approaches the value of the initial shock pressure through damped oscillations. Passage of a shock wave through any foreign layer always increases its rise time.

The step-like loading of thin soft layers can be used to produce quasi-isentropic compression with a minimum increase of the temperature. These wave processes also occur in embedded gauges and control their response time.

The behavior of complicated systems is usually investigated using simple models. A simple model of a composite material can be presented as a plate composed of alternating flat layers of two different materials that lie perpendicular to the direction of shock propagation. Shock-wave phenomena in laminated composites were discussed by Barker, 1971, Oved et al., 1978, and Kanel et al., 1995. In this case the first shock wave decays very rapidly due to numerous wave reflections at interfaces. It has been shown that the wave front induces a large amount of ringing as it passes through the layers of the composite. Obviously, the final shock-compressed state has to correspond to the Hugoniot of the mixture.

Figure 1.13 shows results of computer simulation of shock compression of a laminated plate consisting of copper and polyethylene layers. The pressure and particle velocity distributions at several moments of time show the resonance behavior of a periodic one-dimensional composite. The pressure (and compression) oscillations are concentrated mainly inside the soft polyethylene layers. For this reason, one can expect the main dissipation to be localized in the soft material. The velocity oscillations are concentrated mainly in the heavier copper layers. The oscillations of pressure and velocity do not come to resonance if the thickness of pairs of layers is not constant.

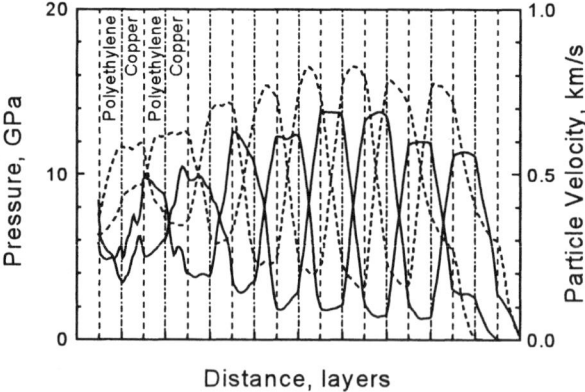

Figure 1.13. Pressure (solid lines) and particle velocity (dashed lines) oscillations in a one-dimensional model of a composite material.

1. Theoretical Background and Experimental Methods of Shock Physics   17

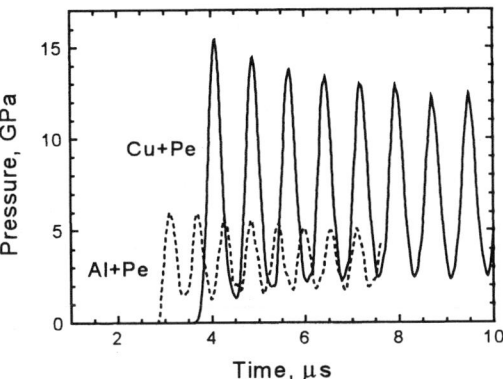

Figure 1.14. Pressure history in the middle sections of two layered composites composed of 20 metal layers and 20 polyethylene layers 0.5 mm in thickness, each driven by a piston moving at a velocity of 0.5 km/s.

Figure 1.14 shows pressure histories in the middle sections of copper–polyethylene and aluminum–polyethylene targets. One can see that the pressure oscillations have an approximately constant period and are near harmonic in form. The period of the oscillations corresponds approximately to the sound propagation time through two adjacent layers of the composite. Wave reflections between layers smear the wave front, but its average propagation velocity is practically equal to the shock-front velocity in the mixture calculated for a given boundary velocity. The average pressure also corresponds to the Hugoniot of the mixture. The rise time of the first compression wave is approximately equal to the period of oscillation. The pressure oscillations can pass into a homogeneous barrier placed behind of the layered target if the dynamic impedance of the barrier is high enough.

The reality of the predicted oscillations has been examined experimentally. With this goal, the pressure profile at the interface between a target, consisting of 10 copper foils 0.2 mm thick and 10 polyethylene films 0.2 mm thick, and a copper plate, was measured with a manganin pressure gauge. The initial load pulse was triangular in form with a total duration about 20 μs. The resulting measurement, presented in Fig. 1.15, confirms the appearance of an oscillation. The relatively small amplitude of the measured oscillations and their relatively fast decay is explained by the viscosity of the soft polyethylene layers in which most of the deformation occurs.

1.5. Generation of Negative Pressures During Reflection of a Compressive Pulse From a Plate Surface

Figure 1.16 shows time–distance, $t-x$, and pressure–particle velocity, $p-u$, diagrams that illustrate the dynamics of reflection of a triangular shock pulse

18 Shock-Wave Phenomena and the Properties of Condensed Matter

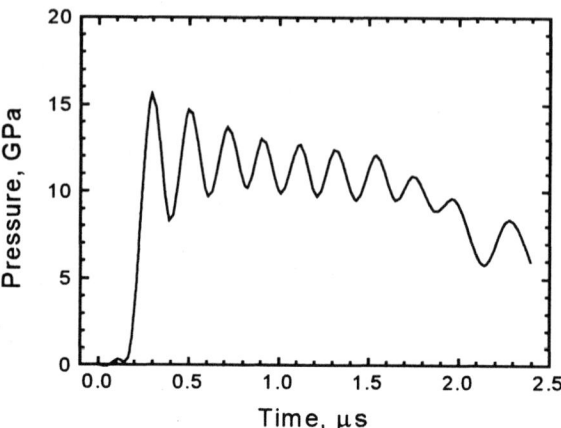

Figure 1.15. Results of a pressure profile measurement with a manganin gauge placed at the interface between the layered copper–polyethylene composite and a copper plate.

when it encounters a stress-free surface of a body. In the time–distance diagram, the shock front trajectory is described by the line 00′. A family of C_+ characteristics represents the unloading wave overtaking the shock front. When the shock front reaches the free surface, the velocity of the latter undergoes a jump from zero up to $u_0 = 2u_p$, where u_p is the particle velocity behind the shock front. The following unloading decreases the free-surface velocity. Reflection of the shock wave by the free surface produces a centered rarefaction wave which is described by a fan of C_- characteristics. The state of the particles must satisfy conditions on both the C_+ and C_- characteristics and is determined in the p–u diagram by the intersection of Riemann's trajectories describing states of matter along the C_- and C_+ characteristics which pass through a given particle at some given moment in time. One can see that some of these intersections lie in the negative pressure region. In other words, tensile stresses are generated in the reflected wave. When the tensile stress becomes high enough, it causes tensile fracture inside the body. This phenomenon is known as a "spall fracture" or "spallation."

The maximum tensile stress is reached at each particle as it is traversed by the terminal characteristic of the centered rarefaction wave. So, the peak tensile stress in the candidate spall plane just before the fracture corresponds to the intersection of trajectories 0′K and 2K in the p–u plane of Fig. 1.16. The line 0′K describes the change of state along the tail C_- characteristic of the centered rarefaction wave; 2K is the trajectory of the change of state along the last of the C_+ characteristics of the incident wave crossing the spall plane before the fracture.

1. Theoretical Background and Experimental Methods of Shock Physics   19

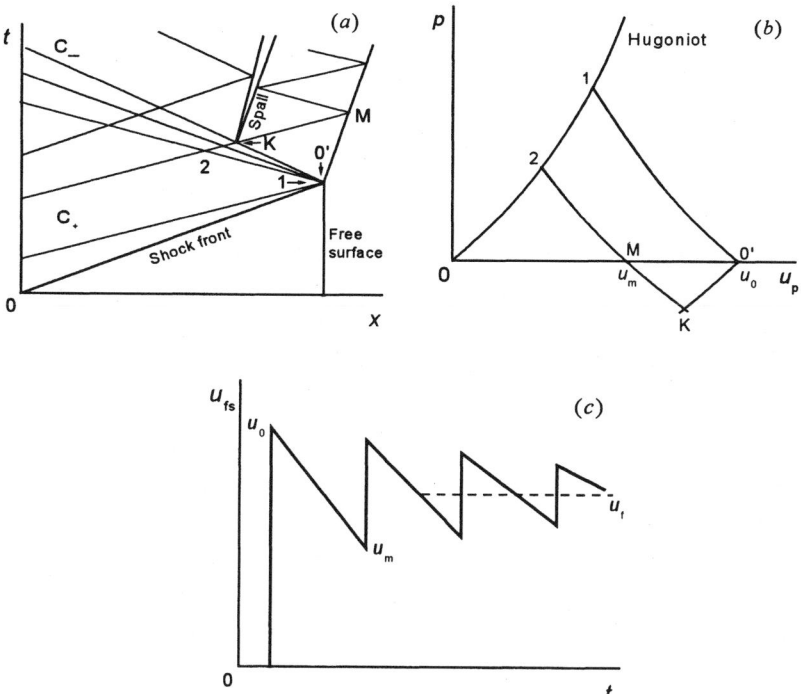

Figure 1.16. The wave dynamics at reflection of a shock pulse from the free surface.

Fracture of material allows the tensile stress to decrease rapidly to zero. As a result, a compression wave appears in stretched material adjacent to the spall plane and propagates to the stress-free rear surface. This compression wave appears and produces a so-called spall pulse in the free surface velocity profile, $u_{fs}(t)$. Subsequent reverberations of the spall pulse between the free surface and spall plane are accompanied by damped oscillations of $u_{fs}(t)$.

The peak velocity, u_0, and the free-surface velocity, u_m, just ahead of the spall pulse, are determined directly from the free surface velocity profile. The tensile stress value just before spalling is then determined by intersection of Riemann's trajectories passing through point $p=0, u=u_0$ for C_- and $p=0$, $u=u_m$ for C_+. Within the acoustic approach, the linear approximation

$$\sigma^* = \tfrac{1}{2}\rho_0 c_0 \Delta u_{fs} \tag{1.12}$$

is used (Novikov et al., 1966), where $\Delta u_{fs} = u_0 - u_m$ is the so-called "velocity pullback." Non-linearity of material compressibility has to be taken into account in the case of high tensile stresses. This can be done by extrapolation of the material isentrope in the $p-u$ plane to the negative pressure region. This correction does not exceed 10% in practical cases. In the elastic–plastic material, the spall pulse front propagates at the longitudinal elastic wave velocity whereas the incident rarefaction plastic wave ahead of it propagates at the bulk sound speed, which is less than the longitudinal sound speed. As a result, the measured velocity pullback becomes less than it would be in a liquid-like material having the same spall strength. A way to introduce necessary corrections was discussed by Kanel, 2001.

1.6. Generation and Measurement of Shock Waves in Condensed Matter

In order to investigate shock-wave phenomena in condensed matter we have to be able to create plane shock pulses in our samples and to measure the evolution of these pulses inside the sample. In this section, we shall discuss methods of producing and recording intense load pulses in condensed media. The shock-wave techniques are well documented in original papers, monographs (for example, Caldirola and Knoepfel, 1971, Graham, 1993), and reviews (for example, Al'tshuler, 1965, Graham and Asay, 1978, and Chhabildas and Graham, 1987). In this section we just mention the methods of generation of shock waves and measurement of their parameters that were used in the experiments discussed in following chapters.

Plane shock waves are created in condensed matter by impacting the sample to be studied with a flyer plate or by detonating an explosive plane wave generator in contact with it. Measurements in the range of shortest load duration are performed using laser or particle beams as shock-wave generators. To plan shock-wave experiments and correctly interpret the results obtained, it is important to understand all details of the loading history for each individual case. Let us to consider the wave dynamics under these conditions.

Figure 1.17a shows the time–distance diagram of the wave process after impact of a flyer plate upon the plane target. At the instant of collision, shock waves are produced in the impactor and the target, and propagate away from the collision surface. All possible values of the pressure and particle velocity in the shock-compressed target have to be described by the Hugoniot of the target material, H_t, in p–u coordinates (Fig. 1.17b). The shock wave in the impactor propagates in the opposite direction and particles of impactor, which have the initial velocity u_i, are decelerated by the shock wave. Possible states of the impactor material behind the shock wave are described by the "Hugoniot of deceleration," H_i, of the impactor material. Because the pressure and particle velocity values have to be the same on both sides of the impact surface, these parameters have to correspond to the intersection, in the p–u plane, of the Hugoniots H_t and H_i. If the target and the impactor are of the same material, then, owing to the symmetry of the Hugoniots, the particle velocity of the shock-compressed material is exactly half of the initial impactor velocity.

1. Theoretical Background and Experimental Methods of Shock Physics   21

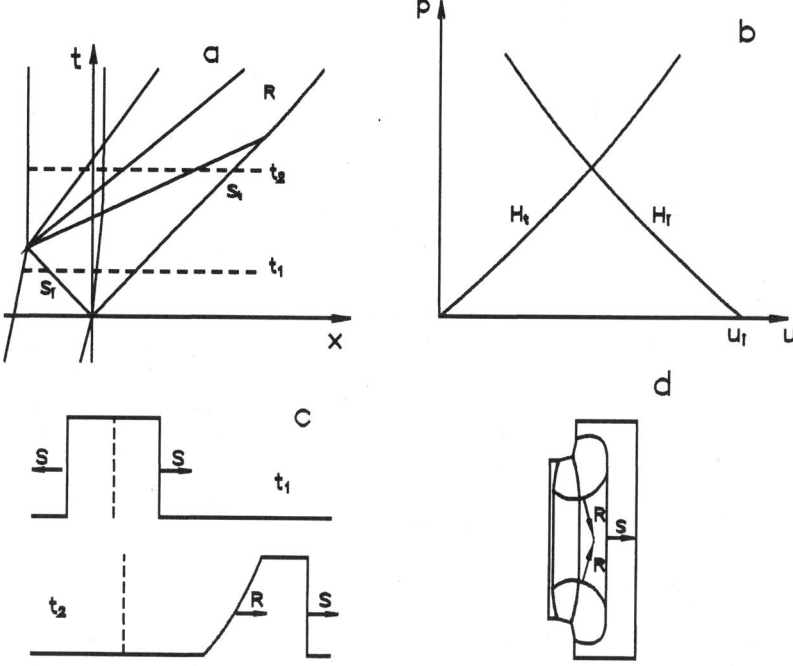

Figure 1.17. Creation of a compression pulse by the impact of a flyer plate on a target plate.

When the shock wave is reflected by the rear surface of the impactor, a centered rarefaction wave appears and propagates toward the sample (Fig. 1.17c). Thus, a constant pressure is supported on the impact surface during the time of wave reverberation in the impactor. Because the rarefaction front, which propagates with the sound velocity in shock-compressed matter, is faster, it overtakes the shock front and causes its attenuation. The transverse dimensions of the impactor and target must be large enough to ensure one-dimensional motion of the medium during the period of time required for the measurements to be completed. This condition is violated by the radial release wave propagating from the edges of the impactor and target (Fig. 1.17d). Longer recording time requires an impactor of larger diameter. Plane impactors are launched by a detonating explosive or by ballistic devices such as powder or gas guns.-

Experiments conducted with explosives are compact and rather inexpensive, so this technique has been widely used in shock-wave physics from the time of World War II until now. The ways of maintaining the planarity of the flyer plates and their integrity have been well developed. By varying the composition and density of the high explosives, the velocity of metal or polymer flyer plates 0.2 to 10 mm in thickness can be easily varied over a range of ~0.5 to ~6 km/s. The flyer velocity may be increased up to ~12 km/s or more using multi-stage acceleration. However, experiments with explosions are destructive and require the deployment of safety measures. They need special explosion chambers or test areas, and the availability of the technology for manufacturing suitably-shaped high-grade explosive charges. The alternative is to use smooth-bore ballistic gas guns or powder guns. This has the obvious advantage of permitting continuous variation of the impact velocity. Owing to the smooth long-time acceleration, the state of impactor is not changed by the launch forces and the accuracy of the tests is increased.

Some experiments presented herein have been done using the pulsed high-power proton beam at KALIF, the Karlsruhe Light Ion Facility (Baumung et al., 1996a) as a shock wave generator. The ablative pressure pulse generated by the direct interaction of the 0.15 TW/cm^2 pulsed proton beam creates compression waves in solid targets with peak pressures of up to several tens GPa and pulse durations of approximately 40–60 ns.

Modern methods for making time-resolved measurements of particle velocity histories include capacitor gauges, the magnetoelectric method, and laser Doppler interferometric techniques.

The capacitor gauges (Ivanov and Novikov, 1963) are used to record the movement of the surface of a metal body. The measuring capacitor consists of a flat electrode and the sample surface with a distance x_0 between them. The external voltage is applied to the capacitor via a resistor R_i having a resistance low enough to ensure that the time constant $t_c = R_i C$ is much less than the characteristic time of the measured process. As soon as the sample surface starts moving toward the electrode, the capacitance begins to increase, and an electric current flows in the gauge circuit. This current is proportional to the rate of increase of the capacitance, dC/dt, i.e., the velocity of the sample surface.

The capacitor gauge method provides a non-contact measurement, so that, in principle, its time resolution is limited only by the tilt of the shock wave relative to the sample surface in the region monitored by the gauge. Depending on the required resolution and duration of the process being measured, the gauge diameter and its initial distance from the sample surface are varied within 5 to 25 mm and 1 to 6 mm, respectively. The actual time resolution of a capacitor gauge having an electrode 5 mm in diameter is ~10–20 ns. With a supply voltage of 3 kV, the signal typically is in the range 1–100 mV. As a result of the low level of the useful signal, the capacitor method is sensitive to electrical noise that restricts its applications.

Laser velocimeters use Doppler-shifted light reflected from the target surface. Since the Doppler shift is very small (for velocities of ~1000 m/s the wavelength shift is ~10^{-2}Å), it has to be recorded using interferometer techniques. The VISAR-type laser Doppler velocimeter (Barker and Hollenbach, 1972; Asay and Barker, 1974) is the best known and most widely used instrument for recording the velocity histories of shocked samples. Owing to the apparent optical symmetry of the interferometer, the technique can operate with both specular and diffuse reflecting surfaces. The accuracy of the velocity measurements with VISAR is ~1–2% or better and the time resolution is ~2 ns or better. The optically recording velocity interferometer system (ORVIS) uses a high-speed electronic streak camera to record the interference and this improves the time resolution to ~200 ps (Bloomquist and Sheffield, 1983). The Doppler shift causes the interference fringes to shift. A streak record of the fringe pattern that is changing position in time directly yields the surface-velocity history. The accuracy of the velocity measurement is somewhat lower in this case. Baumung et al. (1996b) have modified the optical scheme of the ORVIS velocimeter to permit illumination of a line on the target surface so that the velocity history at points along this line can be measured with a standard argon-ion laser and a streak camera.

In contrast to measurement of velocity, all methods of pressure measurement require calibration of gauges. The manganin pressure gauge is now the basic tool for recording the stress or pressure history in the interior of a body. It was first used by Bridgman, 1940, to measure pressures under static conditions. Bridgman found that the resistivity of the manganin alloy increased significantly with increasing ambient pressure but was only slightly affected by temperature changes. Since 1964 (Fuller and Price, Bernstein and Keough) manganin gauges have been used for pressure measurements in plane shock waves. The gauge, in the form of a plane zigzag strip 10–30-μm thick, is usually embedded in the sample such that the plane of the active gauge element is normal to the load direction. It is usually electrically insulated from the sample material by thin polymer films or by mica. A constant electrical current is passed through the gauge. When a shock pulse passes through the gauge plane, the recorded voltage increases with pressure applied to the gauge. In order to increase the precision of the pressure measurements, a resistance bridge is usually used. The manganin alloy has been calibrated in well-controlled shock-wave experiments by measuring the fractional change in resistance, $\Delta R / R_0$, and empirically correlating these results to the pressure. This function is common to all sensors made of manganin of a given grade. Special measurements (Kanel et al., 1978) have shown that, at pressures above 7–10 GPa, the change in the resistivity of manganin is reversible and does not depend on whether dynamic compression occurs by single or multiple shocks, or if it is quasi-isentropic. The release to zero pressure produces a slight hysteresis in the gauge resistance. The irreversible component is attributed to strain-hardening of the material due to the shock compression.

1.7. Analysis and Interpretation of Measurements

By examining self-similar flows, such as steady shock waves or simple Riemann waves, we can determine the thermodynamic characteristics of a material from its measured kinematic parameters. Experiments conducted with different initial conditions and shock-wave amplitudes define the caloric equation of state $\mathscr{E} = \mathscr{E}(p,V)$ in the region of the $p-V$ plane covered by the Hugoniots and isentropes. Interpretation of the measurements is thus based on general conservation laws and determination of the caloric equation of state is reduced to the measurement of shock and particle velocities, i.e., to the measurement of distance and time intervals.

Measurements of stress or particle-velocity histories are used to study the elastic–plastic and strength properties of condensed materials, as well as polymorphic transformations and phase transitions that occur under shock-wave loading. Measured wave profiles contain information on relaxation processes occurring in the media. By recording the evolution of wave profiles we can analyze the associated variations in pressure or stress and the specific volume at all stages of the shock loading. Direct analysis of wave profiles and computer simulations of the experiments are used to achieve this goal.

When the experimental data are analyzed directly, the profiles of mechanical stress $\sigma_x(t)$ or particle velocity $u_p(t)$ obtained in different cross-sections of the sample are used to recover the fields of pressure, $p(x, t)$, stress, $\sigma_x(x,t)$, or particle velocity, $u_p(x, t)$. The equations of motion of a compressible material, Eqs. (1.3), are not limited to any particular thermophysical and mechanical properties, so we can determine all the parameters of state over these fields of kinematic parameters (see Fowles and Williams, 1970, Seaman, 1974, Kanel, 1977). In this way, the trajectories of variations of the state parameters are recovered. Each point of such a trajectory corresponds to a particular instant of time. The position of this trajectory relative to the metastable and equilibrium compression curves allows determination of the parameter of non-equilibrium at each point of the trajectory. The time dependence of the non-equilibrium parameter (e.g., the concentration of the products of a physical or chemical transformation) can thus be determined for each layer in the sample as a function of time. The simple wave approximation is often used to simplify the procedure.

A computational simulation of a dynamic event (impact, explosion, thermal radiation) is conducted by representing the geometry and materials of all the objects in the event, and the initial conditions of the event. The material behavior represented can include standard stress–strain relations, detonation, melting, yielding, and fracture. The most common solution procedures used are finite difference methods which involve discretization of the domain of the problem. The solution is performed in such a manner as to determine successive states of equilibrium in the material at a sequence of times and at a large number of points distributed over the objects participating in the event. With the solution known at these discrete points in time and space, we can interpolate to determine a complete solution over all locations in the objects and throughout the time interval of interest.

The computations are based on the simultaneous solution of the differential equations representing the conservation of mass, momentum, and energy, plus the constitutive relations. The solution procedure consists of determining the particle velocity, stress, strain, internal energy, temperature, fracture damage, and other quantities at each selected point in time and space. With these quantities known at one time, we can step forward to the next time and solve again for new values of these quantities using the conservation laws and the constitutive relations. The constitutive relations must be provided for each material participating in the event. Hence, the development of the constitutive relations is a crucial step in the solution of problems with the computational methods.

References

Al'tshuler, L.V. (1965). "Application of shock waves in physics of high pressures," *Sov. Phys.–Usp.* **8**, pp. 52–91, (1965) [trans. from: *Usp. Fiz. Nauk* **85(2)**, pp. 197–258 (1965)].

Asay, J.R., and L.M. Barker (1974). "Interferometric measurement of shock-induced internal particle velocity and spatial variations of particle velocity," *J. Appl. Phys.* **45(6)**, pp. 2540–2546.

Barker, L.M. (1971). "A Model for Stress Wave Propagation in Composite Materials," *J. Composite Materials* **5**, p. 140.

Barker, L.M., and R.E. Hollenbach (1974). "Shock Wave Study of Phase Transition in Iron," *J. Appl. Phys.* **5(11)**, pp. 4872–4887.

Baumung, K., H.J. Bluhm, B. Goel, P. Hoppe, H.U. Karow, D. Rush, V.E. Fortov, G.I. Kanel, S.V. Razorenov, A.V. Utkin, and O.Yu. Vorobjev (1996a). "Shock-Wave Physics Experiments with High-Power Proton Beams," *Laser and Particle Beams* **14(2)**, pp. 181–210.

Baumung, K., J. Singer, S.V. Razorenov, and A.V. Utkin (1996b). "Hydrodynamic Proton Beam–Target Interaction Experiments Using an Improved Line-Imaging Velocimeter," in: *Shock Compression of Condensed Matter — 1995* (eds. S.C. Schmidt and W.C. Tao), American Institute of Physics, New York, pp. 1015–1018.

Bernstein, D., and D.D. Keough. "Piezoresistivity of Manganin," *J. Appl. Phys.* **35(5)**, pp. 1471–1474.

Bloomquist, D.D., and S.A. Sheffield (1983). "Optically Recording Interferometer for Velocity Measurements with Subnanosecond Resolution," *J. Appl. Phys.* **54**, p. 1717.

Bridgman, P.W. (1940). "The Measurements of Hydrostatic Pressure to 30 000 kg/cm^3," *Proc. Am. Acad. Arts Sci.* **74**, p. 1.

Bushman, A.V., and V.E. Fortov (1983). "Models of equations of state of a matter," *Sov. Phys.–Usp.* **26(6)**, pp. 465–496 [trans. from: *Usp. Fiz. Nauk* **140(2)**, pp. 177–232 (1983)].

Caldirola, P., and H. Knoepfel (eds.) (1971). *Physics of High Energy Density*, Academic Press, New York.

Chhabildas, L.C., and R.A. Graham (1987). "Development in Measurement Technique for Shock Loaded Solids," in: *Techniques and Theory of Stress Measurements for Shock Wave Applications* (eds. R.R. Stout, F.R. Norwood, and M.E. Fourney), American Society of Mechanical Engineers, New York, pp. 1–18.

Courant, R., and K.O. Friedrichs (1948). *Supersonic Flow and Shock Waves*, Interscience, New York.

Fowles, R., and R.F. Williams (1970). "Plane Stress Wave Propagation in Solids," *J. Appl. Phys.* **41(1)**, pp. 360–363.

Fuller, J.A., and J.H. Price (1964). "Dynamic Pressure Measurements to 300 kbar with a Resistance Transducer," *Brit. J. Appl. Phys.* **15(6)**, pp. 751–758.

Graham, R.A. (1993). *Solids under High-Pressure Shock Compression*, Springer-Verlag, New York.

Graham, R.A., and J.R. Asay (1978). "Measurements of Wave Profiles in Shock-Loaded Solids," *High Temp. High Press.* **10(2)**, pp. 355–390.

Ivanov, A.G., and S.A. Novikov (1963). "The Method of Capacitor Gauge for Registration of Momentary Velocity of Moving Surface," *Apparatus and Experimental Technique* **7(1)**, pp. 135–138 (in Russian).

Kanel, G.I. (1977). "Experimental Determination of the Kinetics of Relaxation Processes During the Shock Compression of Condensing Media," *J. Appl. Mech. Tech. Phys.* **18(5)**, pp. 685–689 [trans. from *Zh. Prikl. Mekh. Tekh. Fiz.* **18(5)**, pp. 117–122 (1977)].

Kanel, G.I. (2001). "Distortion of the Wave Profiles in an Elastoplastic Body Upon Spalling," *J. Appl. Mech. Tech. Phys.* **42(2)**, pp. 358–362 [trans. from *Zh. Prikl. Mekh. Tekh. Fiz.* **42(2)**, pp. 194–198 (2001)].

Kanel, G.I., M.F. Ivanov, and A.N. Parshikov (1995). "Computer Simulation of the Heterogeneous Materials Response to the Impact Loading," *Int. J. Impact Engineering* **17(1–6)**, pp. 455–464.

Kanel, G.I., S.V. Razorenov, and V.E. Fortov (1988). "Viscoelasticity of Aluminum in Rarefaction Waves," *J. Appl. Mech. Tech. Phys.* **29(6)**, pp. 824–826 [trans from *Zh. Prikl. Mekh. Tekh. Fiz.* **29(6)**, pp. 67–70 (1988)].

Kanel, G.I., G.G. Vakhitova, and A.N. Dremin (1978). "Metrological Characteristics of Manganin Pressure Pickups under Conditions of Shock Compression and Unloading," *Comb. Expl. Shock Waves* **14(2)**, pp. 244–248 [trans. from *Fiz. Goreniya Vzryva* **14(2)**, pp. 130–135 (1978)].

Landau, L.D., and E.M. Lifshitz (1959). *Fluid Mechanics*, Pergamon Press, Oxford.

Novikov, S.A., I.I. Divnov, and A.G. Ivanov (1966). "The Study of Fracture of Steel, Aluminum and Copper under Explosive Loading," *Phys. Met. Metall.* **21(4)** pp. 122–128 [trans. from *Fiz. Metall. Metalloved.* **21(4)**, pp. 607–615 (1966)].

Oved, Y., G.E. Luttwak, and Z. Rosenberg (1978). "Shock Wave Propagation in Layered Composites," *J. Composite Materials* **12**, p. 84.

Razorenov, S.V., G.I. Kanel, and V.E. Fortov (1985). "Measurement of the Width of Shock Fronts in Copper," *Sov. Phys.–Tech. Phys.* **30(9)**, pp. 1061–1062 [trans. from *Zh. Tekh. Fiz.* **55(9)**, pp. 1816–1818 (1985)].

Razorenov, S.V., G.I. Kanel, O.R. Osipova, and V.E. Fortov, (1987). "Measurement of the Viscosity of Copper in Shock Loading," *High Temp.* **25(1)**, pp. 57–61 [trans. from *Teplofiz. Vys. Temp.* **25(1)**, pp. 65–69 (1987)].

Seaman, L. (1974). "Lagrangian Analysis for Multiple Stress or Velocity Gauges in Attenuating Waves," *J. Appl. Phys.* **45(10)**, pp. 4303–4314.

Skidmore, I.C., and E. Morris (1962). "Experimental equation-of-state data for uranium and its interpretation in the critical region," in: *Thermodynamics of Nuclear Materials*, International Atomic Energy Agency, Vienna, pp.173–216.

Zel'dovich, Ya. B., and Yu. P. Raizer (1967) *Physics of Shock Waves and High-Temperature Hydrodynamic Phenomena* (eds. W.D. Hayes and R.F. Probstein), Vol. II, Academic Press, New York, Reprinted by Dover Publications, Mineola, New York (2002).

CHAPTER 2

Elastic–Plastic Response of Solids Under Shock-Wave Loading

Because shock-wave and high-strain-rate phenomena are involved in a broad range of technological applications, we are interested in understanding time-dependent mechanical properties of materials subjected to these extreme loading conditions. The shock-wave technique also provides a powerful tool for scientific investigation of material properties at extremely high strain rates. With modern diagnostics, elastic–plastic yielding can be studied by recording and analyzing shock-wave structures. Investigations of the resistance of materials to shock-wave deformation are based on the analysis of elastic precursors in compression and rarefaction waves, of plastic shock-front rise times, on measurements of principal stresses in shock-compressed matter, and on other more sophisticated measurements and analyses. Empirical data are generalized by constitutive relationships that are used, for example, for computer simulations of impact phenomena.

In this chapter we will discuss ways of investigating mechanical properties of solids under shock-wave loading and some results of these investigations. Although a comprehensive discussion of the fundamentals of the shock mechanics of solids can be found, for example, in the books of Zel'dovich and Raizer, 1967, and Graham, 1993, we have found it useful to begin this chapter with a brief account of the main relationships and the simplest model describing elastic–plastic response of materials. In the following sections we will discuss finer details of the structure of waveforms in solids, including the shape and decay of elastic precursor waves, the plastic wave rise time, the behavior of shock-compressed solids undergoing unloading and reloading, hysteresis effects, specific effects seen in free surface velocity histories, and specifics of measurements of principal stresses. The chapter is completed by some examples of constitutive models and relationships.

Despite a quite sufficient general understanding of high-rate yielding phenomena having been developed, experiments, theory, and material models do not agree in many of the details concerning material behavior. Mechanical yielding and strength behavior in shock waves show complexities that are not yet fully understood. One of the goals of this chapter is to attract attention to some unidentified aspects of high-rate deformation of solids under conditions of shock-wave loading.

2.1. The Main Relationships

This section provides a brief account of the background needed for understanding and interpreting the shock-wave data. For more details, see the monographs by McClintock and Argon, 1966, Zel'dovich and Raizer, 1967, and many others.

The state of solids in this case is described by two tensors: the stress, σ_{ik}, and the strain, ε_{ik}, where the subscripts i and k represent the coordinate directions x, y, and z of an orthogonal coordinate frame. The stress component σ_{ik} is the force per unit area in the body along the direction i, acting on an area with normal oriented along the k axis. The components $\sigma_{xx}, \sigma_{yy}, \sigma_{zz}$ are the normal stresses and $\sigma_{xy} = \sigma_{yx}$, $\sigma_{yz} = \sigma_{zy}$, and $\sigma_{xz} = \sigma_{zx}$ are tangential or shear stresses. The normal stresses are also represented by σ_i and shear stresses by τ_{ik}. The strain components $\sigma_{xx}, \sigma_{yy}, \sigma_{zz}$, or ε_k, are the normal strains describing elongation along the axes. Tangential or shear strains are represented by $\varepsilon_{xy}, \varepsilon_{xz}$, and ε_{yz}; the notation $\gamma_{ik} = 2\varepsilon_{ik}$ is also used. It is accepted in shock-wave physics, as in geophysics, that stresses and deformations are positive in states of compression and negative in states of tension.

Three mutually perpendicular planes on which there are no shear stresses can be drawn through each point in a body. At least two of the normal stresses $\sigma_1 \geq \sigma_2 \geq \sigma_3$ that act on these planes, and are called the principal stresses, have the maximum and minimum magnitudes for all possible orientations. The maximum shear stress acts on the area that bisects the angle between the maximum and minimum principal stresses and is equal to half the difference between these two stresses:

$$\tau_{\max} = \tfrac{1}{2}(\sigma_1 - \sigma_3). \tag{2.1}$$

The normal strains $\varepsilon_1 \geq \varepsilon_2 \geq \varepsilon_3$ also have maximum and minimum magnitudes along theses axes. Filaments lying along the principal axes of stress and strain can change their length but do not rotate. The maximum shear occurs in the direction intermediate between the directions of the maximum and minimum normal strains and is given by

$$\gamma_{\max} = \varepsilon_1 - \varepsilon_3. \tag{2.2}$$

For small strains, the relative change in specific volume V is equal to the sum of the relative elongations along the three orthogonal directions drawn through the given point:

$$-\frac{dV}{V} = d\varepsilon_x + d\varepsilon_y + d\varepsilon_z. \tag{2.3}$$

The stress and strain tensors can be separated into spherical and deviatoric components. The spherical component of the stress tensor is equivalent to a hydrostatic pressure, p. The spherical components of the strain tensor are equal to one third of the change in volume. Deviatoric components represent the change in shape of the body. In the linear theory of elasticity, the stress and strain increments are related by Hooke's law. For deviatoric components, Hooke's law takes the form

$$d(\sigma_k - p) = 2G\left(d\varepsilon_k + \frac{1}{3}\frac{dV}{V}\right) \quad (2.4)$$

$$d\tau_{ik} = G\,d\gamma_{ik},$$

where G is the shear modulus. The increments in the spherical components of stress and strain are related by the equation

$$dp = -K\frac{dV}{V}, \quad (2.5)$$

where K is the bulk modulus.

The limiting elastic state is defined in terms of a yield criterion. The purpose of the latter is to use a standard test to define the conditions under which plastic flow occurs for given load conditions. For example, according to the criterion of Tresca, the limiting elastic state is reached at a given point when the maximum shear stress reaches the value corresponding to the yield strength, Y, of a given material under uniaxial stress conditions:

$$|\tau_{\max}| \leq \frac{Y}{2}. \quad (2.6)$$

During the following plastic deformation, the increment of strain along each axis is the sum of elastic and plastic components:

$$d\varepsilon_{ik} = d\varepsilon_{ik}^{\text{el}} + d\varepsilon_{ik}^{\text{pl}}. \quad (2.7)$$

No change in volume results from the accumulation of plastic strain, so

$$d\varepsilon_x^{\text{pl}} + d\varepsilon_y^{\text{pl}} + d\varepsilon_z^{\text{pl}} = 0.$$

Figure 2.1a shows the idealized stress–strain diagram of a standard test under uniaxial stress conditions. In this case, $\sigma_y = \sigma_z = 0$, $-Y \leq \sigma_x \leq Y$, and $\varepsilon_y = \varepsilon_z \neq 0$. Until the yield strength, Y, is reached, the material responds elastically to the load so that $\sigma_x = E\varepsilon_x$, where E is Young's modulus. In the plastic region, $\sigma_x = Y$. When the strain direction is changed, the material again behaves elastically until the yield condition is satisfied in the stress region of reverse sign.

In plane compression and rarefaction waves, the loading conditions are characterized by one-dimensional strain for which $\varepsilon_y = \varepsilon_z = 0$, $\sigma_y = \sigma_z \neq 0$. The increment of longitudinal strain in this case is $d\varepsilon_x = -dV/V$. Figure 2.1b shows the stress–strain diagram for a solid body under such one-dimensional compression and unloading. In the elastic region, the longitudinal compressibility of the material is given by the equation

$$\frac{1}{V}\frac{dV}{d\sigma_x} = -\frac{1}{K+(4/3)G} = -\frac{1}{E'}, \qquad (2.8)$$

where E' is the modulus of longitudinal deformation. In this region, the longitudinal compressibility is less than the bulk compressibility which is characterized by the bulk modulus

$$K = -\frac{dp}{dV}V. \qquad (2.9)$$

The yield condition is satisfied when $|\sigma_x - p| = (2/3)Y$. Thus, above the yield point, the state of an elastic–plastic body deviates from the corresponding hydrostatic pressure $p(V, T)$ by as much as $(2/3)Y$. The longitudinal compressibility of the idealized elastic–plastic body for deformations in the plastic region is equal to the bulk compressibility. The transition from elastic to plastic deformation occurs at

$$\sigma_x = Y\left(\frac{K}{2G} + \frac{2}{3}\right). \qquad (2.10)$$

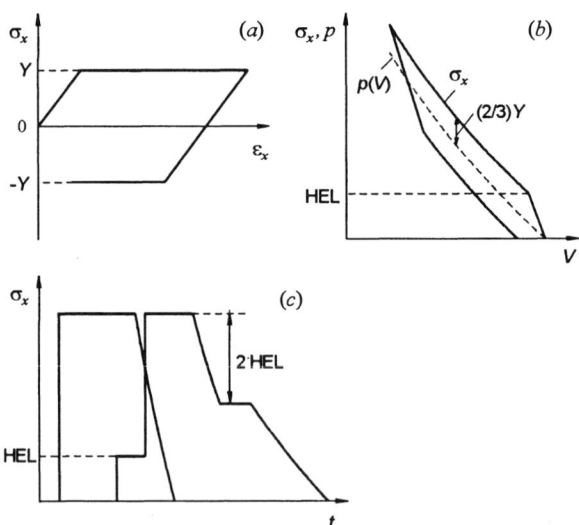

Figure 2.1. Stress–strain diagram of an elastic–plastic body. (a) Uniaxial stress states, (b) uniaxial strain states, and (c) evolution of a compression pulse in an elastic–plastic material.

This transition stress under shock-wave loading is called the Hugoniot elastic limit (HEL). The segment of elastic strain that occurs upon release of the imposed load is $2Y$ under uniaxial stress conditions and twice the HEL under one-dimensional strain conditions because the shear stress first drops to zero in the unloading wave, and this is followed by a change in the sign of the shear stress, τ, and an increase in the magnitude of τ to the limiting value $Y/2$ at further unloading.

Figure 2.1c shows the evolution of an initially square compression pulse in an ideal elastic–plastic material. Since the longitudinal compressibility is different in the elastic and plastic regions, elastic precursors appear in both compression and rarefaction waves. Elastic precursors propagate with the velocity of longitudinal elastic waves,

$$c_\ell = \sqrt{\frac{1}{\rho}\left(K + \frac{4}{3}G\right)}. \tag{2.11}$$

The velocity of wave propagation in the plastic region is the bulk sound velocity

$$c_b = \sqrt{K/\rho}. \tag{2.12}$$

The longitudinal and bulk sound velocities are related through Poisson's ratio, ν:

$$\frac{c_\ell}{c_b} = \sqrt{\frac{3(1-\nu)}{1+\nu}}. \tag{2.13}$$

Several equivalent relationships can be used for calculating the stress at the HEL:

$$\sigma_{HEL} = Y\frac{1-\nu}{1-2\nu} = \frac{2}{3}Y\left(1 - \frac{c_b^2}{c_\ell^2}\right)^{-1} = \frac{1}{2}Y\frac{c_\ell^2}{c_s^2} \tag{2.14}$$

where $c_s = \sqrt{(G/\rho)}$ is the shear sound speed.

The HEL characterizes the dynamic yield strength of the material. This parameter has been measured for many metals and alloys, ceramics, minerals, high explosives, etc. However, elastic–plastic waveforms contain much more information about the behavior of the material. The observed wave profiles exhibit various shapes of the elastic precursors in homogeneous (prior to the experiment) materials characterized by different strain and strain-rate effects. Beginning with the early works of Taylor, 1965, and Ahrens and Duvall, 1966, numerous measurements of the elastic precursor decay have been carried out to evaluate the stress relaxation at high strain rates. Some initial nonuniformity of the properties of the tested samples should also influence the wave structure.

The elastic–plastic model is the simplest idealization of mechanical response of solids. Real materials exhibit strain hardening and strain-rate effects and the yield stress varies with the temperature, pressure, etc. A relationship, or set of relationships, that describes the resistance of a material to inelastic deformation as a function of strain, strain rate, and other parameters is called a "mechanical equation of state" or "constitutive relationship." Usually the material properties above the HEL are recovered from the measured stress, particle velocity, or free surface velocity history of the sample by means of computer simulations of the experiments using assumed constitutive relationships. It is supposed that an agreement between the simulated and the measured profile justifies the physical description of the material properties employed in the computations. To make the computations described, one must begin with preliminary estimates of the material response inferred directly from the measured wave profile.

In the following sections we'll discuss quantitative contributions of the strain hardening, the yield strength gradients, and the stress relaxation to the shock-wave structures formed in solids.

2.2. Rate-Independent Strain Hardening Effects

A constant flow stress governed by a constant yield stress, Y, is only a rough approximation of the behavior of most materials. Strain hardening following yielding is much more typical behavior although some steels, single crystals, and ceramics demonstrate partial softening after the yield stress is reached. Figure 2.2 shows three idealized stress–strain diagrams that correspond to perfect plasticity with constant Y, plasticity with strain hardening, and a response with temporal softening. Curves for both uniaxial stress and uniaxial strain are shown for each type of behavior.

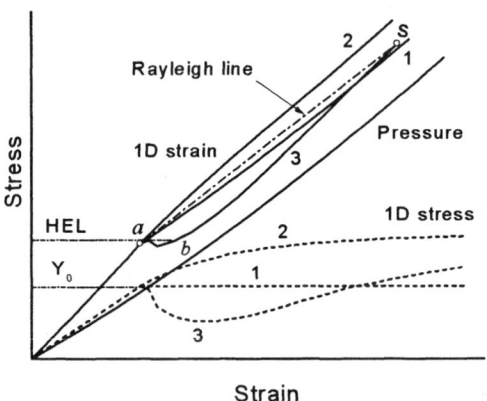

Figure 2.2. Simplified stress–strain diagrams of elastic–plastic compression for materials with perfect plasticity (curves 1), strain hardening (curves 2), and initial softening (curves 3). The dashed lines show diagrams for uniaxial stress conditions and the solids lines are for uniaxial strain conditions. The line aS is an example of the Rayleigh line describing the states in a plastic shock compression wave.

In the cases of a perfect elastic–plastic material and a material with initial softening, curves 1 and 3 in Fig. 2.2, the initial state for the Rayleigh line is the HEL point. It is important to note that the Rayleigh line cannot have intermediate intersections with the stress–strain curve and may have only positive slope in $\sigma - \varepsilon$ coordinates.

States on the stress–strain diagram of the softening material that lie below the horizontal line ab, which corresponds to zero velocity of a plastic compression wave, cannot be reached in a compressive wave. If we consider shocks with various peak stresses in such a material, we'll see that the propagation velocity of a plastic wave decreases to zero with decreasing peak stress. In other words, the fraction of a sample subjected to plastic deformation per unit time decreases to zero. For further decrease of peak stress, plastic deformation will not occur and only an elastic compressive wave will propagate. Thus, equilibrium softening cannot exhibit itself in the shape of the elastic precursor wave but it may appear in the subsonic velocity of plastic shock wave.

The strain hardening behavior of solids is described by various empirical re-relationships that express the flow stress, Y, as a function of plastic strain, γ_p. Some simple examples of strain hardening curves are illustrated in Fig. 2.3. The smoothness of the dependence of the velocities of propagation of a stress perturbation on the deformation is important for analyzing the wave structures in strain hardening materials. Any discontinuity in the sound speed results in the appearance of a flow region with constant parameters. In other words, the start of the plastic deformation accompanying a kink of the stress–strain curve results in a rectangular shape of the elastic precursor wave. Such behavior also occurs for materials with temporal softening having a stress–strain diagram of type 3 in Fig. 2.2. Smooth strain hardening produces a region of smoothly decreasing slope of the stress–strain curve of the material, as described by curves 2 in Figs. 2.2 and 2.3. This region corresponds to a continuous decrease of the wave speed $c_\sigma = V(d\sigma/d\varepsilon)^{1/2}$ starting from c_ℓ. As a result, wave dispersion occurs immediately behind the front of the elastic precursor wave and only the part of the stress–strain diagram with a slope $d\sigma/d\varepsilon > \rho_a U_s^2$ is represented in the precursor. In the case of a strain hardening material, the Rayleigh line from its initial point must be tangent to the stress–strain curve. This results in an increase of the stress σ_a ahead the plastic wave when the peak stress σ_s behind the plastic compression wave decreases.

Figures 2.4 and 2.5 present examples of free surface velocity histories obtained after planar impact loading of Ti-6Al-4V alloy (Razorenov et al., 2000) and quenched steel 40 Kh (Razorenov et al., 1997). The profile for the titanium alloy contains an elastic precursor wave with a distinct shock front followed by a plateau, a gradual rise, and a plastic shock wave. The profile for the quenched steel sample has another shape of the elastic precursor wave which consists of the initial velocity jump followed by a velocity rise with gradually decreasing slope. Such waveforms are typical for strain hardening materials.

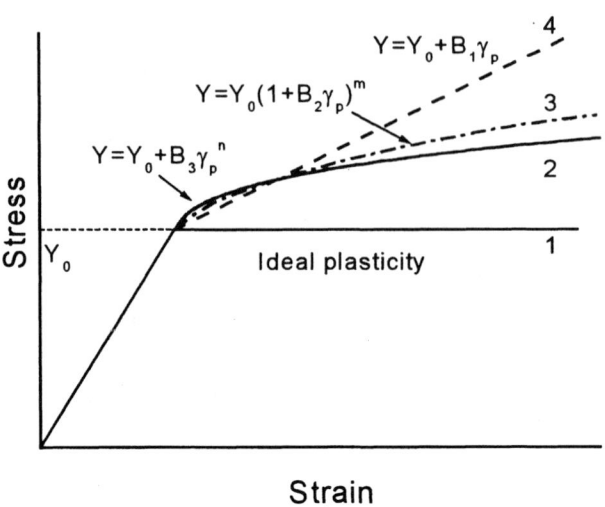

Figure 2.3. Strain hardening curves as they are described by different relationships.

In order to analyze hardening effects, let us assume that the strain hardening is described by the empirical relationship

$$\tau = \tau_0 + B\gamma_p^n. \qquad (2.15)$$

In this equation, $n < 1$ is a constant exponent and γ_p is related to the plastic part of the shear strain. For simplicity of the analysis, the dependence of the elastic moduli and sound speeds on stress is neglected.

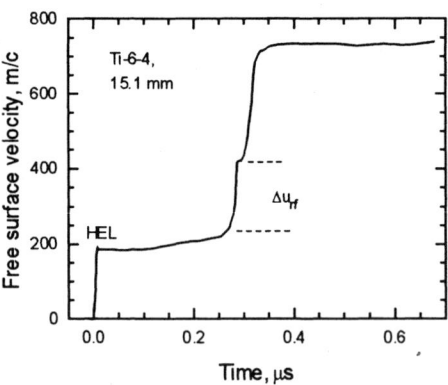

Figure 2.4. A free surface velocity history measured with a VISAR velocimeter for a 15.1-mm thick sample of Ti-6Al-4V alloy.

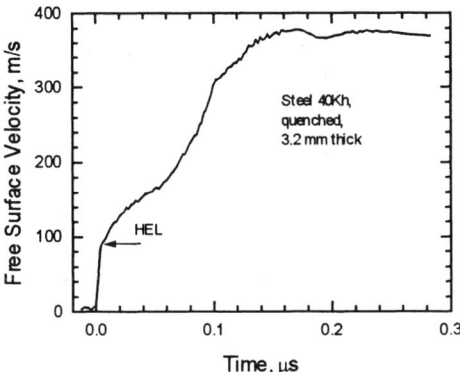

Figure 2.5. A free surface velocity history for a 3.2-mm thick sample of quenched steel 40 Kh.

It will be shown below that Eq. (2.15) provides a smooth transition from elastic to elastic–plastic deformation and, consequently, a continuous decrease, under uniaxial strain conditions, of the wave speed starting from c_ℓ. In this case, a part of the compression wave is a centered simple wave, which is described by a fan of characteristics immediately behind the front of the elastic precursor wave as shown in Fig. 2.6. For this part of the flow,

$$d\sigma = \rho_0 \, a_\sigma^2 \, d\varepsilon_x \qquad (2.16)$$

and

$$d\tau = \frac{3}{4}\left(1 - \frac{a_b^2}{a_\sigma^2}\right) d\sigma, \qquad (2.17)$$

where a_σ is the Lagrangian velocity of propagation of stress perturbations (Fowles and Williams, 1970), and $a_b = c_b V_0/V$ is the bulk sound speed in Lagrangian coordinates. In this case of uniaxial strain, $\varepsilon_y = \varepsilon_z = 0$, the increment of plastic strain is $d\gamma = d\varepsilon_x$, and Eqs. (2.7), (2.16), and (2.17) yield the plastic strain derivatives

$$\frac{d\gamma_p}{d\tau} = \frac{d\varepsilon_x}{d\sigma}\frac{d\sigma}{d\tau} - \frac{1}{G} = \frac{4}{3}\frac{1}{\rho_0(a_\sigma^2 - a_b^2)} - \frac{1}{G} \qquad (2.18)$$

and

$$\frac{d^2\gamma_p}{d\tau^2} = -\frac{8}{3}\frac{a_\sigma}{(a_\sigma^2 - a_b^2)^2}\frac{da_\sigma}{d\tau} = -\frac{32}{9}\frac{a_\sigma^3}{\rho_0(a_\sigma^2 - a_b^2)}\frac{da_\sigma}{d\sigma}. \qquad (2.19)$$

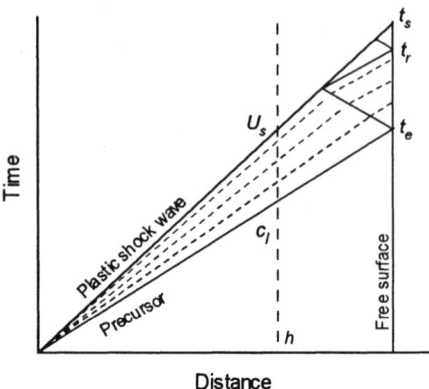

Figure 2.6. Characteristics of an elastic–plastic compression wave in a material with strain hardening.

Differentiating the strain hardening relationship (2.5) gives

$$\frac{d\gamma_p}{d\tau} = \frac{\gamma_p^{1-n}}{Bn} \tag{2.20}$$

and

$$\frac{d^2\gamma_p}{d\tau^2} = \frac{1-n}{B^2 n^2} \gamma_p^{1-2n}. \tag{2.21}$$

Equating the plastic strain derivative of Eq. (2.18) with that of Eq. (2.19) and of Eq. (2.20) with that of Eq. (2.21), we obtain

$$a_\sigma^2 - a_b^2 = \frac{4}{3} \frac{G}{\rho_0} \frac{Bn/G}{\gamma_p^{1-n} + (Bn/G)}, \tag{2.22}$$

and

$$\frac{da_\sigma}{d\sigma} = -\frac{9}{32} \frac{\rho_0}{B^2} \frac{(a_\sigma^2 - a_b^2)^3}{a_\sigma^3} \frac{1-n}{n^2} \gamma_p^{1-2n}. \tag{2.23}$$

Inspecting Eqs. (2.22) and (2.23) shows that, for $n \leq 1/2$, the phase speed, a_σ, gradually decreases with increasing plastic strain, γ_p, from the initial value $a_\sigma = c_\ell$ at $\gamma_p = 0$ to the limit $a_\sigma \to a_b$ as $\gamma_p \to \infty$. In the case that $1/2 < n < 1$, the decrease of phase speed passes through a downward jump at the HEL. If the compression wave between the elastic precursor front and the plastic shock wave is a centered simple wave, the phase velocity is

$$a_\sigma = \frac{h}{(h/c_\ell) + t(\sigma)}, \qquad (2.24)$$

where h is the Lagrangian coordinate of the point at which the stress history $\sigma(t)$ of the compression wave is analyzed, and $t(\sigma)$ is the time interval after passage of the elastic precursor front. Differentiating Eq. (2.24) yields

$$\frac{da_\sigma}{d\sigma} = \left(\frac{\partial a_\sigma}{\partial \sigma}\right)_h = -\frac{a_\sigma^2}{h \cdot (\partial \sigma/\partial t)_h}. \qquad (2.25)$$

It follows from Eqs. (2.23) and (2.25) that, if the hardening exponent $n < 1/2$, the time derivative of stress, $\partial\sigma/\partial t$, is infinite immediately behind the initial discontinuity of the elastic precursor wave (where $\gamma_p = 0$), and it decreases smoothly with the increasing plastic strain, as is shown for the quenched steel in Fig. 2.5. If $n > 1/2$, the time derivative $\partial\sigma/\partial t$ is equal to zero immediately behind the precursor front, and it increases smoothly with plastic deformation. If $n = 1/2$, the derivative $\partial\sigma/\partial t = \text{const.} \neq 0$ behind the elastic precursor front. In this case, the coefficient B of Eq. (2.15) is determined from the wave profile using the relationship

$$B^2 = -\frac{9}{16} \frac{(a_\sigma^2 - a_b^2)^3 \rho_0}{a_\sigma^3 (da_\sigma/d\sigma)}. \qquad (2.26)$$

For strain hardening materials, it is sometimes difficult to accurately determine the HEL point corresponding to the beginning of inelastic deformation in a measured wave profile. Probably it is more reasonable to characterize such materials by a yield stress at some fixed plastic strain. This parameter, which is analogous to the 0.2% yield strength usually reported for quasistatic measurements, may be relatively easy to evaluate from a measured wave profile.

2.3. Development of Elastic Precursor Waves in Relaxing Materials

Since the velocity of carriers of plastic deformation (dislocations) depends on the resolved stress and the number of carriers is limited, the flow stress increases with increasing strain rate. The dependence of flow stress on the strain rate is treated phenomenologically as a manifestation of the viscosity or stress relaxation in the material. The dynamics of deformation of relaxing media is described by a variety of models of elastic–viscous–plastic bodies. The simplest of these is the Maxwell model that involves, in tandem, an elastic element G and a viscous element η (Fig. 2.7). The total strain, γ, in this model is the sum of the elastic component, γ_e, and the viscous (plastic) component, γ_p

$$\gamma = \gamma_e + \gamma_p.$$

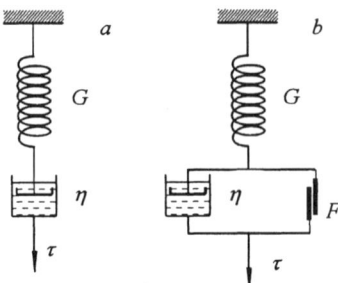

Figure 2.7. Schemes of the (a) Maxwell and (b) Schwedoff–Bingham rheological models.

The elastic strain component is related to the stress by Hooke's law, $\gamma_e = \tau/G$, and the viscous strain rate obeys Newton's law of viscosity, $d\gamma_p/dt = \tau/\eta$, where η is the viscosity coefficient. As a result, the equation for the total strain rate is

$$\frac{d\gamma}{dt} = \frac{d\gamma_e}{dt} + \frac{d\gamma_p}{dt} = \frac{1}{G}\frac{d\tau}{dt} + \frac{\tau}{\eta}. \tag{2.27}$$

When the load is applied instantaneously, the strain is at first localized in the elastic element. After that, viscous deformation accumulates and this process is accompanied by stress relaxation. At fixed total strain ($\gamma = \text{const.}$ or $d\gamma/dt = 0$) the stress relaxation is described by the equations

$$\frac{1}{\tau}\frac{d\tau}{dt} = -\frac{G}{\eta}$$

$$\tau = \tau_0 \exp\left(-\frac{G}{\eta}t\right). \tag{2.28}$$

The ratio of the viscosity coefficient to the shear modulus, η/G, within a given model, is defined as the relaxation time, a parameter that is frequently used in descriptions of the motion of viscoelastic media.

In the other limit case, that in which the strain rate is fixed, $d\gamma/dt = \text{const.}$, and the stress gradually increases to its limiting value, which is determined by the strain rate and the viscosity according to the equation

$$\tau = \eta\frac{d\gamma}{dt}\left[1 - \exp\left(-\frac{G}{\eta}t\right)\right]. \tag{2.29}$$

The viscoelastic, strain-rate-dependent character of deformations results in a specific form of shock-load pulses in solids. The stress relaxation exhibits itself in decay of the precursor wave, in the shape of its profile, and in an increased rise time of the plastic shock wave. Regions with high gradients of the parameters are followed by relaxation zones in which final states are reached asymptotically.

Although fluid-dynamic effects in relaxing media are rather complicated for an exhaustive analysis, some useful information on evolution the precursor front can be obtained in a relatively simple way. Accounting for the conservation laws, we may write the equations

$$\left.\frac{d\sigma}{dt}\right|_{\text{HEL}} = \dot{\sigma} - \rho_0 c_\ell \dot{u}$$

$$\left.\frac{du}{dt}\right|_{\text{HEL}} = \dot{u} + \rho_0 c_\ell \dot{V} \qquad (2.30)$$

for stress and particle velocity at the precursor front. In these equations, a dot over a function $f(h, t)$ indicates the partial time derivative: $\dot{f} = \partial f / \partial t$. From Eqs. (2.30), accounting for the Hugoniot relationship, $\sigma = \rho_0 c_\ell u$, we obtain

$$2\left.\frac{d\sigma}{dt}\right|_{\text{HEL}} = \dot{\sigma} + \rho_0^2 c_\ell^2 \dot{V}. \qquad (2.31)$$

Let relaxation of the shear stress, τ, be described by some function $F = G\dot{\gamma}_p$. Then

$$\dot{\tau} = -G\rho_0 \dot{V} - F$$

$$\dot{\sigma} = -\rho_0 E' \dot{V} - \frac{4}{3}F, \qquad (2.32)$$

and, after substitution of $\dot{\sigma}$ from Eq. (2.32) into Eq. (2.31), we get the known (Ahrens and Duvall, 1966) relationship,

$$\left.\frac{d\sigma}{dt}\right|_{\text{HEL}} = -\frac{2}{3}F, \qquad (2.33)$$

for the precursor decay within the assumption of linear compressibility of the material.

Equation (2.33) does not contain any flow parameters except the stress at the elastic front. Therefore, the precursor wave decays independently of whether the stress increases or decreases behind its front. Asay et al., 1972, analyzed an additional contribution of non-linear compressibility to the precursor decay. In some cases, the non-linearity may be important for more accurate quantitative interpretation of experimental data.

Some materials exhibit a spike-like shape of the front part of the elastic precursor wave as illustrated by the example shown in Fig. 2.8. This spike may form as a result of stress relaxation under some conditions.

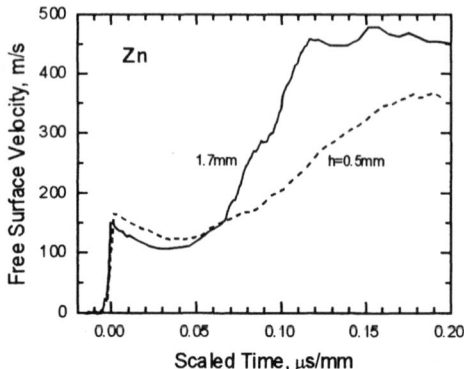

Figure 2.8. Free surface velocity profiles for zinc single crystal samples impacted upon the ($10\bar{1}0$) plane. Data by Razorenov et al., 1998. The normalized times scale used is t/h_s where h_s is the sample thickness.

Let us consider in more detail the process of establishing the shape of the precursor wave formed after collision of two plates of the same material. As a result of the symmetry of such an impact, the velocity of the impact surface is maintained constant, i.e., $\dot{u} = 0$. It follows from Eqs. (2.30) and (2.31) that, under these conditions,

$$\dot{\sigma} = \left.\frac{d\sigma}{dt}\right|_{\text{HEL}} = -\frac{2}{3}F, \quad \text{and} \quad \dot{V} = \frac{\dot{\sigma}}{\rho_0 E'}. \tag{2.34}$$

In order to make clear the direction of further change of the wave profile, let us consider the equation

$$\left.\frac{d\dot{\sigma}}{dt}\right|_{\text{HEL}} = \frac{\partial \dot{\sigma}}{\partial t} - \rho_0 c_\ell \frac{\partial \dot{u}}{\partial t} \tag{2.35}$$

for evolution of the slopes of the σ(t) and u(t) profiles right behind the precursor front.

In the case of symmetric impact under consideration, u = const. at the impact surface, $\partial u / \partial t = 0$ and $\partial \dot{u} / \partial t = 0$, so

$$\frac{\partial \dot{\sigma}}{\partial t} = -\frac{2}{3}\dot{F} \quad \text{and} \quad \left.\frac{d\dot{\sigma}}{dt}\right|_{\text{HEL}} = -\frac{2}{3}\dot{F}. \tag{2.36}$$

Thus, the initial slope of the spike on the elastic precursor front has to be maintained if the strain rate is maintained constant ($\dot{F} = 0$), has to increase in its absolute value if the deformation occurs with acceleration ($\dot{F} > 0$), and has to decrease in the case of decelerating plastic flow. It follows from Eq. (2.32) that, since in this case of $\dot{u} = 0$ the specific volume decreases behind the precursor front with a rate of

$$\dot{V} = \dot{\sigma}/(\rho_0 E'),$$

the shear stress decreases proportionally to the plastic strain rate in accord with the equation

$$\dot{\tau} = -F\frac{(E'+K)}{2E'}. \tag{2.37}$$

For an elastic–viscous material, the rate of inelastic deformation is proportional to the current value of the shear stress and does not depend on the prior history. For such a material, the stress spike on the elastic precursor front has to decrease, and eventually disappear, as it propagates. Since the shear stress may remain constant or increase only when

$$\dot{\sigma} \geq \frac{4}{3}\frac{K}{E'-K}F,$$

precursor decay occurs in these materials even when the stress is increasing behind its front.

The theory of dislocation dynamics describes the plastic strain rate as a function of the average dislocation velocity, $\bar{v}$, and the mobile dislocation density, N_m, using the well known Orowan equation (Gilman, 1969):

$$\dot{\gamma}_p = N_m b \bar{v}. \tag{2.38}$$

The dislocation velocity increases with increasing shear stress. The mobile dislocation density increases with plastic deformation due to processes of multiplication and heterogeneous nucleation (Gupta et al., 1975), and decreases due to their immobilization and annihilation. The growth of dislocation density may compensate for the effect of decreasing shear stress as the plastic strain rate decreases.

Even if the high-rate plastic deformation that occurs under shock-wave loading involves more complicated mechanisms, it seems reasonable to present the relaxation function as $F = F(\tau, N_m)$ and the acceleration of relaxation as

$$\dot{F} = \frac{\partial F}{\partial \tau}\dot{\tau} + \frac{\partial F}{\partial N_m}\dot{N}_m. \tag{2.39}$$

Thus, the initial slope of the spike at the elastic precursor front is maintained or increases in absolute value if

$$\frac{\partial F}{\partial N_m}\dot{N}_m \geq -\frac{\partial F}{\partial \tau}\dot{\tau} \quad \text{or} \quad \dot{N}_{m,i} \geq F(\tau_i, N_0)\frac{(E'+K)}{2E'}\left(\frac{\partial F}{\partial \tau} \bigg/ \frac{\partial F}{\partial N_m}\right)_i, \qquad (2.40)$$

where the subscript i signifies a value at the elastic wave front at the moment stress relaxation begins. Obviously, if the effect of multiplication $\partial F / \partial N_m$ exceeds the effect of decreasing shear stress, the spike steepness, $\dot{\sigma}$, will increase in absolute value until it approaches the equality in Eq. (2.40).

Figure 2.9 shows an example of simulation of a compression wave in a relaxing elastic–plastic material. The calculations have been done assuming that the multiplication rate increases proportionally to the density of carriers of plastic deformation. Increase in the rate of plastic deformation in this example results in growth of the spike steepness and in increasing the depth of the valley between the elastic and plastic waves. Although the wave, in general, is compressive, the stress relaxation occurs so rapidly that it causes rarefaction behind the elastic front. As a result of faster relaxation of stress, the threshold yield conditions will be reached earlier in the valley than at the precursor front. After this, the spike will continue to decay asymptotically to the quasistatic elastic limit. In other words, the elastic spike has a transient nature and may only appear during the process leading to establishment of the wave. Note also that the slope of the Rayleigh line describing the states in the plastic compression wave is smaller in the initial phase and corresponds to a subsonic ($< c_b$) propagation speed. Subsonic plastic waves were recorded simultaneously with a spike-like precursor wave in iron (Arnold, 1992), sapphire (Mashimo et al., 1988), and other materials. The observations of subsonic plastic waves were treated as evidence of softening of the material in the shock wave but, actually, that may not be entirely correct.

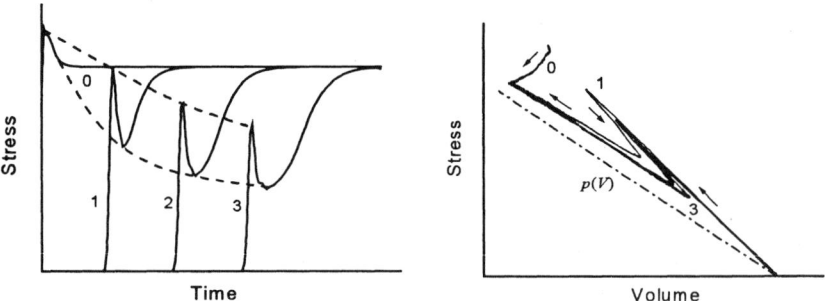

Figure 2.9. An example of a computer simulation of shock compression of a relaxing elastic–plastic material when the multiplication rate increases proportionally to the density of carriers of plastic deformation. The numbers indicate the distance from the impact surface, $p(V)$ is the bulk compression curve, and arrows show the direction of the change of state.

2.4. Interpreting Free Surface Velocity Histories

Most advanced diagnostics used to study shock compression of condensed matter are based on measuring the history of the free surface velocity of a shocked plate. The VISAR velocimeters provide very high resolution of time and space and high accuracy of the velocity data. However, interaction of an elastic–plastic compression wave with the plate surface creates a series of reflections that distort the recorded waveform. It is useful to understand the consequences of these reflections in order to correctly interpret the free surface velocity histories and, at least, to understand the limits to which the direct and indirect assumptions used for interpreting the data are satisfied.

Figure 2.10 shows a free surface velocity history obtained by computer simulation of shock compression of an elastic–plastic material. One can see that, being generally similar to the stress waveform, the free surface velocity history contains some additional peculiarities. Clearly seen is an intermediate step 2 in the plastic shock and one or more steps on the plateau behind it. The stress–particle velocity and time–distance diagrams in Figs. 2.11 and 2.12 will help us to understand details of the wave reflections.

As it has to be, the arrival of an elastic precursor wave at the stress-free rear surface of the plate sets it in motion with the velocity $u_{fs,1}$, that is essentially twice the particle velocity at the HEL: $u_{fs,1} = 2u_{p,HEL}$, and causes the appearance of a reflected elastic unloading wave. When the reflected unloading wave meets the plastic shock, a new elastic compression wave is formed in the unloaded layer. This can be interpreted as a reflection of the elastic wave at the plastic shock front. This reflection forms an intermediate step in the free surface velocity profile, designated point 2 in Fig. 2.10. The intermediate step is seen also in

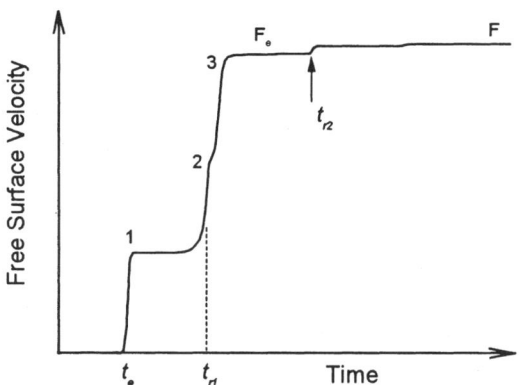

Figure 2.10. Simulated free surface velocity history of a shocked elastic–plastic plate.

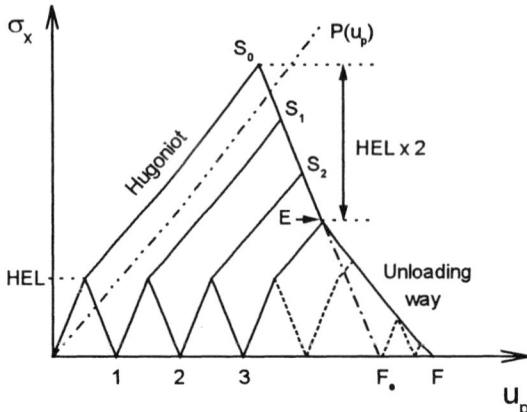

Figure 2.11. Stress–particle velocity diagram of interaction of an elastic–plastic compression wave with a plate surface. The numbers 1, 2, and 3 indicate free surface velocities at the points marked in Fig. 2.10. The symbols S_0, S_1, and S_2 indicate the initial shocked state and the states after interaction of the original elastic–plastic wave with reflected unloading waves. Points F and F_e correspond to those marked on Fig. 2.10.

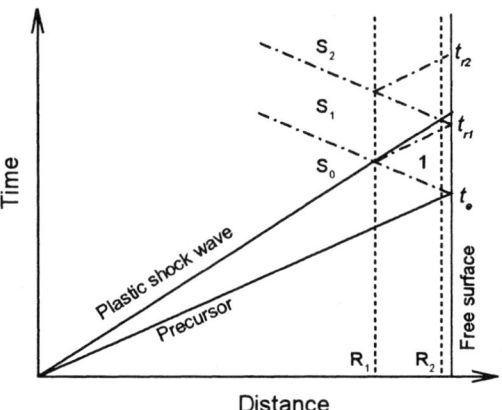

Figure 2.12. Time–distance diagram of the reflection of an elastic–plastic compression wave from a free surface. The symbols S_0, S_1, and S_2 correspond to states thus marked on Fig. 2.11.

the experimental wave profile shown in Fig. 2.4. Obviously, for an elastic–plastic material, the magnitude of the free surface velocity at this intermediate step should be double the value of the free surface velocity at the HEL: $u_{fs,2} = 2u_{fs,1}$. For real materials this magnitude is also affected by strain hardening effects and rate-dependent material properties.

Let the initial shock-compressed state correspond to the point S_0 on the original Hugoniot in Fig. 2.11. After interaction of the reflected elastic unloading wave with the original plastic shock wave, further propagation of the latter is governed by the Hugoniot $1-S_1$ shifted by $2u_{p,\mathrm{HEL}}$ relative to the original Hugoniot. It is essential that unloading of shock-compressed matter from point S_0 to S_1 occurs elastically. Next, reverberation of the elastic wave between the free surface and the plastic shock causes further attenuation from point S_1 to S_2 and, again, this occurs elastically because the shocked state S_1 on the shifted Hugoniot is above the hydrostat and corresponds to the yield condition. Reverberations of the elastic wave between the plate surface and the plastic front will stop only when the latter disappears.

As a result of this reverberation process, attenuation of the elastic–plastic compression wave occurs from the initial shocked state S_0 down to point F_e in Fig. 2.11, along the elastic unloading trajectory S_0-F_e. A thin surface layer deforms only elastically. Meanwhile, deeper layers (on the left side of line R_1 in Fig. 2.12) have been shocked much above the HEL and have to unload in an elastic–plastic way along the trajectory S_0-E-F. As a result of this unloading, the magnitude of the free surface velocity must be larger and correspond to the point F. The mismatch of average impedances of different sections of the shocked plate is resolved by means of new wave reverberations between the plate surface and the surfaces R_1 and R_2 (which is much closer to the surface than R_1), etc. (Fig. 2.12), of interactions of reflected elastic unloading waves with the plastic shock. This process explains the appearance of long steps on the top of the free surface velocity profile. Thus, some part of an elastic–plastic free surface velocity history always looks inclined even for rate-independent materials which do not produce this feature on waveforms inside a plate.

When a free surface velocity history $u_{\mathrm{fs}}(t)$ is analyzed instead a stress history $\sigma(t)$, the approximation $u_{\mathrm{fs}}(t) = 2u_p(t)$ is usually used. For the phase speeds a_σ which were used in Eqs. (2.16)–(2.26), this assumption gives

$$\frac{da_\sigma}{d\sigma} = -\frac{2a_\sigma}{\rho_0 h (\partial u_{\mathrm{fs}}/\partial t)}. \tag{2.41}$$

A more detailed analysis accounting for interaction between the incident compression wave and the reflected unloading gives

$$a_\sigma = c_\ell \frac{2h - c_\ell t(\sigma)}{2h + c_\ell t(\sigma)}$$

$$\frac{d\sigma}{du_{\mathrm{fs}}} = \rho_0 \frac{c_\ell a_\sigma (c_\ell + a_\sigma)}{c_\ell (c_\ell + c_\sigma) + 2c_\sigma^2} \tag{2.42}$$

$$\frac{da_\sigma}{d\sigma} = -\frac{c_\ell^2 + c_\ell a_\sigma + 2a_\sigma^2}{\rho_0 a_\sigma (2h + c_\ell t) \cdot \partial u_{\mathrm{fs}}/\partial t}.$$

Hysteresis of a path of elastic–plastic loading and subsequent unloading precludes estimating the particle velocity as one-half of the free surface velocity. For example, in the case of an ideal elastic–plastic material, the elastic part of unloading is twice that of the compression and this, as it is clear from Fig. 2.11, results in

$$u_{\text{fs,F}} = 2u_{\text{p,S}_0} - \frac{\sigma_{\text{HEL}}}{\rho}\left(\frac{1}{c_b} - \frac{1}{c_\ell}\right). \qquad (2.43)$$

For analyzing the free surface velocity history of a relaxing material, we have to account for the variations of $\dot{\sigma}$ and $\dot{u}$ behind the precursor front and for the precursor decay. At the time t_c when the precursor front arrives at the free surface having the coordinate h_{fs}, the velocity of the latter becomes equal to twice the particle velocity:

$$u_{\text{fs}}(t_c) = u(h_{\text{fs}}, t_c) + \frac{1}{\rho_0 c_\ell}\sigma(x_{\text{fs}}, t_c). \qquad (2.44)$$

Reflection of the precursor front from the free surface creates an unloading wave that propagates back into the sample with the longitudinal sound speed, c_ℓ, along a C_- characteristic line in the time–distance diagram shown in Fig. 2.13. During the time interval dt, the unloading front propagates back a distance $c_\ell dt$ from the sample surface. The unloading obviously stops the stress relaxation process so the states along the corresponding C_+ characteristic after intersection with the unloading wave front are described by the Riemann invariant line in the stress–particle-velocity diagram. When the characteristic comes to the sample surface at the time $t_c + 2dt$, the surface velocity becomes

$$u_{\text{fs}}(t_c + 2dt) = u(h_{\text{fs}} - c_\ell dt, t_c + dt) + \frac{1}{\rho_0 c_\ell}\sigma(h_{\text{fs}} - c_\ell dt, t_c + dt). \qquad (2.45)$$

The particle velocity and the stress appearing in right member of Eq. (2.45) may be found from the relationships

$$u(h_{\text{fs}} - c_\ell dt, t_c + dt) = u(h_{\text{fs}}, t_c) - \left.\frac{du}{dt}\right|_{\text{HEL}} dt + 2\dot{u}dt \qquad (2.46)$$

and

$$\sigma(h_{\text{fs}} - c_\ell dt, t_c + dt) = \sigma(h_{\text{fs}}, t) - \left.\frac{d\sigma}{dt}\right|_{\text{HEL}} dt + 2\dot{\sigma}dt. \qquad (2.47)$$

Substitution of $u(h_{\text{fs}} - c_\ell dt, t_c + dt)$ from Eq. (2.46) and $\sigma(h_{\text{fs}} - c_\ell dt, t_c + dt)$ from Eq. (2.47) into the relationship (2.45) gives us the value of the velocity increment during the time interval $2dt$:

$$u_{\text{fs}}(t_c + 2dt) - u_{\text{fs}}(t) = 2\dot{u}dt + \frac{2}{\rho_0 c_\ell}\dot{\sigma}dt - \left.\frac{du}{dt}\right|_{\text{HEL}} dt - \frac{1}{\rho_0 c_\ell}\left.\frac{d\sigma}{dt}\right|_{\text{HEL}} dt. \qquad (2.48)$$

2. Elastic–Plastic Response of Solids Under Shock-Wave Loading   49

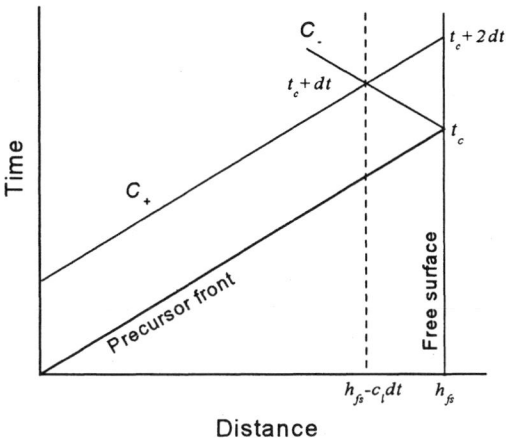

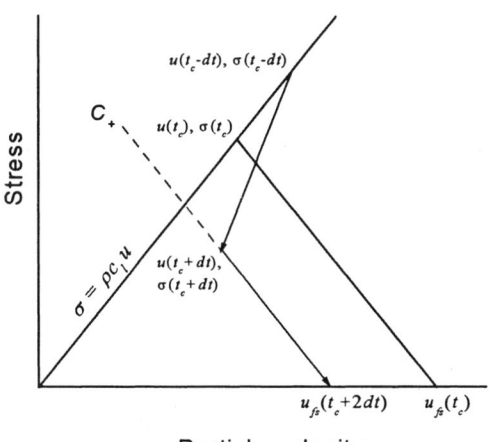

Figure 2.13. Time–distance and stress–particle-velocity diagrams explaining the slope of the free surface velocity profile behind the precursor front.

Dividing Eq. (2.48) by $2\,dt$ and using the Rankine–Hugoniot relationships for the elastic front, we obtain the relationships

$$\frac{\partial u_{\mathrm{fs}}}{\partial t} = \dot{u} + \frac{1}{\rho_0 c_\ell}\dot{\sigma} - \frac{1}{2}\left(\frac{du}{dt} + \frac{1}{\rho_0 c_\ell}\frac{d\sigma}{dt}\right)_{\mathrm{HEL}}$$

or (2.49)

$$\frac{\partial u_{\mathrm{fs}}}{\partial t} = \dot{u} + \frac{1}{\rho_0 c_\ell}\dot{\sigma} - \frac{1}{\rho_0 c_\ell}\frac{d\sigma}{dt}\bigg|_{\mathrm{HEL}}$$

for the initial slope of the free surface velocity profile. Finally, accounting for Eq. (2.30), we arrive at the expected but not obvious relationship

$$\frac{\partial u_{fs}}{\partial t} = 2\dot{u}, \qquad (2.50)$$

which means that the initial part of free surface velocity history is the doubled particle velocity profile.

2.5. Hugoniot Elastic Limits and Dynamic Yield Stresses of some Solid Materials

After development of methods for recording stress and particle velocity histories of waves in shock-loaded materials, Hugoniot elastic limits (HEL) have been measured for many metals and alloys, ceramics, minerals, high explosives, etc. Some results of these measurements for metals and alloys with 4–10-mm-thick samples are presented in Table 2.1.

For many metals and alloys the dynamic yield stress is 1.5–2 times the value obtained in quasistatic tests. This difference is less for high-strength materials. Figure 2.14 shows results of measurements by Taylor and Rice, 1963, of the Hugoniot elastic limit of Armco iron as a function of the wave propagation distance. The decay of the HEL is evidence of stress relaxation behind the precursor front.

The data shown in Fig. 2.14 are approximated by the relationship

$$\sigma_{HEL} = \bar{\sigma} - g\ln(x), \qquad (2.51)$$

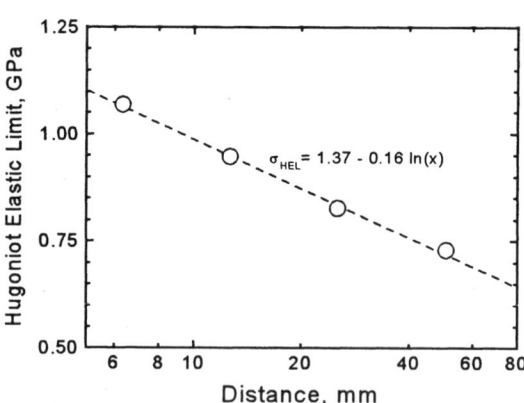

Figure 2.14. Decay of the elastic precursor wave in Armco iron. Data by Taylor and Rice, 1963.

Table 2.1. Properties of some metals and alloys.

Material	Density ρ_0, g/cm^3	Bulk sound velocity, c_b, km/s	Longitudinal sound velocity, c_ℓ, km/s	HEL, GPa	Dynamic yield strength, Y_d, GPa	Spall strength, GPa	Reference
Aluminum 2024	2.78	5.25	6.36	0.6	0.29	1.15	Morris, 1982
Aluminum AMg6M	2.61	5.3	6.4	0.38	0.18	0.8–1.1	Kanel et al., 1996
Magnesium Ma1	1.75	4.5	5.6	0.2	0.1	0.8	Kanel et al., 1996
Titanium Ti-6Al-4V	4.45	5.11	6.15	2.0	0.92	3.6–4.4	Kanel et al., 1996
Tungsten	19.27	4.02	5.22	4.0	2.4	0.7	Morris, 1982
Armco iron	7.8	4.65	5.97	0.95	0.56	1.4–1.7	Kanel et al., 1996
Steel 1018	7.86	4.63	5.92	1.4	0.82		Morris, 1982
KhVG doped tool steel	7.95	4.65	5.85	1.8			Kanel et al., 1996
Fe-Cr-Ni-Mo steel 35Kh3NM, 46–49 HRC	7.76		5.9	2.2	1.25	3.4–3.9	Gluzman et al., 1985
As received 40Kh chromium-doped structural steel, 17–19 HRC	7.82		5.66	1.7	0.83	2.3	Razorenov et al., 1997
Quenched 40Kh steel, 45–54 HRC	7.78		5.50	2.15	0.92	4.2	Razorenov et al., 1997

where $\bar{\sigma}$ and g are empirical constants. Using Eq. (2.33), we may obtain the relaxation function, $F = G\dot{\gamma}_p$, and the average plastic strain rate, $\dot{\gamma}_p$, from the empirical relationship (2.51):

$$F = -\frac{3}{2}c_\ell \frac{d\sigma}{dx}\bigg|_{\text{HEL}} = \frac{3}{2}c_\ell \frac{g}{x}$$

$$\dot{\gamma}_p = \frac{3}{2}\frac{g}{G}\frac{c_\ell}{x}.$$

The data shown in Fig. 2.14 correspond to plastic strain rate decreasing from ~3000 s^{-1} to ~300 s^{-1} with increasing propagation distance from 6 to 60 mm.

2.6. Structure of Plastic Waves

Although shock waves are often considered to be discontinuities of infinitesimally small thickness, the thickness of shock waves of moderate strength in solids is quite resolvable. As examples, Figs. 2.15 and 2.16 show results of high-resolution measurements of waveforms in plates of aluminum and titanium alloys. The thickness of plastic shock waves or the so-called "shock front rise time" is controlled by the stress relaxation time at the high strain rate within the shock wave—a quantity closely related to material viscosity.

Figure 2.17 presents results of measurements of strain rates within shock waves in aluminum and copper. Swegle and Grady, 1985, observed that the maximum strain rate $\dot{\varepsilon}_m$ in the plastic wave and the stress increment $\Delta\sigma$ in the plastic wave for different condensed materials are related by the equation

$$\dot{\varepsilon}_m = A(\Delta\sigma)^4, \tag{2.52}$$

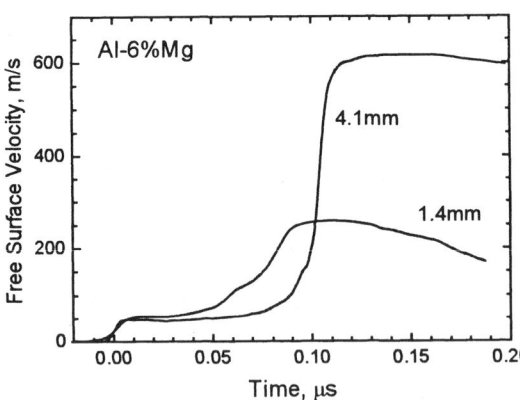

Figure 2.15. Free surface velocity histories of Al-6Mg plates shocked to different peak stresses.

where A is a material constant, and $\Delta\sigma$ is the stress increment from the precursor level to the peak stress of the steady wave.

Swegle and Grady, 1985, assumed that the rate dependence is associated only with the deviator stresses. If the wave is steady, evolution of deviatoric stresses during compression is completely determined by the relative position of the Hugoniot and the Rayleigh line, as illustrated in Fig. 2.18. The compression rate is established automatically in response to the corresponding shear stresses within the steady compression wave.

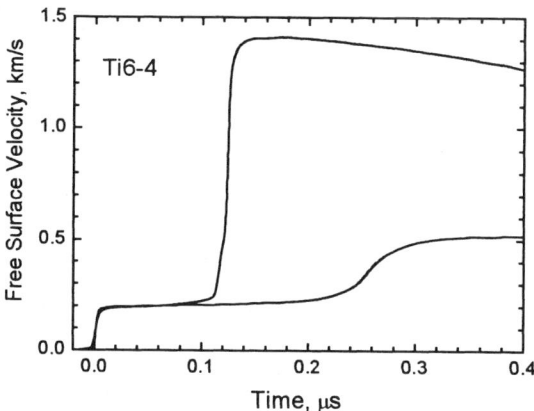

Figure 2.16. Free surface velocity histories for 10-mm-thick Ti-6Al-4V plates. The data at different peak stresses are from Razorenov et al., 2000.

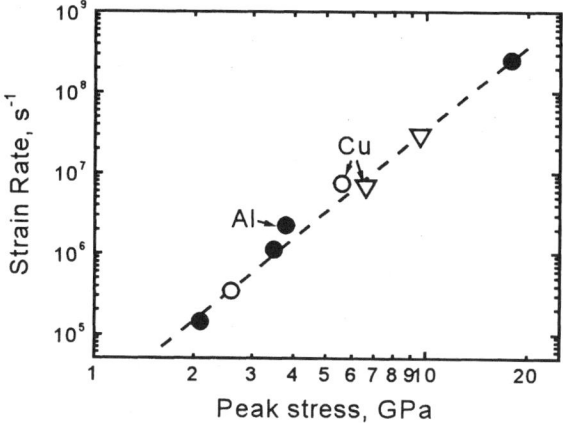

Figure 2.17. Results of measurements of strain rates within shock waves in aluminum (solid circles) and copper (open symbols) using different experimental methods (Johnson and Barker, 1969, Swegle and Grady, 1985, Razorenov et al., 1985, 1987, Kesler et al., (1994).

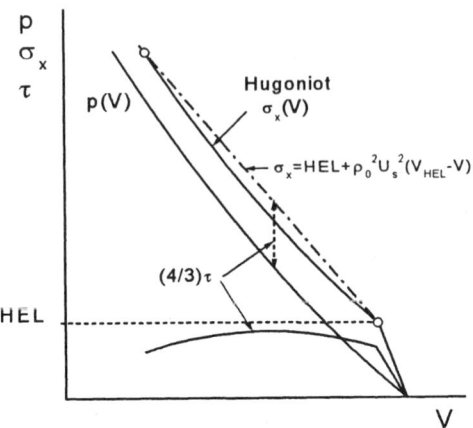

Figure 2.18. Evolution of stresses within elastic–plastic compression wave.

It can be shown that the maximum shear stress within a shock wave is approximately proportional to the squared value of the peak compressive stress. In accordance with Eqs. (2.27) and (2.32), at maximum shear stress ($\dot{\tau}=0$) the plastic strain rate $\dot{\gamma}_p$ is equal to the total strain rate $\dot{\varepsilon}_m$. In this case, the empirical relationship (2.52) is transformed to

$$\dot{\gamma}_p = A'\left(\tau - \frac{1}{2}Y\right)^2. \tag{2.53}$$

The value of the constant parameter A' is approximately 10^8 GPa^{-2}s^{-1} for aluminum, 5×10^7 GPa^{-2}s^{-1} for beryllium, 6×10^8 GPa^{-2}s^{-1} for bismuth, 3×10^8 GPa^{-2}s^{-1} for copper, and 3×10^7 GPa^{-2}s^{-1} for iron (Swegle and Grady, 1985). This observation emphasizes that the viscosity is not constant, but decreases as the strain rate increases.

The rise time of a plastic shock wave decreases rapidly with increasing peak stress and becomes less than the time resolution of modern waveform diagnostics. Analysis of light refraction at a shock front in transparent dielectric crystals (see Kormer, 1968) has shown that the thickness of strong shock waves may be less than 0.1 μm (corresponding to a rise time $\leq 10^{-11}$ s) that corresponds to the viscosity $\eta \leq 1$ to 10 Pa·s.

Dunn and Grady, 1986, have shown that, in contrast to the unique determination of stress-dependent relaxation, a strain-dependent viscosity (as a result, for example, of multiplication of mobile dislocations) may also satisfy the empirical relationship (2.52).

An additional factor that governs the rise time of a compression shock propagating in a multicomponent composite material is the acoustic interaction between the matrix and the filler. Wave reflections between components of the material occur as a result of their different dynamic impedances and produce dispersion of a load pulse in the material.

Comparison of the wave propagation characteristics of composite and viscous (i.e., stress relaxing) materials reveals that there are some basic similarities in their wave profiles. Composite and viscous materials appear to sustain steady waveforms. The smooth wave transitions in viscous materials, which are attributable to rate effects, appear to be similar to smooth wave transitions in the composite that are attributable to dispersive effects. In view of this similarity, the gross behavior of a composite can be described by an appropriate viscous material constitutive model, even though the response of the individual constituent materials of the composite may be strain-rate independent. Such a model was discussed by Barker, 1971. Following the Maxwellian model, he has introduced metastable (instantaneous) and equilibrium stress–strain loading paths and stress relaxation from the instantaneous to the equilibrium response. It is necessary to say that determination of the metastable loading path for a composite is a problem. Indeed, computations and available measurements of shock-wave processes in composite materials do not reveal any signs of the ultimate metastable states. Kanel et al., 1995, described the process with an empirical constitutive relationship without determination of metastable paths. Instead of metastable states they assumed that fast bulk compression produces superfluous pressure in the material that depends on the compression rate. Then, establishment of mechanical equilibrium occurs through relaxation of the superfluous pressure to zero. The total pressure was presented as a sum of an equilibrium component determined by the equation of state and a non-equilibrium component that depends on the strain rate and load history. An empirical constitutive relationship that describes the rate of production and relaxation of the superfluous pressure has been fit to observations.

2.7. The Stressed States of Shock-Compressed Solids.

There is a series of published works devoted to direct measurement of stressed states inside shock-compressed solids. The method is illustrated schematically in Fig. 2.19 (Dremin and Kanel, 1976, Kanel et al., 1977). A specimen for these measurements is assembled of several blocks in a configuration having thin plane gaps that are oriented perpendicular and parallel to the direction of shock compression. The gaps are filled with soft polymer films that serve as insulators for manganin pressure gauges placed in the gaps. The polymer films have very low yield stress, so the behavior of the filling is essentially that of a liquid. It is supposed that, after many wave reverberations, pressures in the gaps become equal to the normal stress components acting on the gap surfaces. Thus, a gauge placed in a gap on a plane perpendicular to the load direction records the history of the longitudinal stress $\sigma_x(t)$, whereas a gauge placed in a gap on a plane parallel to the load direction records the history of the transverse stress component, $\sigma_y(t)$. To improve the accuracy with which the stress difference $\sigma_x - \sigma_y$ is measured, the gauges may be connected into a Winston resistance bridge. In this case, both main stress components, $\sigma_x(t)$ and $\sigma_y(t)$, and their small difference, $\sigma_x - \sigma_y$, are directly recorded in one experiment.

This method was effectively applied to the study of ceramics and glasses with high yield stresses. Reasonable data on the shock response of aluminum have been obtained by Dremin and Kanel, 1976. However, results of such experiments with heavy metals like lead or tungsten show unreasonably low transverse stresses. The point is that the data should be consistent with the entire volume of experimental information on the equation of state and the unloading and recompression behavior of shock-compressed solids. Large stress differences in shock-compressed states result in a significant difference between the Hugoniot and the hydrostat of the material. A correspondingly large elastic part of the unloading wave should be observed in the case of a large stress difference. Disappearance of deviatoric stresses at melting should cause a kink on the Hugoniot. However, direct measurements of abnormally large stress differences were never confirmed by these other observations.

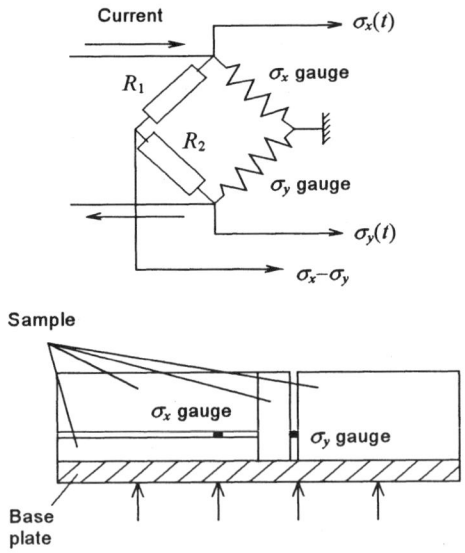

Figure 2.19. Schematics of measurements of normal stress components in shock-compressed solids. The upper part of the picture shows connection of the gauges into the measuring circuit for direct recording of the stress difference. R_1 and R_2 are the balancing resistors.

Direct measurements of stressed states in shock-compressed solids are still accompanied by several sources of uncertainty. Whereas the working conditions of the σ_x gauges are quite clear, there is a lack of knowledge about the flow in a gap on a plane oriented parallel to the direction of shock compression. Feng et al., 1997, performed two-dimensional computer simulations of the shock-wave processes in a silicon carbide ceramic specimen assembled as shown in Fig. 2.19. The results demonstrate that the mechanical states within and near the lateral gauge deviate significantly from uniaxial strain due to perturbations caused by the gap emplacement. The stress equilibration of the σ_y gauge takes much more time than equilibration of the σ_x gauge, so measurement of transverse stress components is possible only behind the compression front at later times. However, the established equilibrium gauge response is a good measure of the far-field, transverse stress in the shocked sample.

Al'tshuler et al., 1999, and Bakhrakh et al., 2001, simulated the flow in the gap coinciding with the shock direction in aluminum, copper, lead, uranium, and other metals. For all metals they found that the pressure within the gap, p_g, never approached the value of the stress σ_y in the shocked metal sample. The difference $p_g - \sigma_y$ may be of different sign: in the case of aluminum $p_g - \sigma_y$ was approximately 0.2 of the yield stress Y whereas, for copper, this difference was negative and reached $-0.8Y$. In other words, measurements of the stress difference in shocked metals may be accompanied by errors of -20% or $+80\%$ due solely to hydrodynamic effects. They also observed that the particle velocity of a low-density liquid-like material inside the gap exceeds the particle velocity of the surrounding shock-compressed metal and insulation is extruded from the gap. The most dramatic disagreement between p_g and σ_y took place for uranium, the heaviest metal in the series of simulations. At peak stresses of ~20–40 GPa, the pressure within the gap for the transverse gauge was lower by 30% than the compressive stress σ_x in the uranium, independently of its yield stress in the range $0 \le Y \le 4$ GPa. The interesting observation was that artificial doubling of the sound speeds in the model equation of state of uranium markedly decreased the difference $p_g - \sigma_y$.

Thus, computer simulations do not confirm the universal validity of presuppositions on which the method of measuring the principal normal stresses in shocked solids is based. Whereas the pressure in a gap on a plane oriented perpendicular to the shock direction does indeed become equal to the longitudinal stress component, the pressure in a gap on a plane oriented parallel to the shock direction may deviate by hundreds of per cent from the transverse stress in a shock-compressed specimen. The flow during shock compression of a solid body with an axial gap filled by another material is more complicated than was expected and requires additional analysis.

Since, as the computer simulations show, deviation of the pressure inside the axial gap from the transverse stress in the surrounding material is smaller when the sound speed in the latter is larger, the following picture may be intuitively guessed. If, at the same peak stress, the shock front velocity in the material filling the gap is larger than that in the specimen material, the compression wave inside the gap should be ahead of shock front in the specimen. In this situation, unloading propagates backward along the gap and decreases the pressure there. Thus, a rough intuitive criterion for choosing the insulating material for direct measurement of tangential stresses is: The Hugoniot of the insulating material must be such that, at the peak stress encountered, the shock front velocity inside the gap is less than that in the tested sample.

2.8. Unloading of Shock-Compressed Solids

Figure 2.20 shows stress histories in AD1 aluminum (analogous to the Al 1100 alloy in the AAS specification) recorded at various distances from the impact surface. The waveforms were recorded by Dremin et al., 1981 with insulated manganin pressure gauges. The peak stresses were much above the HEL. The stress profiles demonstrate elastic–plastic unloading behavior of the material. The elastic part clearly seen at the beginning of unloading propagates faster than the rest of the wave. However, in contrast to the example of the ideal elastic-plastic body shown in Fig. 2.1c, the elastic-to-plastic transition is smooth and diffusive. This makes it difficult to determine the yield stresses before and during unloading directly from the measured stress history.

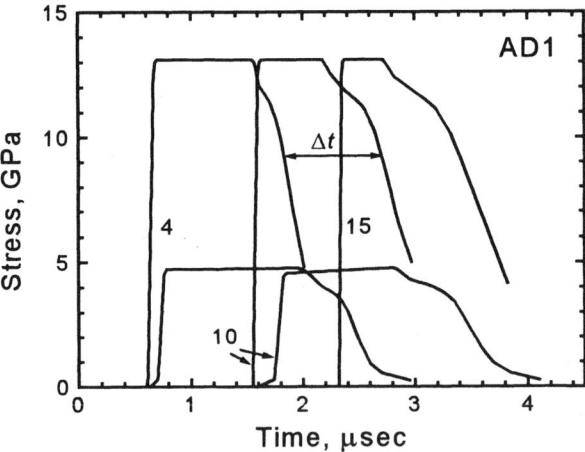

Figure 2.20. Evolution of compression pulses in AD1 aluminum produced by impacts of 5-mm-thick aluminum flyer plates at a velocity of 595 m/s, and 4-mm-thick plates at a velocity of 1500 m/s. The wave profiles were measured by Dremin et al., 1981 with manganin pressure gauges. The numbers indicate the distance (in mm) from the impact surface to the given gauge location.

Variations of the flow stress during unloading may be determined by means of computer simulations. A more direct way of interpreting the stress histories is based on recovery of the stress–strain diagram of the process of compression followed by unloading. Usually, calculation of the trajectory of the changing material state in the stress–strain plane is based on the simple wave approach. For simple waves, the longitudinal stress increments, $d\sigma$, the strain increments, $d\varepsilon_x = -dV/V_0$, the resolved shear stress, τ, and the pressure, p, are related by Eqs. (2.16) and (2.17) through the propagation velocity, a_σ, of the corresponding part of the wave at the longitudinal stress, σ_x, in Lagrangian coordinates. In fact, the stress–strain trajectories are calculated by integrating the dependencies, $a_\sigma(\sigma_x)$, of the phase velocity in the wave with respect to the stress.

If stress histories have been measured at two or more different propagation distances, determination of the function $a_\sigma(\sigma_x)$ is quite obvious. The wave process is illustrated schematically by the time–distance diagram in Fig. 2.21. In the simplest version, the stress histories $\sigma_x(t)$ are recorded simultaneously at two propagation distances in the sample. Since the distance, Δx, between the stress gauges is known and the time, Δt, between arrival of the shock and the rarefaction fronts at the first and second gauges is measured directly from the record, the shock front velocity, U_s, and Lagrangian velocity of the rarefaction front are easily determined just by dividing Δx by Δt. For determining the phase velocity of some intermediate part of the wave at a given level of stress, the time interval Δt is measured at this stress level, as shown in Fig. 2.20. If the velocity of shock waves in the impactor and the sample and the flyer thickness are known precisely, a single $\sigma_x(t)$ history may be sufficient for determination of the sound speed.

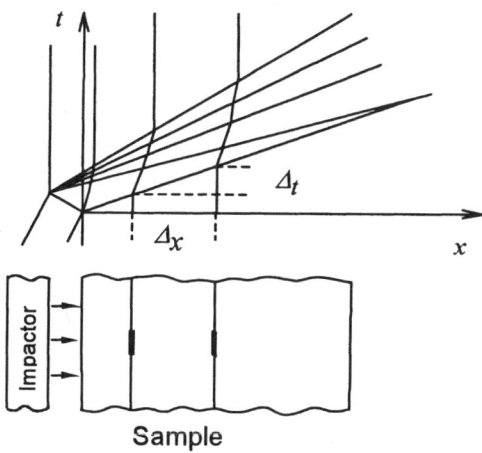

Figure 2.21. Schematics of measurements of the rarefaction wave velocity.

As an example, Fig. 2.22. shows variations of Lagrangian phase speeds, $a_\sigma(\sigma_x)$, of unloading waves measured in experiments with aluminum and the estimated dependence of the longitudinal and bulk sound speeds on the pressure.

An estimate of the bulk sound velocity at elevated pressures was derived using the Hugoniot of the material. In the pressure–particle-velocity plane, the release isentrope of many solids deviates from the Hugoniot by not more than 3% for peak pressures up to at least 50 GPa. If we assume that the Hugoniot and the rarefaction isentrope coincide on the p–u plane, we find that

$$\left(\frac{dp}{du_p}\right)_S = \pm\rho_0 a = \pm\left(\frac{dp}{du_p}\right)_H = \pm\rho_0(c_0 + 2su_p), \qquad (2.54)$$

where c_0 and s are the coefficients of a linear relationship between the shock front velocity, U_s, and the particle velocity, u_p, and the derivatives $(dp/du_p)_S$ and $(dp/du_p)_H$ are evaluated along the isentrope and the Hugoniot, respectively. The Lagrangian sound velocity in a shock-compressed material can now be estimated using Eq. (2.54) in combination with the Hugoniot equation, $U_s = c_0 + su_p$, and the momentum conservation law

$$a = c_0 + 2su = \sqrt{c_0^2 + (4sp/\rho_0)}. \qquad (2.55)$$

The estimate of the longitudinal sound velocity, $c_\ell(\sigma_x)$, shown in Fig. 2.22 was made assuming Poisson's ratio to be constant in Eq. (2.13). In this case

$$\frac{a_\ell(p)}{a_b(p)} = \frac{c_\ell(0)}{c_b(0)} = \sqrt{\frac{3(1-v)}{1+v}}. \qquad (2.56)$$

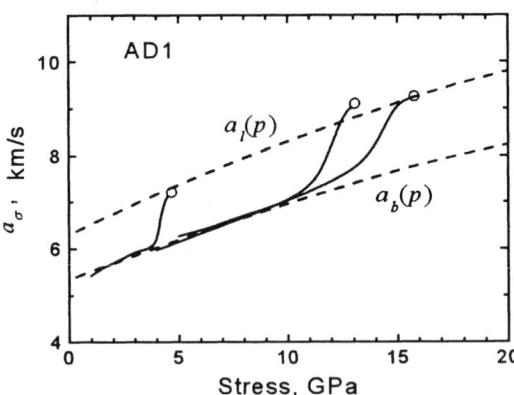

Figure 2.22. Lagrangian phase speeds of unloading waves in AD1 aluminum (solid lines). Points indicate the states behind shock wave. Dashed lines show estimated longitudinal, $a_\ell(\sigma_x)$, and bulk, $a_b(\sigma_x)$, sound speeds.

Figure 2.22 shows that the phase velocities in unloading decrease monotonically from the velocity of longitudinal elastic waves to values corresponding to the bulk compressibility. This means that the unloading wave has a small leading elastic part followed by a purely plastic wave. Using the known dependence of the phase speed on the stress, $a_\sigma(\sigma_x)$, the strain, ε_x, and shear stress, τ, at given σ_x are calculated by integrating Eqs. (2.16) and (2.17):

$$\varepsilon_x(\sigma_x) = \varepsilon_H + \frac{1}{\rho_0} \int_{\sigma_H}^{\sigma_x} \frac{d\sigma_x}{a_\sigma^2(\sigma_x)} \tag{2.57}$$

$$\tau(\sigma_x) = \tau_H + \frac{3}{4} \int_{\sigma_H}^{\sigma_x} \left(1 + \frac{a_b^2}{a_\sigma^2}\right) d\sigma_x, \tag{2.58}$$

where the subscript H designates the shock-compressed state on the Hugoniot of the material.

Figure 2.23 presents the results of calculations of maximum variations of shear stresses in aluminum and aluminum alloys during unloading from shock-compressed states.

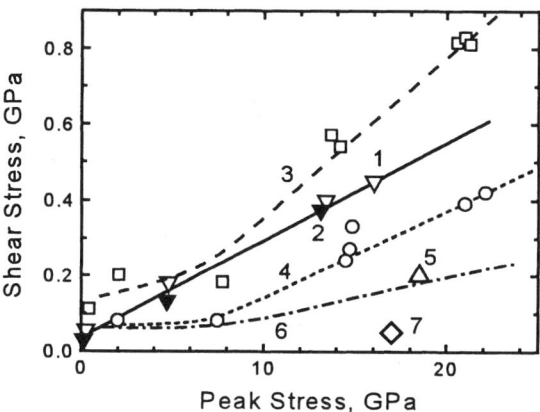

Figure 2.23. Curves 1, 2, and 3 are the results of measurement of the maximum variation (absolute value) of shear stresses during unloading from shock-compressed states (data 1 and 2 by Dremin and Kanel, 1976, and Dremin et al., 1981, for D16 alloy and aluminum AD1, respectively; data 3 by Asay and Chhabildas, 1981, for 6061-T6 aluminum). 4 and 5 are the stress increments in reshock (data 4 by Asay and Chhabildas, 1981, and point 5 by Dremin and Kanel, 1976). Curve 6 is the difference between data 3 and 4; it represents an estimate of the stressed state of shock compressed aluminum. Point 7 is the result of direct measurements of the difference of principal stresses in shock-compressed aluminum by Dremin and Kanel, 1976.

Figures 2.24 and 2.25 show the evolution of compression pulses in steel and Armco iron plates and stress–strain diagrams of the compression–release cycles recovered from these and similar data. Although there are no obvious elastic precursors in the release waves, it is evident that the deformation of the stronger steel is associated with the higher deviatoric stresses over the entire cycle of

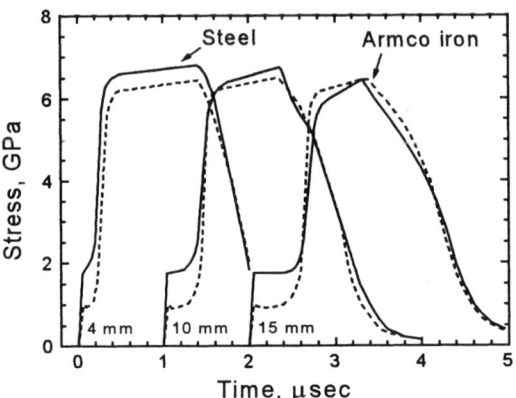

Figure 2.24. Evolution of a compression pulse in steel KhVG and in Armco iron. The samples were impacted by 5-mm-thick aluminum flyer plates at a velocity of 590 m/s. The numbers indicate distance from the impact surface.

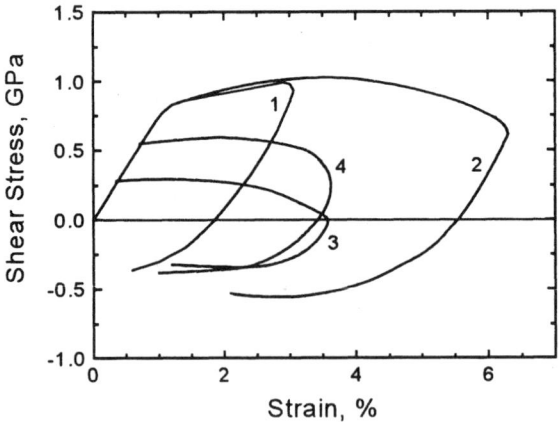

Figure 2.25. Stress–strain diagrams of compression–release cycles for steel 35Kh3NM at two peak stresses (curves 1, and 2), Armco iron (curve 3), and steel KhVG (4). Data by Gluzman et al., 1985.

shock compression and release. The stress–strain diagrams clearly show a strong Bauschinger effect during unloading from shock-compressed states. It is interesting to note that the analysis of wave profiles for iron shows that its shock-compressed state is almost hydrostatic. For harder steels, there is also a trend toward lower deviatoric stresses behind the plastic shock wave.

Figure 2.26 presents free surface velocity histories recorded during shock compression and unloading of metal sample plates. The wave profiles exhibit an obvious hysteresis: After unloading, the free surface velocity does not return to zero but maintains some residual value. The hysteresis of the free surface velocity is a consequence of the hysteresis of elastic–plastic deformation. Because the longitudinal compressibilities in the elastic and plastic deformation ranges are different, the residual particle velocity after elastic–plastic compression and unloading inside the plate is

$$u_{\text{pl}} = \frac{\sigma_{\text{HEL}}}{\rho_0} \frac{c_\ell - c_b}{c_\ell c_b}. \tag{2.59}$$

After reflection of the load pulse from the free surface, the particle under consideration goes through a tensile elastic–plastic cycle, so that its velocity becomes

$$u_{\text{p2}} = u_{\text{pl}} + 2\frac{\sigma_{\text{HEL}}}{\rho_0}\frac{c_\ell - c_b}{c_\ell c_b} = 3\frac{\sigma_{\text{HEL}}}{\rho_0}\frac{c_\ell - c_b}{c_\ell c_b}. \tag{2.60}$$

The residual free surface velocity at the end of the reflection process must be equal to the particle velocity in the interior of the plate, i.e., u_{p2}. Thus, measurement of the residual free surface velocity is a way to evaluate the yield stress of the material immediately after shock loading.

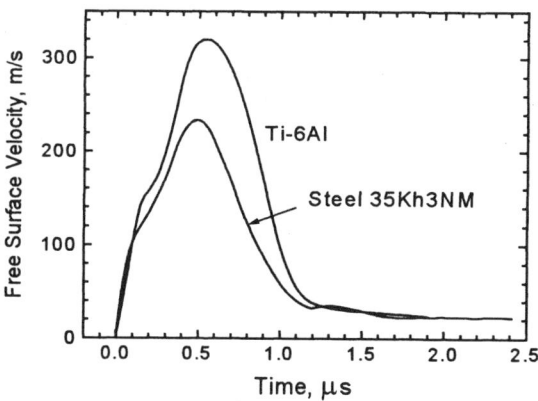

Figure 2.26. The free surface velocity histories of samples of Ti-6Al-4V alloy and 35Kh3NM steel impacted below the spall threshold. Measurements with capacitor gauges. Data by Kanel and Petrova, 1981, and Gluzman et al., 1985, respectively.

Figure 2.27 shows an experimental free surface velocity history and results of computer simulations using the model of an ideal elastic–plastic body. In this case, the compression portion of the load pulse looks like two steep steps with a plateau between them whereas the measured velocity history exhibits a rise right behind the elastic precursor front and a plastic wave with a measurable rise time. It has been shown earlier that strain hardening leads to the stress and particle velocity rise behind the elastic precursor front. The calculations give a large amplitude of the elastic unloading wave, a result that does not agree with the experimental data: Whereas the unloading front arrives at the sample surface at the same time in the calculated and measured free surface velocity profiles, the experimental velocity history shows a slower unloading than the calculated profiles. We may consider this discrepancy as evidence of a strong Bauschinger effect or some softening of the material. The experimental velocity profile during unloading is situated between the free-surface velocity histories for elastic–plastic and liquid-like bodies. It seems that some softening of the material occurs as a result of high-rate deformation in the plastic shock wave.

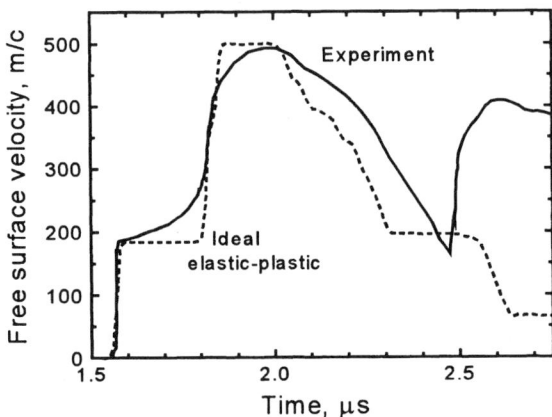

Figure 2.27. Results of computer simulations of the impact loading of a Ti-6Al-4V sample using a model of an ideal elastic–plastic body in comparison with experimental data. The spall strength was assumed to be above the tensile stresses generated under these load conditions.

2.9. Sound Speeds in Shock-Compressed Solids

For analyzing shock waveforms and for calculation of shock-wave phenomena, one needs to know the bulk and longitudinal sound speeds or the corresponding elastic moduli at elevated pressures and temperatures. The bulk sound speed is completely determined by the thermophysical equation of state (EOS) of the material. If the EOS is not known, simple estimates may be good enough in many cases. Some such estimates are discussed below.

At moderate shock pressures, when the Hugoniot deviates from the isentrope by only a small amount, the quasi-acoustic approach for treating the shock-wave speed (Landau and Lifshitz, 1959) is satisfactory. According to this approach, the speed of the shock wave is the average of the sound speed ahead of the shock discontinuity, c_0, and the Lagrangian sound speed behind it, a:

$$U = \tfrac{1}{2}(c_0 + a) = \tfrac{1}{2}(c_0 + c_0 + 2su). \tag{2.61}$$

Hence, we have an estimation procedure for the Lagrangian sound speed, Eq. (2.55), that is consistent with the quasi-acoustic approach for shock-wave computations. For these estimates we need to know only the Hugoniot of the material. Figure 2.28 compares the bulk sound speeds calculated from Eqs. (2.55) or (2.61) with experimental data by Al'tshuler et al., 1960, for aluminum, copper, iron, and lead. Reasonable agreement between measurements and calculations confirms the validity of the quasi-acoustic approximation.

A natural approach for estimation of the longitudinal sound speed is the assumption that Poisson's ratio, ν, is constant. In this case the ratio c_ℓ/c_b is constant and we arrive at the relationship (2.56). Figure 2.29 compares such estimates with experimental data for aluminum. In the experiments, the unloading front velocities were measured as a function of shock pressure. These data were identified as the longitudinal sound speeds in shock-compressed matter. The bulk sound speeds were found from propagation velocities of intermediate parts of the unloading waves. The calculated longitudinal sound velocities are in satisfactory agreement with experimental data for shock pressures of up to 125 GPa. From 125 GPa onward, the rarefaction front velocity is found (McQueen et al.,

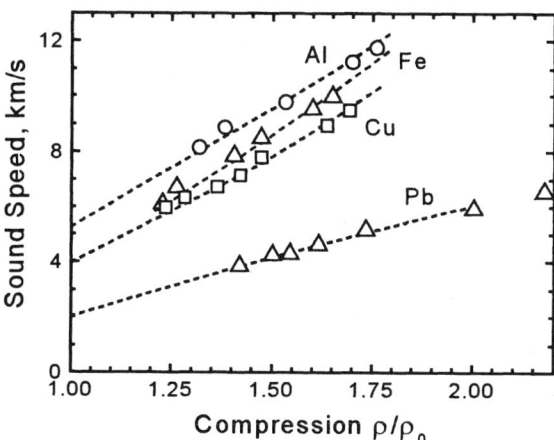

Figure 2.28. Measured bulk sound speeds in shock-compressed metals (Al'tshuler et al., 1960) and those estimated using Eq. (2.55) (Dremin and Kanel, 1970).

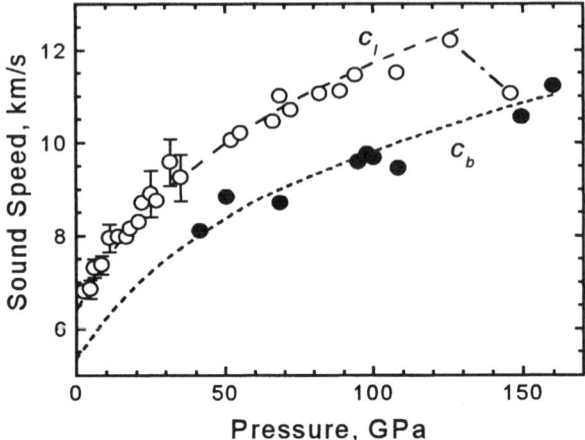

Figure 2.29. Measured longitudinal and bulk sound speeds in shock-compressed aluminum (Kusubov and van Thiel, 1969, Vorobiev et al., 1974, McQueen et al., 1984) and those estimated using Eqs. (2.55) and (2.56).

1984) to decrease and become equal to the bulk sound velocity at ~150 GPa. The reason for this loss of shear rigidity is melting as a result of the temperature increase in the shock-compressed material.

The quasi-acoustic approach coupled with the assumption of a constant Poisson's ratio is in reasonable agreement with known experimental data for aluminum, copper, tantalum, and titanium. For these metals, the discrepancy between estimated and measured sound speeds does not exceed 5% at shock pressures up to 100 GPa. For iron, steel, and other solids whose crystal structure is changed under compression, the quasi-acoustic approach is not applicable and Poisson's ratio is not maintained constant. For such materials the estimates described above may be applied only for the pressure range below the phase transition.

Guinan and Steinberg, 1974, developed another approach for estimation of the shear modulus at elevated pressures and temperatures. They derived an equation to estimate the temperature and pressure derivative of the isotropic polycrystalline shear modulus and have found that estimated values generally compare well with experimental data. In their paper, extensive information on elastic moduli and their derivatives has been tabulated for 65 elements. Being linear, this approach is useful only for relatively low shock pressures. On the other hand, very often the data on the influence of temperature on the elastic properties of metals or alloys are needed but are not available. In such cases the work by Guinan and Steinberg provides a way to solve the problem.

2.10. Behavior of Rubber Under Shock Compression

Elastomers have the ability to undergo large reversible deformations and exhibit a pronounced frequency dependence of the elastic moduli. Kalymykov et al., 1990, and Kanel et al., 1994, presented results of measurements of the Hugoniot and sound speeds for a rubber.

Figure 2.30 shows results of measuring the shock wave velocities and Lagrangian velocities of unloading waves in rubber as a function of pressure. The diagram also presents an estimate of the bulk sound velocity, $a_b(p)$, for shock-compressed rubber made using the quasi-acoustic approximation, and an upper estimate of the sound velocity. The latter has been obtained by differentiation of the Hugoniot in $p-V$ coordinates to obtain

$$a_H = (\rho/\rho_0)\sqrt{(dp/d\rho)_H},$$

whereas the quasi-acoustic approach is based on the assumption of coincidence of the Hugoniot and the unloading isentrope in the $p-u_p$ coordinates. In fact, the upper estimate of the bulk sound speed, a_H, corresponds to a zero coefficient of thermal expansion.

It is certainly seen from the figure that the rarefaction wave front velocity exceeds the upper estimate of the bulk sound speed. As the unloading proceeds, the measured phase velocities come close to the curve $a_b(p)$. It is natural to

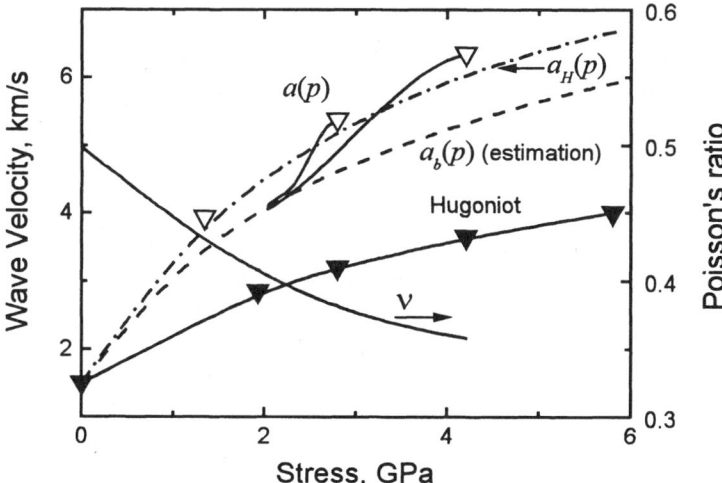

Figure 2.30. Dependence of the shock wave velocity, the estimated and measured Lagrangian sound velocities, and Poisson's ratio upon the pressure for rubber. The points show the velocities of wave fronts. The solid line originated from the light points corresponding to the unloading processes.

connect such behavior of the rarefaction wave velocity with an elastic–plastic response of the shock-compressed rubber. If this assumption is true, the rarefaction front is an elastic wave and Poisson's ratio can be calculated from the relation between the rarefaction front velocity and the bulk sound speed at the same pressure:

$$v = \frac{3(c_b/c_\ell)^2 - 1}{3(c_b/c_\ell)^2 + 1}.$$

Values of Poisson's ratio calculated in this way are presented in Fig. 2.30, which shows that Poisson's ratio decreases from the value of about 0.5 typical for elastomers down to 0.36 at a pressure of 4.2 GPa. This latter value is typical for solids. The low value of Poisson's ratio can be treated in terms of solidification of rubber under the high pressure of shock-wave compression. Earlier, Weaver and Peterson, 1969, observed a glass transition of rubber under static compression at room temperature. Within the framework of the experimental data obtained, it is impossible to exactly identify the mechanism of glassing. Nevertheless, the capability for elastomers to glass in shock waves is demonstrated quite distinctly.

2.11. Behavior of Metals Under Stepwise and Repetitive Shock Compression

It is usually recognized that the mechanical behavior of metals and alloys is well understood and is well described by modern advanced phenomenological and dislocation models. Nevertheless, there is a series of observations that indicates that metals experience specific changes in their properties when subjected to the high rate of strain in shock waves. Let us consider examples.

Figure 2.31 shows stress histories recorded in aluminum plates subjected to step-like shock compression. The elastic–plastic model suggests that the second, compression wave should be purely plastic, whereas experiment shows that the velocity of the second wavefront is equal to the longitudinal sound speed, c_ℓ. In contrast to the prediction of the purely elastic–plastic model, the experimental evidence is that the initial unloading or reloading from the shocked state is elastic in both cases. The results of evaluations of stressed states in second compression waves in aluminum and copper are presented in Fig. 2.32 in terms of deviations of stresses from the corresponding Hugoniots. For both materials, the deviatoric stresses increase in absolute value when the material is subjected to additional compression. Furthermore, the difference between increments in the deviatoric stress in compression and rarefaction waves is less than the magnitude of the increments, i.e., the stressed state of material immediately behind the shock front is almost isotropic.

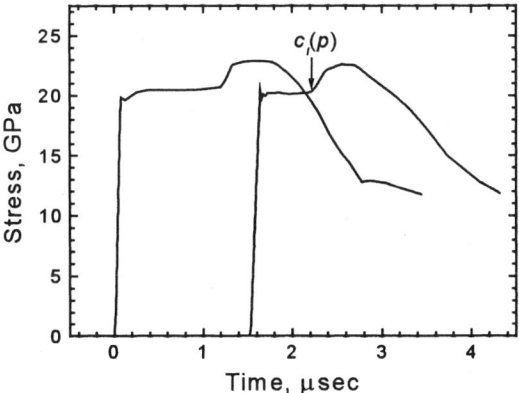

Figure 2.31. The stress histories at two propagation distances in a plate of D16 aluminum alloy (comparable to 2024 aluminum) subjected to step-like shock compression. The measurements were carried out with manganin pressure gauges by Dremin and Kanel, 1976.

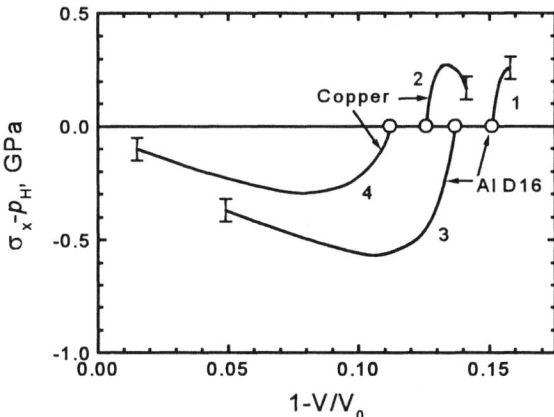

Figure 2.32. Deviations of stresses from the Hugoniots in second compression waves (curves 1 and 2) and in unloading waves (curves 3 and 4). Data by Dremin and Kanel, 1976.

The comparison of increments of deviatoric stresses produced by reshock and release from shock-compressed states has been suggested by Dremin and Kanel, 1976, Asay and Lipkin, 1978, and Asay and Chhabildas, 1981, as a self-consistent way to estimate the stressed states of shock-compressed solids. It is assumed that the difference of principal stresses in the shock-compressed state will be equal to the value of the yield strength in this state. Some such estimates are presented in Fig. 2.23.

On the other hand, the observations may be interpreted in terms of a transition from dominance of rate-independent yielding in unshocked material to dominance of stress relaxation processes in the material behind the shock wave.

Figure 2.33 shows stress histories in Armco iron and steel samples loaded successively by two compression pulses (Kanel et al., 1979). Measurements were done with manganin pressure gauges. The stress history exhibits a change in the character, and a drop in the amplitude, of the elastic compression precursor in the second load pulse. Similar behavior was observed for titanium (Kanel et al., 1979) and molybdenum (Furnish et al., 1992). It looks, indeed, as though materials become softer or undergo a transition from elastic–viscous–plastic to elastic–viscous response after shock compression.

Finally, Fig. 2.34 presents the free surface velocity history of a monocrystalline aluminum plate through which a shock-compression wave, a release wave, and a short tensile pulse passed in sequence. The arrangement of such experiments has been described by Kanel et al., 1988. Since the tensile pulse propagated in an elastic–plastic body after rarefaction to zero stress, the tensile part of this pulse should be purely plastic. However, the measurements demonstrate arrival of its front at the time corresponding to that expected for an elastic wave.

From the whole volume of observations we may conclude that shock waves appear to have a specific effect on materials leading to relatively high mobility of the crystal lattice and permitting fast relaxation of shear stress. Zaretsky et al.,

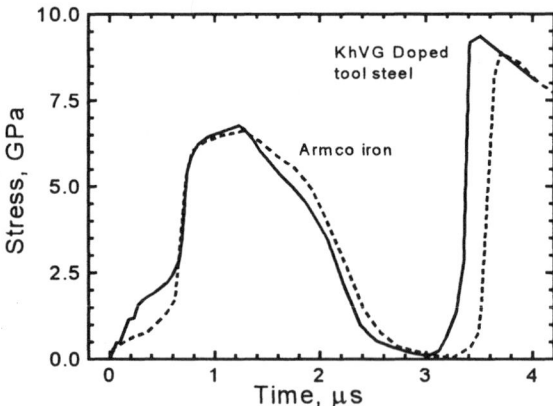

Figure 2.33. Stress profiles in Armco iron and steel samples loaded by two successive compression pulses. Measurements were done with manganin gauges embedded in the samples at a distance of 15 mm from the impact surface. Low time resolution at small stresses in the vicinity of the Hugoniot elastic limit is a result of the gauge insulation.

2. Elastic–Plastic Response of Solids Under Shock-Wave Loading 71

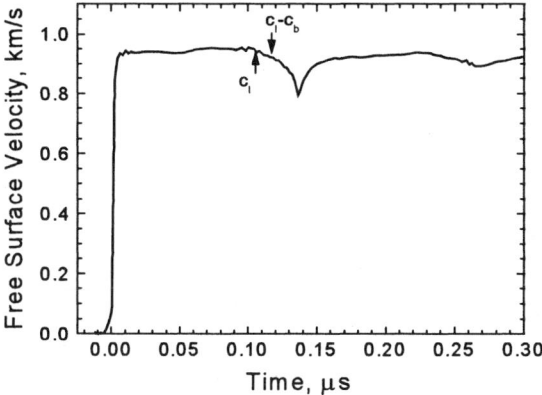

Figure 2.34. A free surface velocity history during wave reverberation in a 0.4-mm-thick Al single crystal. Arrows show pre-estimated times of arrival of elastic and plastic waves.

1991, and Zaretsky, 1992, have found evidence of formation of stacking faults in shock waves. These two-dimensional defects are a breach of the atomic layer stacking sequence. During plastic flow, the stacking faults arise as a result of a splitting of complete dislocations in the (111) planes into two partial dislocations in the same planes. Simulations of the shock-induced plasticity by molecular-dynamic methods (Mogilevsky and Mynkin, 1978, Holian, 1995) confirm the generation of partial dislocations.

At the maximum concentration of stacking faults, the fcc lattice transforms into an hcp lattice, which is not stable. Estimates (Zaretsky, 1992) show that all, or almost all, of the plastic strain has to be produced by the partial dislocations to provide the observed anomalous shift of the diffraction peaks. As a result, the stacking fault concentration reaches few percent in shock waves of several GPa peak stress. The large concentration of stacking faults means that a remarkable amount of the metastable phase is produced during shock compression. Thus, a possible basis for the physical description of the observed time-dependent shear stress response is that the lattice becomes unstable and, due to that, fast stress relaxation becomes possible during the disappearance of the metastable phase.

2.12. On Constitutive Relationships and Models of High-Rate Yielding

The main goal of studies of the behavior of materials under shock-wave loading is to provide a basis for predicting effects of high-velocity impacts, explosions, pulsed laser irradiation, etc. In such calculations, the behavior of the material is represented by the equation of state and constitutive relations that describe yielding of the material. Because of the complexity of physical processes of deformation of real materials, empirical criteria are developed to meet the practical requirements of accurate prediction of yielding and simplicity of use. Since the time when high-resolution methods for monitoring stress and particle velocity histories were developed, numerous investigations of mechanical properties of different classes materials have been conducted. During the same time numerous phenomenological and micromechanical models were developed for describing material behavior under high strain rate and shock-wave loading conditions. Usually not one, but many empirical constitutive relationships can be constructed to fit experimental data. Since different mechanisms of plastic deformation may operate depending on the strain rate, it seems that different material models and corresponding constitutive relationships should be used at different levels of stress. Some examples are considered below.

The basic mechanisms of plastic deformation of crystalline solids involve the motion of microscopic dislocations. There is an extensive literature devoted to dislocation dynamics, published mainly in the 1970s. The specifics of dislocation dynamics at extremely high strain rates in shock waves were summarized by Mogilevsky, 1983. In contrast to low and moderate strain rates, shock-wave loads are accompanied by the activation of additional slip planes, a significant reduction in the dislocation cell size (Arvidson and Eriksson, 1973), and a greater contribution of twinning to deformation, even in those materials that, under normal conditions, are deformed without twinning (see, for example, Smith, 1958). Higher dislocation, point-defect, and stacking-fault densities (as compared with quasi-static deformation) give rise to hardening as a result of the shock-wave treatment. The residual properties of metals were found to depend on the duration of the load pulse (Murr et al., 1978).

The complexity of internal structure forces a statistical approach to the description of plasticity in terms of dislocation theory. This approach relies on a simple statistical treatment that operates with the mean values of the parameters. More advanced approximations that use the distribution functions and their moments (Johnson, 1969) to describe macroscopic strain are probably unjustified because of the lack of experimental data. The phenomenon has so many aspects and governing factors that it is unrealistic to attempt to predict the macroscopic response of real materials over a wide range of loading conditions "from first principles." Simplified models with adjustable empirical parameters are therefore most useful.

Depending on the relationships between the average displacement, $d\bar{s}$, of mobile dislocations during plastic deformation and the increment, dN_m, of their density, N_m, the increment of plastic strain, $d\gamma_p$, is expressed in terms of the dislocation parameters by the equations

$$d\gamma_p = bN_m d\bar{s} \qquad (2.62)$$

or

$$d\gamma_p = b\bar{s}dN_m, \qquad (2.63)$$

where b is the Burgers vector (the elementary displacement caused by unit dislocation, $b \approx 10^{-8}$ to 10^{-7} cm). As a rule, Orowan's equation (2.62) is used to analyze the kinetics of plastic deformation. It follows from Orowan's equation that the strain rate is proportional to the velocity of mobile dislocations. The latter is a function of the resolved shear stress. In the simplest case, the velocity v of steady motion of a dislocation between obstacles is determined by the viscous drag law (Gilman, 1968, 1969)

$$v = b\frac{\tau}{B}, \tag{2.64}$$

where B is the drag coefficient, equal to $\sim 10^{-4}$ Pa·s for pure metals. In reality, as a result of interactions of dislocations with each other, with grain boundaries, with inclusions, and with other obstacles, the average velocity of the dislocations is less than the value given by Eq. (2.64). Moreover, a physical threshold for the maximum dislocation velocity exists because small mechanical disturbances cannot propagate faster than the velocity of sound. Gilman, 1968, suggested an empirical relationship for the average dislocation velocity

$$v_d = c_s\, e^{-d/\tau}, \tag{2.65}$$

where d = const. is a characteristic drag stress. In the deformation process, the dislocation density increases as

$$N \approx N_0 + M\gamma_p, \tag{2.66}$$

where M is the dislocation multiplication factor. The fraction of mobile dislocations decreases during the deformation process because of interactions with different obstacles and blocking. Taking account of Eq. (2.66), we have the equation

$$N_m = (N_0 + M\gamma_p)e^{-\varphi\gamma_p} \tag{2.67}$$

for the density of mobile dislocations. Thus, the rate of plastic deformation, $\dot\gamma_p(\tau, \gamma_p)$, is described by Eqs. (2.62), (2.65), and (2.67) with four empirical constants: d, φ, M, and N_0.

Various modifications of the expressions for the dislocation velocity and their multiplication, and numerous calculations of the evolution of the load pulse based upon them, have been done. However, it should be noted that this approach actually provides only a partial description of the evolution of the compression pulse, such as decay of the elastic precursor wave, or the rise time of a steady plastic shock wave. Under carefully controlled experimental conditions, satisfactory agreement between calculations and measurements is achieved when the initial density, N_0, of mobile dislocations is 1–3 orders of magnitude higher than the measured total dislocation density. The rigid connection between dislocation density and strain when one ignores heterogeneous nucleation of dislocations on stress concentrators limits the capability of description of such specific phenomena as the accelerated stress relaxation behind the shock-wave front.

Mogilevsky et al., 1978, compared the hardening of copper single crystals as a result of treatment by a shock wave and a relatively smooth quasi-isentropic compression wave of the same peak stress and, correspondingly, the same strain. They observed that transformation of the quasi-isentropic compression wave to a shock wave was accompanied by a sharp increase in residual hardness of the metal. The hardening is a result of high dislocation density. The particular feature of the shock is that its wave front produces higher shear stresses than the smooth quasi-isentropic wave. We come to the conclusion that the elevated residual dislocation density is a consequence of higher shear stresses acting during shock compression as compared to that during quasi-isentropic compression. This shows that the dependence of the rate of multiplication and/or nucleation of plastic deformation carriers, i.e., dislocations, on the acting shear stress must be taken into account.

Krasovsky, 1980, and Kanel, 1982, described the kinetics of plastic deformation by a two-term relation that includes both Eqs. (2.62) and (2.63):

$$\dot{\gamma}_p = b\dot{N}\bar{s} + bN_m v . \qquad (2.68)$$

It is assumed that, for great overstressing for which the dislocation velocities are high and the mean time of their movement between the obstacles is short, the deformation is controlled by the processes of nucleation and multiplication of dislocations. For low stresses, the deformation is controlled by the velocity of the dislocations. On this basis, it is possible to describe in full the behavior of a metal when subjected to impulsive loading.

It should be noted, however, that the application of dislocation models is often impeded by difficulties in obtaining values of the numerous material constants. The microscopic models are probably best confined to one-dimensional problems with short load duration, for which the structure and attenuation of shock waves have to be accurately calculated. A simpler approach to stress relaxation relies on the use of phenomenological models.

In order to simulate the wave profiles observed in experiments, phenomenological models have to account for both strain hardening and strain softening and must include strain-rate effects. In this sense, the structural model of Mazing is very promising (see Gokhfeld and Sadakov, 1984). Mazing's model reflects nonuniformity of the structure of polycrystalline solids and is effectively used for calculations of repetitive loading. The model represents each elementary volume as a series of parallel elastic–viscous–plastic subelements as shown in Fig. 2.35. Owing to the parallelism, the strains in the subelements are the same but the stresses may be different. The sub-elements have equal elastic moduli,

E_0, but different yield strengths, Y_k. The deviatoric stress in an elementary volume is determined from the equation

$$\sigma = \sum_{k=1}^{N} \sigma_k g_k, \qquad (2.69)$$

where N is number of subelements, σ_k is the stress in the sub-element, and g_k is its weight factor. The weight factor may be interpreted as a relative cross section of the subelement.

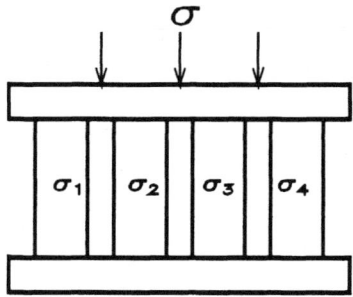

Figure 2.35. Multielement Mazing model.

Figure 2.36 shows the stress–strain diagram for Mazing's model. Under loading, an inelastic strain component appears when the yield stress of the weakest subelement is reached. After that, the average stress increases only due to the resistance to deformation of the remaining subelements. When a subelement yields, the effective modulus, $d\sigma/d\varepsilon$, decreases. The modulus decreases further as the yield criterion is met in additional subelements. When n subelements have been loaded to their yield stress, the average modulus of the whole system becomes

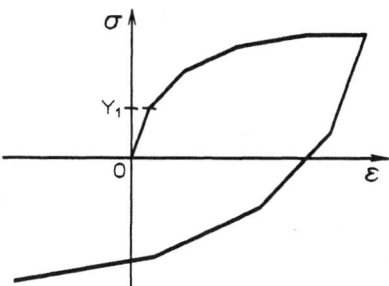

Figure 2.36. The stress–strain diagram for the Mazing model.

$$\overline{E}_{n+1} = E_0\left(1 - \sum_{k=1}^{n} g_k\right). \tag{2.70}$$

The ultimate stress in the system is

$$\overline{\sigma} = \sum_{k=1}^{N} Y_k g_k. \tag{2.71}$$

The initial phase of unloading is elastic for all subelements. Again, the yield criterion is first reached in the weakest subelement and yielding of that element results in a decrease of the modulus, $d\sigma/d\varepsilon$. As a result, plastic deformation of the whole system during unloading may begin before the average stress has decreased to zero. Later, the stress in the next subelement reaches the yield strength, etc. The resulting stress–strain diagram corresponds to one exhibiting a Bauschinger effect in many of its details.

In the range of high strain rates, a constitutive model has to account for viscous properties of the material. According to observations by Swegle and Grady, 1985, the strain rates in plastic shock waves are proportional to the squared shear stresses, Eq. (2.53). To account for the effect of the plastic strain rate of the i-th subelement, Eq. (2.53) was transformed by Kanel, 1988, to the relationship

$$\dot{\gamma}_p = A'\left[\frac{2\overline{\tau}}{Y_i}\left(\tau_i - \frac{Y_i}{2}\right)\right]^2. \tag{2.72}$$

The descriptive capabilities of the model are significantly increased by introducing strain hardening and strain softening of the subelements. As an example, Fig. 2.37 presents results of computer simulations of shock loading of aluminum D16 alloy using Mazing's model with two subelements. In these calculations the yield stress of one subelement increased with strain whereas it decreased in other sub-element. One can see that the model provides quite reasonable agreement between the measured and simulated stress histories.

The model becomes even more flexible if the weight factors of the sub-elements are allowed to vary with the strain. Figure 2.38 shows the result of a computer simulation of a shock-wave experiment with Ti-6Al-4V alloy where the weight factor of the second component increases as a result of inelastic deformation of the elementary volume as whole:

$$\dot{g}_2 = k_g(1-g_2)\dot{\varepsilon}_p, \quad \text{where} \quad \varepsilon_p = g_1\varepsilon_{1p} + g_2\varepsilon_{2p} \quad \text{and} \quad g_1 = 1 - g_2. \tag{2.73}$$

The second subelement was viscous according to the Swegle–Grady relationship, whereas the first one was strain-hardening according to the equation

$$\sigma_1 = Y_0 + Y'\varepsilon_{1p}^n. \tag{2.74}$$

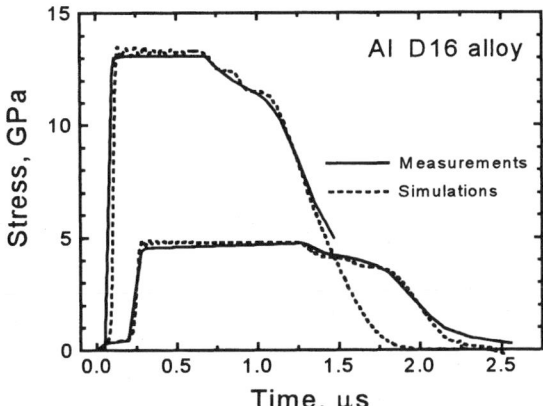

Figure 2.37. Example of use of a two-element model in simulations of shock loading of aluminum D16 alloy.

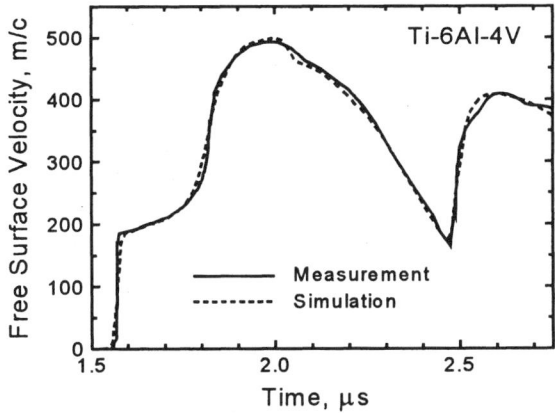

Figure 2.38. Result of computer simulation of a shock-wave experiment with Ti-6Al-4V alloy.

Figure 2.38 shows a result of calculations made using the material parameters $Y_0 = 1.0$ GPa, $Y' = 2.25$ GPa, $n = 0.35$, $Y_2 = 0.075$ GPa, $A' = 10^{-11}$ Pa^{-2}s^{-1}, and $k_g = 4$.

Investigations of material properties under shock-wave loading are carried out with the main goal of predicting the response of these materials to high-velocity impact, explosion, or exposure to pulsed laser or particle beams. In this regard, it seems that relatively simple empirical or semiempirical constitutive models are quite sufficient for most practical applications. Nevertheless, more careful examination of accumulated experimental data displays the complexities of physical and mechanical processes active in the shock-wave event. The unexplained shock-wave properties of solids provide a fascinating field of research activity and can be expected to challenge the experimenters and theorists for some time to come.

References

Ahrens, T.J., and G.E. Duvall (1966). "Stress relaxation behind elastic shock wave in rocks," *J. Geophys. Res.* **71(18)**, pp. 4349–4360.

Al'tshuler, L.V., M.N. Pavlovsky, V.V. Komissarov, and P.V. Makarov (1999). "On shear strength of aluminum in shock waves," *Comb., Expl. Shock Waves* **35(1)**, pp. 92–96. [trans. from *Fiz. Goreniya Vzryva* **35(1)**, pp. 102–107 (1999).]

Al'tshuler, L.V., S.B. Kormer, M.I. Brazhnik, L.A. Vladimirov, M.P. Speranskaya, and A.I. Funtikov (1960). "Isentropic compressibility of aluminum, copper, lead, and iron at high pressures," *Sov. Phys.–JETP* **11**, pp. 766–775 (1960). [trans. from *Zh. Eksp. Teor. Fiz.* **38(4)**, pp. 1061–1072 (1960).]

Arnold, W. (1992). "Influence of twinning on the elasto-plastic behavior of Armco iron," in: *Shock Compression of Condensed Matter — 1991* (eds. S.C. Schmidt, R.D. Dick, J.W. Forbes, and D.G. Tasker), Elsevier, Amsterdam, pp. 539–542.

Arvidson, T., and L. Eriksson (1973). Fragmentation, structure and mechanical properties of some steels and pure aluminum after shock loading," in: *Metallurgical Effects at High Strain Rates* (eds. R.W. Rohde, B.M. Butcher, J.R. Holland, and C.H. Karnes) Plenum Press, New York, pp. 605–613.

Asay, J.R., and L.C. Chhabildas (1981). "Determination of the shear strength of shock-compressed 6061-T6 aluminum," in *Shock Waves and High-Strain-Rate Phenomena in Metals* (eds. M.A. Meyers and L.E. Murr) Plenum, New York, pp. 417–431.

Asay, J.R., G.R. Fowles, and Y. Gupta (1972). "Determination of material relaxation properties from measurements on decaying elastic shock fronts," *J. Appl. Phys.* **43(2)**, pp. 744–746.

Asay, J.R., and J. Lipkin (1978). "A self-consistent technique for estimating the dynamic yield strength of a shock-loaded material," *J. Appl. Phys.* **49(7)**, pp. 4242–4247.

Bakhrakh, S.M., V.N. Knyazev, P.N. Nizovtzev, V.A. Raevsky, and E.V. Shuvalova (2001). "Computational and theoretical analysis of the method of main stresses," *Problems of Atomic Science and Technology. Series: Theoretical and Applied Physics*, pp. 13–17. (in Russian)

Barker, L.M. (1971). "A Model for Stress Wave Propagation in Composite Materials," *J. Composite Materials*, **5**, p. 140.

Dremin, A.N., and G.I. Kanel (1970). "Refraction of oblique shock wave at interface with less rigid media," *J. Appl. Mech. Tech. Phys.* **11(3)**, pp. 488–492 (1970) [trans. from *Zh. Prikl. Mekh. Tekh. Fiz.* **11(3)**, pp. 140–144 (1970)].

Dremin, A.N., and G.I. Kanel (1976). "Compression and rarefaction waves in shock-compressed metals," *J. Appl. Mech. Tech. Phys.* **17(2)**, pp. 263–267 [trans. from *Zh. Prikl. Mekh. Tekh. Fiz.* **17(2)**, pp. 146–153 (1976)].

Dremin, A.N., G.I. Kanel, and O.B. Chernikova (1981). "The resistance to plastic deformation of aluminum AD1 and duralumin D16 to plastic deformation," *J. Appl. Mech. Tech. Phys.* **22(4)**, pp. 558–562 [trans. from *Zh. Prikl. Mekh. Tekh. Fiz.* **22(4)**, pp. 132–138 (1981)].

Dunn, J.E., and D.E. Grady (1986). "Strain rate dependence in steady plastic shock waves," in *Shock Waves in Condensed Matter* (ed. Y.M. Gupta) Plenum Press, New York, pp. 359–364.

Feng, R., Y.M. Gupta, and M.K.W. Wong (1997). "Dynamic analysis of the response of lateral piezoresistance gauges in shocked ceramics," *J. Appl. Phys.* **82(6)**, pp. 2845–2854.

Fowles, R., and R.F. Williams (1970). "Plane stress wave propagation in solids," *J. Appl. Phys.* **41(1)**, pp. 360–363.

Furnish, M.D., L.C. Chhabildas, D.J. Steinberg, and G.T. Gray III (1992). "Dynamic behavior of fully dense molybdenum," in *Shock Compression of Condensed Matter—1991* (eds. S.C. Schmidt, R.D. Dick, J.W. Forbes, and D.G. Tasker), North-Holland, Amsterdam, pp. 419–422.

Gilman, J.J. (1968). "Dislocation dynamics and response of materials to impact," *Appl. Mech. Rev.* **21(8)**, pp. 767–783.

Gilman, J.J. (1969). *Micromechanics of Flow in Solids*. McGraw-Hill, New York.

Gluzman, V.D., G.I. Kanel, V.F. Loskutov, V.E. Fortov, and I.E. Khorev (1985). "Resistance to deformation and fracture of 35Kh3NM steel under conditions of shock loading," *Strength of Materials* **17(8)**, pp. 1093–1098 [trans. from *Problemy Prochnosti* **17(8)**, pp. 52–57 (1985)].

Gokhfeld, D.A., and O.S. Sadakov (1984). *Plasticity of Structural Elements Under Repeated Loads*, Mashinostroenie, Moscow. (in Russian)

Graham, R.A. (1993). *Solids Under High-Pressure Shock Compression*, Springer-Verlag, New York.

Guinan, M.W., and D.J. Steinberg (1974). "Pressure and temperature derivatives of the isotropic polycrystalline shear modulus for 65 elements," *J. Phys. Chem. Solids* **35**, pp. 1501–1512.

Gupta, Y.M., G.E. Duvall, and G.R. Fowles (1975). "Dislocation mechanisms for stress relaxation in shocked LiF," *J. Appl. Phys.* **46(2)**, pp. 532–546.

Holian, B.L. (1995). "Atomistic computer simulations of shock waves," *Shock Waves* **5**, pp. 149–157.

Johnson, J.N. (1969). "Constitutive relation for rate-dependent plastic flow in polycrystalline metals," *J. Appl. Phys.* **40(5)**, pp. 2287–2293.

Johnson, J.N., and L.M. Barker (1969). "Dislocation dynamics and steady plastic wave profiles in 6061-T6 aluminum," *J. Appl. Phys.* **40(11)**, pp. 4321–4334.

Kalymykov, Yu.B., G.I. Kanel, I.P. Parkhomenko, A.V. Utkin, and V.E. Fortov (1990). "Behavior of rubber in shock waves and rarefaction waves," *J. Appl. Mech. Tech. Phys.* **31(1)**, pp. 116–120 [trans. from *Zh. Prikl. Mekh. Tekh. Fiz.* **31(1)**, pp. 126–130 (1990)].

Kanel, G.I. (1982). "Model of the kinetics of metal plastic deformation under shock-wave loading conditions," *J. Appl. Mech. Tech. Phys.* **23(2)**, pp. 256–260 [trans. from *Zh. Prikl. Mekh. Tekh. Fiz.* **23(2)**, pp. 105–110 (1982)].

Kanel, G.I. (1988). "Calculation of strains and failure of steel in shock waves," *Problemy Prochnosti* **20(9)**, pp. 55–58 (in Russian)

Kanel, G.I., M.F. Ivanov, and A.N. Parshikov (1995). "Computer Simulation of the Heterogeneous Materials Response to the Impact Loading," *Int. J. Impact Engineering*, **17(1–6)**, pp. 455–464.

Kanel, G.I., A.M. Molodets, and A.N. Dremin (1977). "Investigation of singularities of glass strain under compression waves," *Comb. Expl. Shock Waves* **13(6)**, pp. 772–779 [trans. from *Fiz. Goreniya Vzryva* **13(6)**, pp. 906–912 (1977)].

Kanel, G.I., A.M. Molodets, and A.N. Dremin (1979). "Variation in the strength of metals under the influence of shock waves," *Phys. Met. Metall.* **46(1)**, pp. 175–177 [trans. from *Fiz. Metall. Metalloved.* **46(1)**, pp. 200–202 (1978)].

Kanel, G.I., and E.N. Petrova (1981) "The strength of titanium VT6 under conditions of shock-wave loading," in *Detonation*, Inst. Chem. Phys., Chernogolovka, pp. 136–141. (in Russian)

Kanel, G.I., S.V. Razorenov, and V.E. Fortov (1988). "Viscoelasticity of aluminum in rarefaction waves," *J. Appl. Mech. Tech. Phys.* **29(6)**, pp. 824–826 [trans. from *Zh. Prikl. Mekh. Tekh. Fiz.* **29(6)**, pp. 67–70].

Kanel, G.I., S.V. Razorenov, A.V. Utkin, and V.E. Fortov (1996). *Shock-Wave Phenomena in Condensed media*, Janus-K, Moscow. (in Russian)

Kanel, G.I., and V.V. Scherban (1980). "Plastic deformation and cleavage rupture of Armco iron in a shock wave," *Comb. Expl. Shock Waves* **16(4)**, pp. 439–446 (1980) [trans. from *Fiz. Goreniya Vzryva* **16(4)**, pp. 93–103].

Kanel, G.I., A.V.Utkin, and Z.G.Tolstikova (1994). "Response of the High-Filled Elastomers to Shock-Wave Loading," *High-Pressure Science and Technology — 1993* (eds. S.C. Schmidt, J.W. Shaner, G.A. Samara, and M. Ross), American Institute of Physics, New York, pp. 1123–1126.

Kesler, G., H.U. Karow, K. Baumung, V.E. Fortov, G.I. Kanel, and V. Licht (1994). "High-Power Light Ion Beams and Intense Shock Waves," *High-Pressure Science and Technology — 1993* (eds. S.C. Schmidt, J.W. Shaner, G.A. Samara, and M. Ross), American Institute of Physics, New York, pp. 1887–1890.

Kormer, S.B. (1968). "Optical investigations of properties of shock-compressed condensed dielectrics," *Sov. Phys.–Usp.* **11(4)**, pp. 229–254. [trans. from *Usp. Fiz. Nauk* **94(4)**, pp. 641–687 (1968).]

Krasovsky, A.Ya. (1980). *Brittleness of Metals at Low Temperatures*, Naukova Dumka, Kiev. (in Russian)

Kusubov, A.S., and M. van Thiel (1969). "Dynamic yield strength of 2024-T4 aluminum at 313 kbar," *J. Appl. Phys.* **40(2)**, pp. 893–899.

Landau L.D., and E.M. Lifshitz (1959). *Fluid Mechanics*, Pergamon Press, Oxford.

Mashimo, T., Y. Hanaoka, and K. Nagayama (1988). "Elastoplastic properties under shock compression of Al_2O_3 single crystal and polycrystal," *J. Appl. Phys.* **63(2)**, pp. 327–336.

McClintock, F.A., and A.S. Argon (1966). *Mechanical Behavior of Materials*, Addison-Wesley, Reading, MA.

McQueen, R.G., J.N. Fritz, and C.E. Morris (1984). "The velocity of sound behind strong shock waves in 2024 Al," in *Shock Compression of Condensed Matter—1983* (eds. J.R. Asay, R.A. Graham, and G.K. Straub) North-Holland, Amsterdam, pp. 95–98.

Mogilevsky, M.A. (1983). "Mechanisms of deformation under shock loading," *Physics Reports (Review Section of Phys. Letters)* **97(6)**, pp. 357–393.

Mogilevsky, M.A., S.A. Bordzilovsky, and N.N. Gorshkov (1978). "Effect of the width of the front with quasi-isentropic compression on the hardening of some metals," *Comb. Expl. Shock Waves* **14(6)**, pp. 794–798 [trans. from *Fiz. Goreniya Vzryva* **14(6)**, pp. 110–116 (1978)].

Mogilevsky, M.A., and I.O. Mynkin (1978). "Effect of the point defects on one-dimensional compression of a lattice," *Comb. Expl. Shock Waves* **14(5)**, pp. 680–682. [trans. from *Fiz. Goreniya Vzryva* **14(5)**, pp. 159–163 (1978).]

Morris, C.E. (ed.) (1982). *Los Alamos Shock Wave Profile Data*, University of California Press, Berkeley, CA.

Murr, L.E., E. Moin, K. Wongwiwat, and K.P. Standhammer (1978). "Effect of peak pressure and pressure-pulse duration on crystallographic transformations in shock-loaded metals and alloys," *Scripta Met.* **12(5)**, pp. 425–431.

Razorenov, S.V., A.A. Bogach, and G.I. Kanel (1997). "Influence of heat treatment and polymorphous transformations on the dynamic rupture resistance of 40X steel," *Phys Met. Metall.* **83(1)**, pp. 100–103 [trans. from *Fiz. Metall. Metalloved.* **83(1)**, pp. 147–152 (1997)].

Razorenov, S.V., A.A. Bogach, G.I. Kanel, A.V. Utkin, V.E. Fortov, and D.E. Grady (1998). "Elastic-Plastic Deformation and Spall Fracture of Metals at High Temperatures," in *Shock Compression of Condensed Matter—1997* (eds. S.C. Schmidt, D.D. Dandekar, and J.W. Forbes) American Institute of Physics, New York, pp. 447–480.

Razorenov, S.V., G.I. Kanel, and V.E. Fortov (1985). "Measurement of the width of shock fronts in copper," *Sov. Phys.–Tech. Phys.* **30(9)**, pp. 1061–1062 [trans. from *Zh. Tekh. Fiz.* **55(9)**, pp. 1816–1818].

Razorenov, S.V., G.I. Kanel, O.R. Osipova, and V.E. Fortov (1987). "Measurement of the viscosity of copper in shock loading," *High Temp.* **25(1)**, pp. 57–61 [trans. from *Teplofiz. Vys. Temp.* **25(1)**, pp. 65–69 (1987)].

Razorenov, S.V., G.I. Kanel, A.V. Utkin, A.A. Bogach, M. Burkins, and W.A. Gooch (2000). "Dynamic strength and edge effects at spall fracture for titanium alloys of varying oxygen content," in: *Shock Compression of Condensed Matter—1999* (eds. M.D. Furnish, L.C. Chhabildas, and R.S. Hixson) American Institute of Physics, New York, pp. 415–418.

Smith, C.S. (1958). "Metallographic studies of metals after explosive loading," *Trans. AIME* **214**, pp. 574–586.

Swegle, J.W., and D.E. Grady (1985). "Shock viscosity and the prediction of shock wave rise times," *J. Appl. Phys.* **58**, pp. 692–701.

Taylor, J.W. (1965). "Dislocation dynamics and dynamic yielding," *J. Appl, Phys.* **36(10)**, pp. 3146–3150.

Taylor, J.W., and M.H. Rice (1963). "Elastic-plastic properties of iron," *J. Appl. Phys.* **34**, pp. 364–371.

Vorob'ev, A.A., A.N. Dremin, and G.I. Kanel (1974). "Dependence of the coefficients of elasticity of Al on the degree of compression in the shock wave," *J. Appl. Mech. Tech. Phys.* **15(5)**, pp. 661–665. [trans. from *Zh. Prikl. Mekh. Tekh. Fiz.* **15(5)**, pp. 94–100 (1974).]

Weaver, C.W., and M.S. Paterson (1969). "Stress-strain properties of rubber at pressures above the glass transition pressure," *J. Polym. Sci.*, **7** pt. A-2, № 3, pp. 587–592.

Zaretsky, E.B. (1992). "X-Ray diffraction evidence for the role of stacking faults in plastic deformation of solids under shock loading," *Shock Waves* **2**, pp. 113–116.

Zaretsky, E.B., P.A. Mogilevsky, G.I. Kanel, and V.E. Fortov (1991). "Device for investigating X-ray diffraction studies on shock-compressed materials," *High Temp.* **29(5)**, pp. 805–811 [trans. from *Teplofiz. Vys. Temp.* **29(5)**, pp. 1002–1008 (1991)].

Zel'dovich, Ya.B. and Yu.P. Raizer (1967). *Physics of Shock Waves and High-Temperature Hydrodynamic Phenomena*, Academic Press, New York. Reissued in 2002 by Dover Publications, Mineola, NY.

CHAPTER 3

Yield and Strength Properties of Metals and Alloys at Elevated Temperatures

The main driving force behind high-strain-rate testing is the need to obtain values for the parameters in the various material constitutive models used in the numerical simulation of the impact or shock response of materials and structures. Information about the high-strain-rate properties of materials at elevated temperatures is important for problems such as penetration and high-rate metallurgical treatment by cutting or forging, since the transient loading processes are accompanied by irreversible heating. The knowledge of the temperature dependencies of dynamic responses of materials would also help us better understand the nature of thermomechanical instabilities as manifested by the formation of shear bands. The existing theory on formation of shear bands is based on a competition between the strain hardening and the thermal softening of material during an adiabatic deformation process. There is a general agreement that the tendency to form adiabatic shear bands increases when the strain hardening decreases and the thermal softening increases. Since high strain rates create the adiabatic conditions for the induced deformation, shear banding is usually associated with impact loading.

Expected temperature effects upon the elastic–plastic and strength properties of solids at very high strain rates are not trivial and are not yet entirely clear. It is well known that, under normal conditions, both the yield strength and the tensile strength of materials are strong functions of the temperature and decrease with heating. However, at very high strain rates other temperature dependencies of the strength properties of polycrystalline metals and metal single crystals have been revealed when the temperature was introduced as a varied parameter in shock-wave tests (Bogach et al., 1998, Kanel et al., 1996, Razorenov et al., 2002).

It is known that crystalline solids exhibit rate-dependent stress–strain behavior. The effect of strain rate on the main mechanisms and the resistance to plastic flow of metals has been successfully rationalized in terms of the dynamics of dislocations (Kumar and Kumble, 1969). For low rates of mechanical loading, the dislocation motion is aided by thermal fluctuations. Dislocation motion is impeded at barriers and a combination of thermal agitation and applied stress is required to activate dislocations over the obstacles. At some point (10^3–10^4 s^{-1}) between intermediate strain rates and the higher strain rates, the plastic deformation process undergoes a transition from being dominated by stress-assisted thermal activation to being controlled by the time it takes a mobile dislocation to move from one barrier to the next; this latter time depends on the effective stress and the viscous-drag force provided by the perfect lattice. At very high strain rates the applied stress is high enough to overcome instantaneously the usual dislocation barriers without any aid from thermal fluctuations so viscous phonon drag becomes dominant. Clifton, 1971, has given the earliest and most complete discussion of transition from thermal activation to dislocation drag in shock-loaded solids; the modern state of the question was recently discussed by Johnson and Tonks, 1992 (see also Follansbee and Weertman, 1982). Sakino, 2000, suggested that this transition in the rate-controlling mechanism should result in a change of sign of the temperature dependence of the flow stress. Since the phonon viscosity is proportional to the temperature, an increase of the flow stress with increasing temperature may be expected at the highest strain rates. In experiments with copper, Sakino reached a point of temperature independence. Bhate et al., 2002, performed atomistic simulations of the dislocation motion for temperatures ranging from 10K to 200K. The results show that the dislocation velocity decreases with increasing temperature and confirms that the drag is due to thermal phonons. The simulations also confirmed the existence of a limiting velocity of dislocations that is less than the shear sound speed. The stress needed to approach this speed is approximately proportional to the temperature.

A possible influence of the temperature upon the bulk tensile strength of materials at very high strain rates is also complicated. As a result of overstressing, the fracture process becomes scattered in nature and consists of nucleation, growth, and coalescence of numerous pores or cracks (Curran et al., 1987, Antoun et al., 2003). Temperature may obviously influence the nucleation and growth processes. Spontaneous nucleation of vacancies and microvoids by thermal fluctuations should result in a decrease of strength with increasing temperature. On the other hand, the resistance to growth of pores is determined by the yield stress and by the viscosity of the surrounding material (Johnson, 1981), so it has to change with the temperature proportionally to the change of flow stress. If the flow stress increases with the temperature as a result of phonon drag at high strain rates, the total resistance to fracture should also increase. It should be possible to reveal the crucial elementary fracture process by simultaneously measuring the flow stress and the fracture stress at different temperatures.

3.1. Spall Strength at Melting

There have been only a few observations of mechanical yielding and strength behavior in shock waves at elevated temperatures. In this chapter we'll begin discussion of temperature effects upon the yield and strength properties of metals and alloys under shock-wave loading from the data for aluminum single crystals obtained by Kanel et al., 2001. Experiments with single crystals exclude effects of grain boundaries and other structural nonuniformities that make material response more complex. The measurements were done over the temperature

3. Yield and Strength of Metals and Alloys at Elevated Temperatures

range from 15 °C to 650 °C. Note that the melting temperature of aluminum is 660.1 °C. Resistive heaters of sufficient power to heat the test specimen to the melting point of aluminum within 10 minutes were used to vary the temperature.

Figures 3.1 and 3.2 present typical free surface velocity histories measured for shock loading by impact of a thin flyer plate or by a pulsed ion beam. The wave profiles show a significant increase in the precursor wave amplitude with increased temperature. At room temperature, the measurements show a ramped profile of the precursor wave with a very small initial jump, but a front spike appears as the temperature is increased. The precursor waveforms with spikes are associated with accelerating stress relaxation. The measurement is not able to

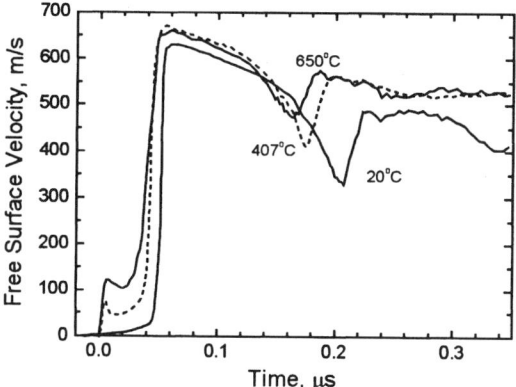

Figure 3.1. Free surface velocity histories of 2.90 ± 0.05-mm-thick aluminum single crystals impacted by 0.4-mm-thick aluminum flyer plates. The temperatures at which the tests were conducted are noted by the profiles.

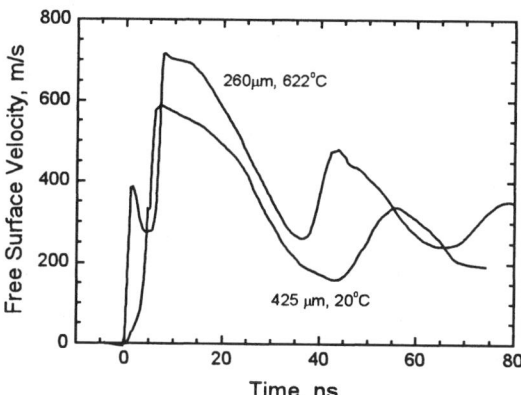

Figure 3.2. Free surface velocity histories of aluminum single crystals shock loaded with an ion beam at 20°C and 622°C. The sample thickness was 425 μm and 260 μm in the tests at normal and elevated temperatures, respectively.

resolve the rise time in the spike-like precursor wave front. Experiments with samples of greater thickness exhibited an increase in the rise time of the plastic shock wave (as measured from 10 to 90% of its peak amplitude) from 4–6 ns at the room temperature to 12–16 ns near the melting temperature.

The free surface velocity profiles show that the velocity pullback, a measure of the spall fracture stress, decreases slightly with increasing temperature, but remains high up to the maximum of 650 °C. The shape of the spall signal varies depending on the shape of the incident shock pulse. Figure 3.3 presents free surface velocity histories measured for single crystals and polycrystalline aluminum at room temperature with an ion beam used as a shock-wave generator. Figure 3.4 illustrates variations of the elastic precursor wave as a function of the propagation distance and peak stress at the elevated temperature.

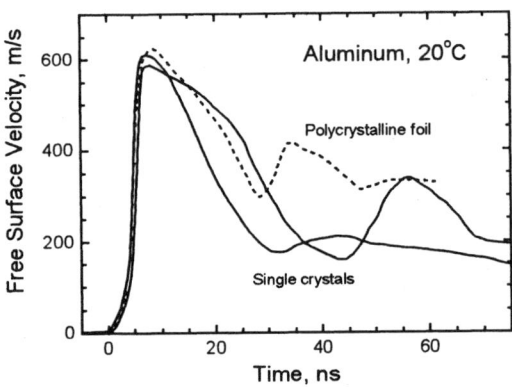

Figure 3.3. Free surface velocity histories for aluminum single crystals and polycrystalline aluminum measured at room temperature.

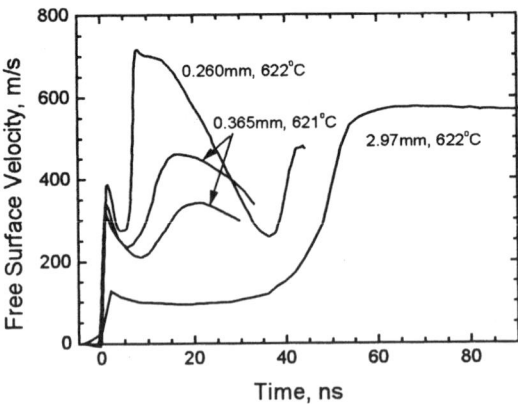

Figure 3.4. Variations of the elastic precursor wave as a function of the propagation distance and the peak stress in aluminum single crystals at 622 °C temperature.

In analyzing free surface velocity profiles, account must be taken of the temperature dependencies of elastic constants (and, hence, of the sound speeds). The effect is illustrated on Fig. 3.5 using data by Tallon and Wolfeden, 1979. The weak temperature sensitivity of the bulk sound speed is rather common for many solids.

Figure 3.6 shows the pressure–temperature diagram for some of the high-temperature spall experiments. The diagram has been calculated using a complete equation of state developed by Asay and Hayes, 1975, based on semi-empirical free-energy functions of liquid and solid aluminum. The $p–T$ state curves for shock compression and subsequent isentropic rarefaction were calculated for solid phase aluminum, assuming that no melting occurs.

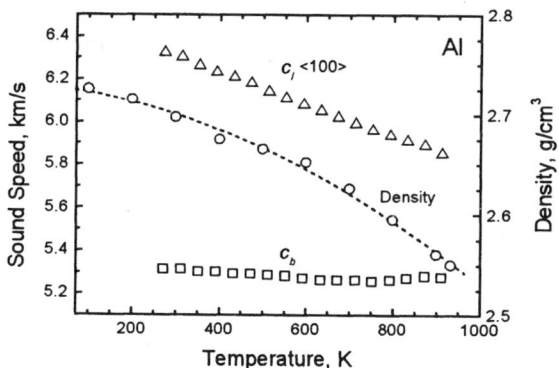

Figure 3.5. Longitudinal sound speed, c_l, bulk sound speed, c_b, and density of an aluminum single crystal as functions of the temperature. Data by Tallon and Wolfeden, 1979.

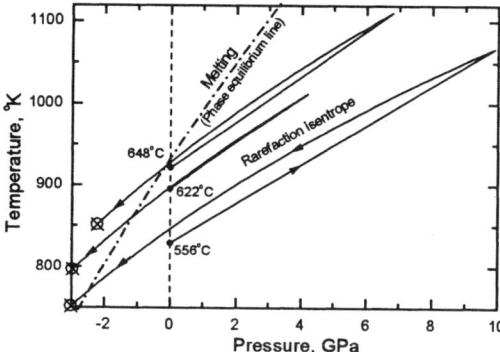

Figure 3.6. Pressure–temperature diagrams for shock compression followed by rarefaction of aluminum. The solid points indicate initial states and the open points correspond to spall conditions.

The lowest points on the curves correspond to the spall strength values. The melting temperature increases with increasing the pressure and, correspondingly, decreases when the pressure turns negative. The relationships between slopes of the melting curves, $T_m(p)$, and isentropes, $T_S(p)$, of solids is such that adiabatic expansion shifts the state of the solid matter toward the melting curve and, at a certain tension, the melting curve is crossed at a temperature $T < T_{m0}$, where T_{m0} is the melting temperature at zero pressure. Since spall fracture of pre-heated crystals occurred after intersection of these curves, we may conclude that the single crystals maintained a high resistance to spall fracture at the point when melting should begin.

Figure 3.7 presents the pressure–volume diagram of the compression–tension process for pre-heated aluminum. It is interesting to note that the specific volume of the stretched crystal at its ultimate tensile stress never reached even the zero-pressure volume of molten aluminum.

Information about melting should appear in the free surface velocity histories. It was expected that, even if the beginning of melting does not result in a sharp decrease of the spall strength, it should increase the compressibility of the material and decrease the yield stress. Both these effects must produce kinks in the wave profiles. According to this criterion, melting was observed in the unloading history of shock compressed tin by Mabire and Hereil, 2000. However, no kinks are recorded in the free surface velocity histories for aluminum single crystals.

Figures 3.8 and 3.9 present examples of free surface velocity profiles for zinc single crystals (Bogach et al., 1998) and for polycrystalline commercial grade aluminum AD1 (Kanel et al., 1996). The data show that HELs of metals increase, or at least do not decrease, with increasing temperature. Single crystals maintain high spall strength whereas the spall strength of polycrystalline metals drops to zero as their temperature approaches the melting point.

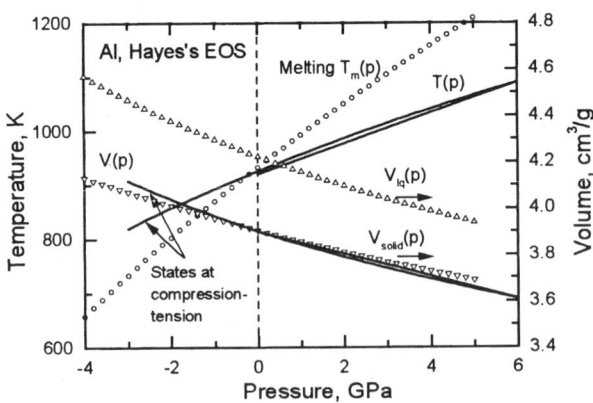

Figure 3.7. Calculated states of a pre-heated aluminum crystal in tension.

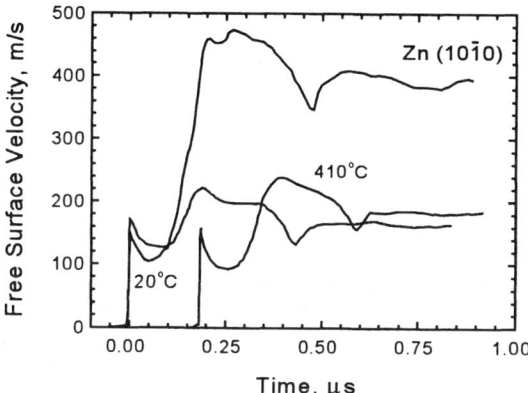

Figure 3.8. Free surface velocity histories of shock-compressed zinc single crystals 1.7–1.8-mm thick at normal and elevated temperatures and different peak stresses. Data by Bogach et al., 1998. The melting temperature of zinc is 419.5 °C.

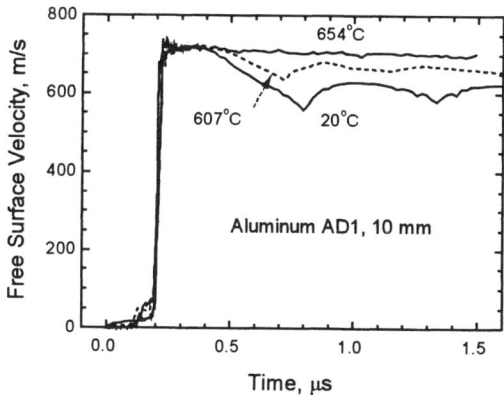

Figure 3.9. Examples of free surface velocity histories of shock-compressed aluminum AD1 at normal and elevated temperatures. Data by Kanel et al., 1996.

For further analysis of melting under tension, it is convenient to use a linear estimate of the pressure at which the isentrope of a solid intersects its melting curve (Kanel, 2000). Neglecting the material viscosity and the yield strength, it is believed that the rarefaction of shock-compressed material occurs along its isentrope. The temperature decrease along a rarefaction isentrope occurs more slowly than along the melting curve:

$$\left(\frac{\partial T}{\partial p}\right)_S < \frac{dT_m}{dp}. \tag{3.1}$$

If the temperature is close enough to the melting temperature, the state of the material under tension must cross the melting boundary.

It is necessary to mention that all states of condensed matter at negative pressures are below the sublimation curve and are metastable with respect to two separate pieces of the material or to mixtures of the condensed and vapor phases. From this point of view, melting under tension, if it may be observed, is a transformation of a metastable solid phase into a metastable liquid phase. On the other hand, equality of the Gibbs potentials of metastable solid and liquid phases at the same pressure and temperature means that the phases are in equilibrium and determines the position of the phase boundary, as in the region of positive pressures. In this sense, states on the phase boundary are in equilibrium even at negative pressures.

Let us now estimate the pressure at which an isentrope of a solid intersects its melting curve $T_m(p)$. Along the melting curve

$$\frac{dV}{dp} = \left(\frac{\partial V}{\partial T}\right)_p \frac{dT_m}{dp} + \left(\frac{\partial V}{\partial p}\right)_T. \tag{3.2}$$

A linear equation of state for the isentrope of a solid with an initial (at $p=0$) temperature T_0 may be written in a form

$$V = V(T_{m0}) + (T_0 - T_{m0})\left(\frac{\partial V}{\partial T}\right)_{p=0} + \left(\frac{\partial V}{\partial p}\right)_S p. \tag{3.3}$$

Equations (3.2) and (3.3) give the pressure at the point of intersection of the isentrope and the melting curve (or, more precisely, the intersection with the solidus line)

$$p\alpha \frac{dT_m}{dp} - \frac{p}{K_T} = \alpha(T_0 - T_{m0}) - \frac{p}{K_S}, \tag{3.4}$$

where

$$\alpha = \frac{1}{V}\left(\frac{dV}{dT}\right)_p$$

is the thermal expansion coefficient, T_0 is the initial test temperature, and

$$K_T = -V\left(\frac{dp}{dV}\right)_T \quad \text{and} \quad K_S = -V\left(\frac{dp}{dV}\right)_S$$

are isothermal and isentropic bulk moduli, respectively.

Estimates of the pressure at the points of intersection of the melting curve and the isentropes as a function of T_0 are shown in Fig. 3.10 for single crystals of aluminum and zinc and in Fig. 3.11 for polycrystalline aluminum and magnesium. When the pressure in states of tension intersects the melting boundary, the solid states cease to be in thermodynamically stable and melting should start. It would be natural to expect a sharp decrease in tensile strength with beginning of melting. As a result, the lines in Figs. 3.10 and 3.11 should describe melting thresholds which limit the high-temperature strength of the material. However, whereas all experimental data for polycrystalline metals are below the estimated melting lines, part of the high-temperature data for single crystals are above these lines. In other words, the strength of polycrystalline metals drops when the material begins to melt whereas single crystals maintain a high resistance to spall fracture even above the point at which melting should start.

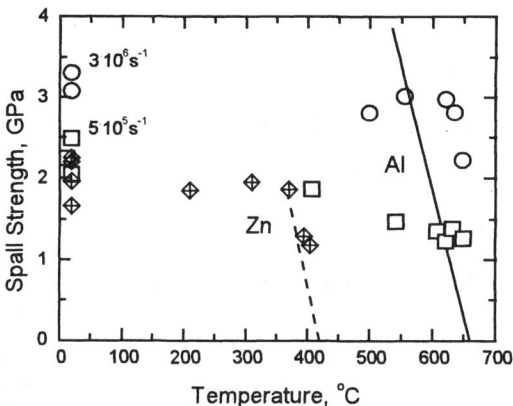

Figure 3.10. The relationship between spall strength (points) and melting thresholds (lines) for zinc and single crystals of aluminum at two different strain rates.

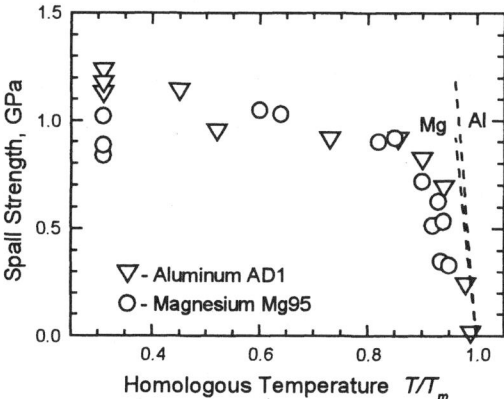

Figure 3.11. The relationship between spall strength and melting thresholds for polycrystalline aluminum and magnesium.

In polycrystalline solids, melting may start along grain boundaries at temperatures below the melting temperature of the crystal. The effect is caused by the disorder and by the larger concentration of impurities in boundary layers of grains (Ubbelohde, 1965; Dash, 1999). Several computational works have been devoted to detailed study of the role of inter-granular surfaces in the melting process at temperatures below the bulk melting point. Lu and Szpunar, 1995, investigated a grain boundary disordering transition of the melting type in a twist boundary of an aluminum bicrystal. They have found that partial melting of the grain boundary had occurred at temperatures above 0.9375 T_m. Besold and Mouritsen, 1994, performed Monte Carlo simulations of grain-boundary melting of a bicrystal. They reported that a disordered liquid-like layer gradually emerged at the grain boundary well below the bulk melting temperature T_m. Monte Carlo data over an extended temperature range indicated a logarithmic divergence of the width w of the disordered layer going approximately as $w(T) = -\ln(T_m - T)$. Very likely, this grain-boundary effect contributed to the precipitous drop in spall strength near, but below, the melting temperature. Besides this, hot spots may be formed under shock compression of polycrystalline materials as a result of partial localization of shock-wave energy at structural imperfections. In preheated samples, the material may melt locally in these hot spots, which, in turn, should reduce the strength of the material.

If molten spots appear in the volume of a single crystal, the crystal is no longer homogeneous and should show a spall strength close to that of polycrystalline aluminum. However, even at the highest temperatures, the single crystals demonstrate a higher strength than polycrystalline materials at room temperature and at the same strain rate. It seems more likely that the crystals did not melt and the spall data in all cases represent the strength of the solid crystals.

If melting did not occur, one has to conclude that superheated solid states were realized in the crystals under tension. It is known that, unlike the case of liquids, superheating of crystalline solids is impossible under normal conditions. It is assumed that the crystal surface plays a crucial role in the melting process. Melting of a uniformly heated crystal always begins on its surface. However, superheated states may be reached inside the crystal body if its surface is below the melting temperature. This condition was realized in the spall experiments discussed. The magnitude of superheating of aluminum crystals reached 60–65 °C at the shortest load durations.

Understanding of the melting of a solid at the microscopic level is an outstanding problem in condensed matter physics. There exist a number of theoretical models to describe how a collection of atoms in a crystalline arrangement becomes unstable against a liquid configuration. Lindemann treated melting as a vibrational instability (see Ubbelohde, 1965). Later, it was suggested that the transition is caused by a lattice-shear instability (rigidity catastrophe, see Boyer, 1985), the catastrophic generation of dislocations (Cotterill, 1980), or the presence of thermal vacancies and other point defects (see Cahn, 1986).

There have been several attempts to estimate the upper limit for the superheating of crystals. Figure 3.12 compares different criteria. Fecht and Johnson, 1988, proposed a thermodynamic stability limit for the superheated crystal in terms of an entropy catastrophe that occurs at a critical temperature T_m^S when the entropy of the superheated crystal equals that of the liquid phase. Later, Tallon, 1989, suggested another inner instability point T_m^g where the entropy for a superheated crystal becomes equal to that for a glass (a diffusionless liquid) rather than that for a liquid. This instability point is slightly lower than T_m^S. It was demonstrated also that the upper limiting temperature for crystals due to the rigidity catastrophe T_m^r is below the instability limits defined by the volume or isochoric catastrophe (T_m^V, at which the volume of the crystal equals that of the liquid) and the entropy catastrophe (T_m^S). These proposed instability limits range from $1.3\,T_m$ to $2.0\,T_m$. Lu and Li, 1998, analyzed homogeneous nucleation kinetics for melting in superheated crystals in order to derive a kinetic stability limit for the lattice. It was found that at a critical temperature (T_m^k, which is about $1.2\,T_m$ for various elemental metals) a massive homogeneous nucleation of melting occurs in the superheated crystal. Iwamatsu, 1999, determined that the catastrophic homogeneous nucleation occurs at $T \sim 1.11\,T_m$. Ma and Lip, 2000, have determined a superheating limit for massive heterogeneous nucleation on dislocations, which is evidently lower than the limit for homogeneous melting. Even for a crystal with a low dislocation density of 10^2 cm^{-2}, the associated threshold temperatures are only $1.068\,T_m$ to $1.095\,T_m$ for various metals.

Thus, the problem of superheating is presently receiving renewed interest. Melting of crystalline materials has been proven to be nucleated at heteroge-

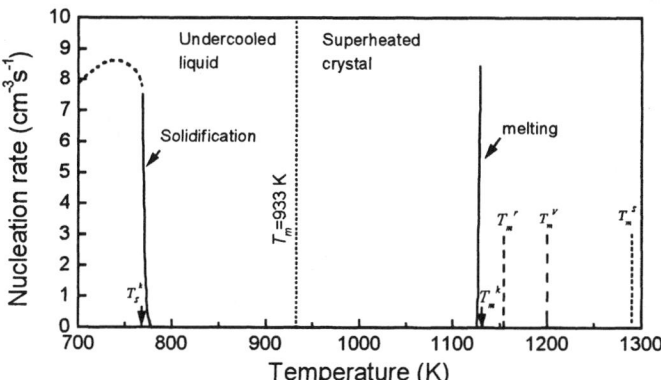

Figure 3.12. Theoretical thresholds for melting and solidification in aluminum. T_m^k is the temperature of a massive homogeneous nucleation catastrophe, T_m^r is the temperature of a rigidity catastrophe, T_m^V is the temperature at which the volume of the crystal equals that of the liquid, and T_m^S is the temperature at which the entropy of the crystal equals that of the liquid. Data by Lu and Li, 1998.

ous nucleation sites such as grain boundaries or free surfaces. Providing heterogeneous nucleation can be avoided for melting, crystals can be superheated above their equilibrium melting points. However, known experimental data are very poor and are not sufficient to validate theoretical predictions. Whereas superheated solid states are allowed by theory, nothing is known about either mechanical or thermophysical properties of matter in these exotic metastable states, or about real temperature and time limits for their existence.

3.2. High-Temperature Yielding

For many metals, the strain rate sensitivity of the flow stress increases steeply above a strain rate of $\sim 10^3 - 10^4$ s^{-1}. This is interpreted as a transition in the rate-controlling mechanism of dislocation motion. At very high strain rates the applied stress is high enough that the usual dislocation barriers are overcome instantaneously without the aid of thermal fluctuations, and other drag mechanisms (such as the phonon viscosity, internal stresses generated by other dislocations and point defects, etc.) become dominant. Since contributions of some of them are proportional to the temperature, an increase of the flow stress with increased temperature may be expected at higher strain rates (Sakino, 2000), as shown schematically in Fig. 3.13. The strong dependence of the flow stress on the strain rate should result in the rapid decay of the elastic precursor wave that was actually observed in the experiments at elevated temperatures (Fig. 3.14).

The results of measurements of dynamic yield strength in plate impact experiments are summarized in Fig. 3.15, where longitudinal stresses at the elastic precursor spike and in the minimum behind the elastic precursor front and corresponding yield stress values are presented as functions of the temperature. The points at room temperature represent an upper estimate of the values because the precursor wave front was not resolved by the measurements. The VISAR data of

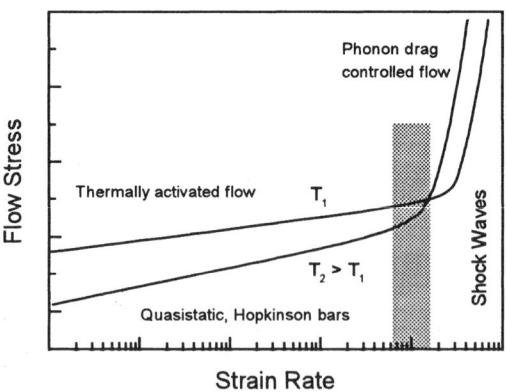

Figure 3.13. Regions of operation of different mechanisms of plastic flow depending on the strain rate.

3. Yield and Strength of Metals and Alloys at Elevated Temperatures 95

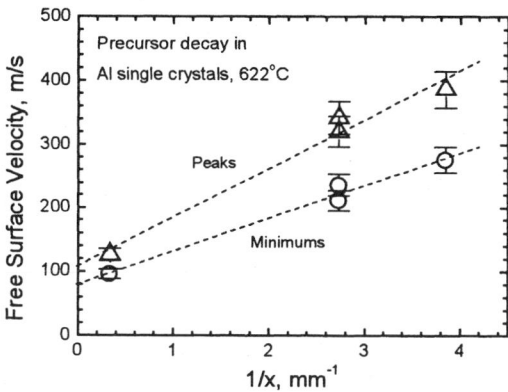

Figure 3.14. Precursor decay in aluminum single crystals at 622°C.

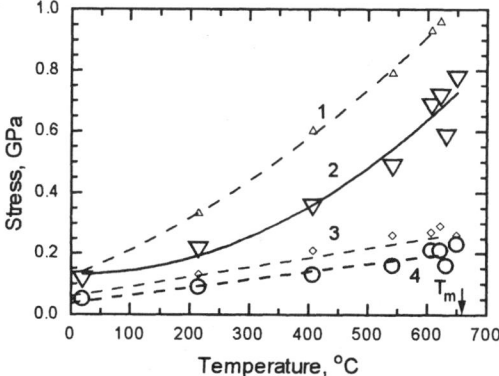

Figure 3.15. Longitudinal stresses at the elastic precursor spike (curve 1) and at the minimum behind the elastic precursor front (curve 2). Data labeled as 3 and 4 are the corresponding values of the yield stress. T_m is the melting temperature.

plate impact experiments may underestimate their peak values because the duration of the elastic peak is close to the time resolution of the measurements. Since Poisson's ratio increases with temperature, the yield strength is less sensitive to the temperature than is the Hugoniot elastic limit. The strong decay of the elastic precursor wave at elevated temperature results in an increased time interval between the elastic and plastic fronts as compared to values calculated for steady waves of the measured amplitudes. For thick samples this discrepancy reached 2–8 ns, depending on the temperature. Obviously, the velocity of the elastic precursor front is larger and the velocity of the plastic shock is smaller near the impact surface than these values far from the impact surface.

The results of measurements show that the dynamic yield strength increases linearly with increasing temperature. In plate impact experiments with thicker samples, the yield strength near the melting temperature exceeds its value at room temperature by a factor of at least four and the ratio of the yield strength to the shear modulus increases by an order of magnitude. For comparison, we note that at low strain rates this ratio for aluminum single crystals is maintained constant over the temperature range of 300–600 K and thereafter decreases to half of this value as the temperature is increased to 900 K (Berner and Kronmuller, 1965).

For the following discussion, let us consider the relationship between the resolved shear stress, τ, the plastic shear strain rate, $\dot{\gamma}$, and the mobile dislocation density, N_m:

$$\tau = \frac{B}{b^2 N_m} \dot{\gamma}, \tag{3.5}$$

where b is Burgers' vector and B is the drag coefficient. The initial density of mobile dislocations, N_m, in single crystal samples obviously does not depend on the temperature or, at least, does not decrease with heating. The observed compressive strain rate obviously does not increase with increasing temperature. Under these circumstances we have to conclude that the observed increase in the yield stress, Y, and, correspondingly, in the resolved shear stress, τ, is determined by an increase in the drag coefficient B.

The over-barrier motion of dislocations at high temperatures is decelerated by various obstacles and by friction forces due to phonons. The interaction of moving dislocations with electrons is essential only at low temperatures. The phonon drag coefficient B_p increases linearly with temperature (Ninomura, 1974):

$$B_p = \frac{k_B T \omega_D^2}{\pi^2 c^3}, \tag{3.6}$$

where k_B is the Boltzmann constant, ω_D is the Debye frequency, and c is the sound speed. The drag forces created by obstacles are obviously proportional to their concentration in the crystal structure. In particular, the equilibrium concentration of point defects grows exponentially with the temperature (Cheremskoy et al., 1990):

$$c_v = \exp\left(\frac{S_F}{k_B}\right) \exp\left(-\frac{H_F}{k_B T}\right) = A \exp\left(-\frac{H_F}{k_B T}\right), \tag{3.7}$$

where H_F is the defect formation enthalpy.

The measurements show a nearly linear increase of the dynamic yield stress with the temperature: Fig. 3.15 shows an increase in Y by a factor of approximately four when the absolute temperature increases 3.2 fold. Such behavior agrees reasonably well with Eq. (3.6) for the phonon friction coefficient. Thus, it seems very probable that the dislocation drag at high strain rates is connected mainly with thermal oscillations of atoms in the crystal lattice. A decrease in steepness of the plastic shock wave with heating evidences an increase in the material viscosity as a result of a smaller dislocation velocity at the resolved shear stresses prevailing in a steady shock wave.

It is interesting to note that the spall strength of aluminum decreases whereas the yield stress increases as the temperature is increased. It is known that the spall fracture consists essentially of processes of nucleation, growth, and coalescence of microvoids. Since the resistance to growth of pores is determined by the yield stress of the surrounding material and should increase with the latter, we may conclude that the growth of voids does not contribute very much to the stress relaxation at spalling and that the measured fracture stresses are controlled mostly by the nucleation processes. Probably a coalescence of vacancies at elevated temperature produces microvoids and thereby additional damage nucleation sites and decreases the total resistance to high-rate fracture by spalling. However, even in this case, it is not quite clear why the temperature sensitivity of the spall strength is less at higher strain rates.

3.3. Shock Yielding and Fracture of Some Alloys

Investigations of the strength properties of alloys, besides their practical importance, may provide additional information for analyzing the governing mechanisms of plastic deformation and fracture at the highest strain rates. A general trend of the mechanical behavior of shock-wave loaded metals is that, whereas the yield strength of pure metals under these conditions does not depend on the temperature or even abnormally increases with the temperature, alloys may exhibit normal behavior with decreasing yield strength as the temperature increases. Rohde, 1969, studied the dynamic yield behavior of iron over the temperature range of 76 to 573 K. The dynamic yield stress at a strain rate of $\sim 10^5$ s^{-1} was found to be independent of the temperature, in contrast to the highly temperature-sensitive quasi-static yield stress which decreased by a factor of 2.5 between 76 and 298 K. It was concluded that the motion of twinning dislocations is the mechanism of dynamic yielding in iron. Similar results have been obtained by Asay, 1974, for bismuth. Unlike the case of pure metals, experiments with stainless steel demonstrated a decrease in both the dynamic yield strength and the spall strength at the temperature of 980 K to half of their values measured at room temperature (Gu and Jin, 1998).

Figure 3.16 presents results of tests of Al-6%Mg alloy at normal and elevated temperature. These experiments are especially interesting because the room temperature yield stress of Al-6%Mg alloy is close to that found for aluminum single crystals at the highest temperatures. The wave profiles in Fig. 3.16

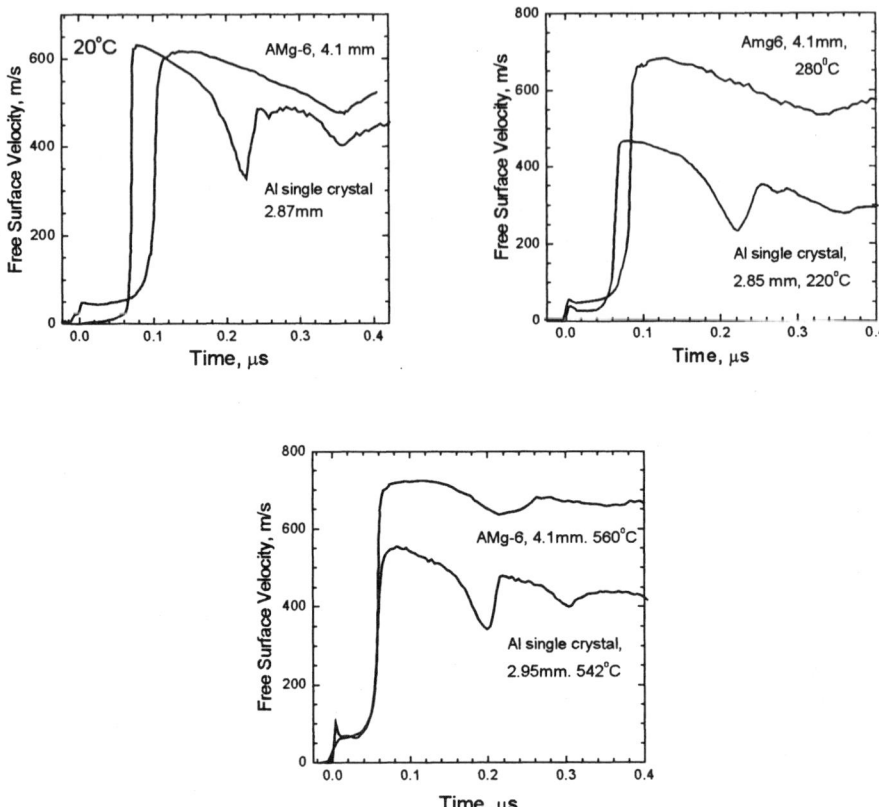

Figure 3.16. Free surface velocity histories of samples of Al-6%Mg alloy shocked at normal and elevated temperatures in comparison with the data for aluminum single crystals.

show that, whereas at room temperature the peak stress in the elastic precursor wave for the alloy is much larger than for single crystals, they gradually become equal as the temperature increases. The observation is in agreement with the hypothesis that phonon friction is the dominant mechanism at high strain rates and temperatures.

Figure 3.17 presents examples of free surface velocity profiles for the $\alpha-\beta$ titanium alloy Ti-6-22-22S (Krüger et al., 2002). An ingot of this alloy with the chemical composition (in wt. %): Al (5.75), Sn (1.6), Zr (1.99), Mo (2.15), Cr (2.10), Si (0.13), Fe (0.04), O (0.082), N (0.006), C (0.009) was prepared in a vacuum arc furnace and was diffusion annealed at 1100°C for 20 hours. After that, a swaging process at 900°C in the $\alpha+\beta$ region was performed. Finally, the material was solution annealed and aged.

3. Yield and Strength of Metals and Alloys at Elevated Temperatures   99

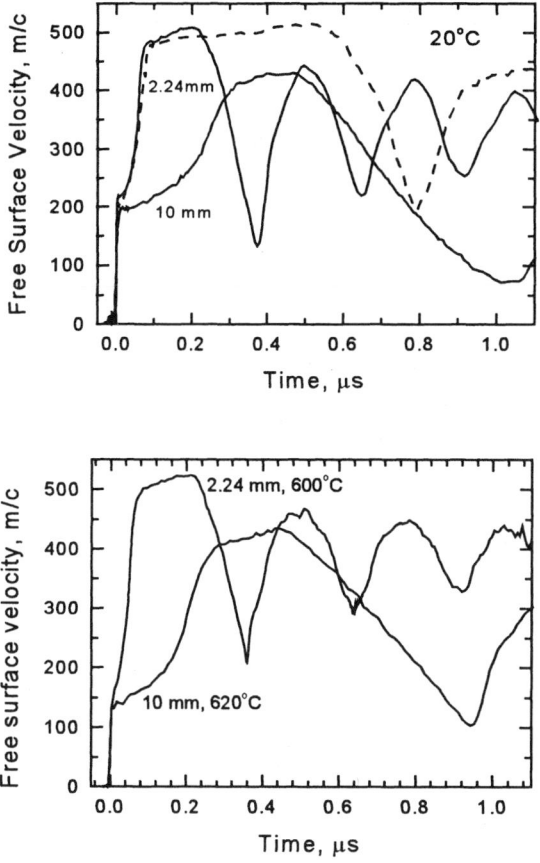

Figure 3.17. Free surface velocity histories of the Ti-6-22-22S titanium alloy samples at room and elevated temperatures.

High-temperature data have been analyzed with account taken of temperature derivatives of the shear modulus according to Guinan and Steinberg, 1974. The elastic–plastic waveforms contain information about the initial yield strength and the subsequent strain hardening. In order to get this information, the stress–strain diagrams have been recovered from the compressive parts of free surface velocity histories. The estimates have been done using the simple wave approach. The compression wave was considered as a centered simple wave, which is described by a fan of characteristics immediately behind the front of the elastic precursor wave. Figure 3.18 shows examples of recovered stress and strain histories at 20°C. The plastic strain, γ, was calculated according to the relationship

$$d\gamma = d\varepsilon_x - d\tau/G.$$

The initial dynamic yield strength values and the flow stresses at 0.2% plastic strain evaluated from the measured free surface velocity histories are presented in Fig. 3.19.

Figure 3.20 summarizes the room-temperature yield strength data evaluated from the shock-wave tests under uniaxial strain conditions and from the uniaxial stress Hopkinson bar tests at lower strain rates. The yield stresses at 0.2% plastic deformation are plotted because it is difficult to determine precisely the initial Y values from the Hopkinson bar tests. The strain rates during shock compression have been estimated as average compression rates in the middle sections of the samples. Figure 3.21 summarizes both the spall strength data over the temperature range from –170°C to 620°C and the yield strength data. The spall strength

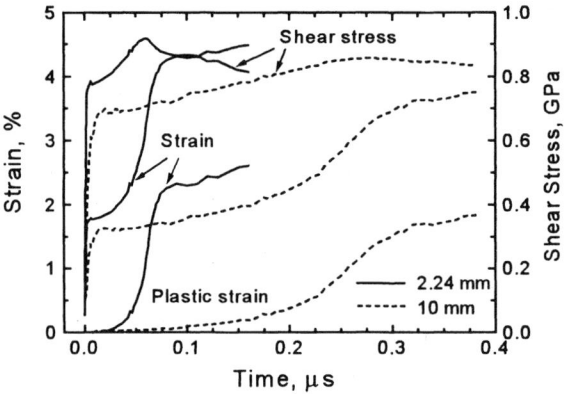

Figure 3.18. Examples of recovered stress and strain histories at 20°C.

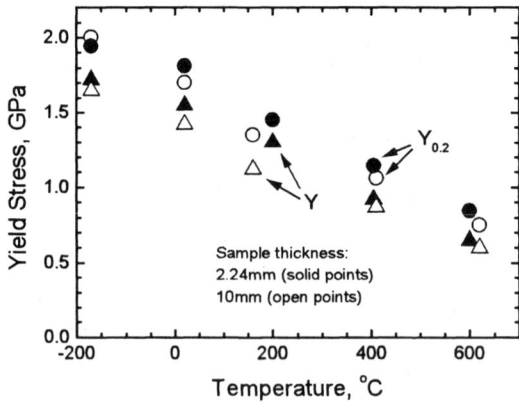

Figure 3.19. Dynamic yield data as a function of temperature for different impact conditions.

data do not vary as much as the yield strength and exhibit a maximum tensile strength near room temperature.

The strain-rate and the temperature dependencies of the yield stress obtained from the uniaxial stress tests and from the shock-wave experiments are in good agreement and demonstrate, in general, a logarithmic dependence over the strain rate range of 10^{-4} s^{-1} to 10^6 s^{-1}. This indicates that the thermal activation mechanism of plastic deformation of the alloy is maintained.

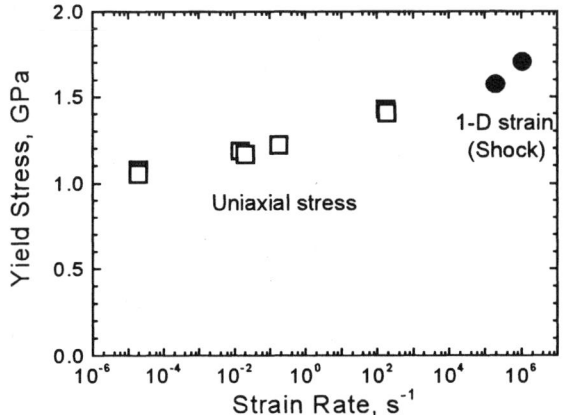

Figure 3.20. Yield strength at 0.2% plastic strain of Ti-6-22-22S alloy as a function of the strain rate.

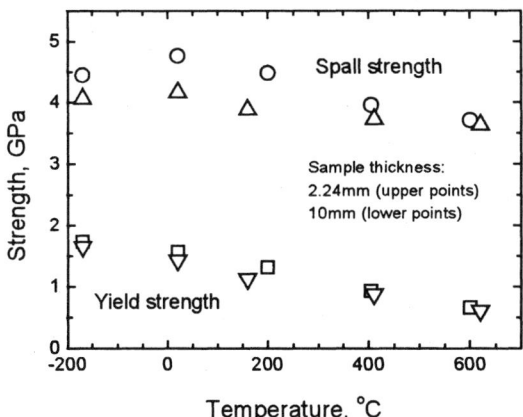

Figure 3.21. Spall strength of Ti-6-22-22S alloy as a function of temperature at two load durations in comparison with the yield strength data.

Figure 3.22 compares yield strength data for the Ti-6-22-22S alloy, commercial grade titanium, and titanium of 99.99% purity. The commercial titanium contained (in wt.%) O_2: 0.15, Fe: 0.10, Cr: 0.018, Ni: 0.015, C: 0.016, Al < 0.02, and Cu, Zr, V, Mn <0.01 each. The 2-mm-thick samples were cut from a textured rolled sheet. The specimens of pure titanium were cut from a rod of high-purity titanium produced by the electron-beam zone melting technique. Both the commercial and the high-purity titanium of were in the α phase.

It is well known that, under normal conditions, the yield stress and tensile strength of titanium increase with increasing the oxygen content (see, for example, Zwicker, 1974). As a result, at room temperature the yield strength of pure titanium is much lower than that of commercial titanium with high oxygen content. However, with increasing the temperature this difference decreases. Whereas the yield strength of commercial titanium decreases with increasing temperature, pure titanium demonstrates the opposite trend. As in the case of aluminum, pure titanium shows an anomalous increase of the yield strength with increasing temperature.

In general, the different behavior of pure titanium and titanium alloys does not contradict the assumption about a transition in the rate-controlling mechanism. The flow stress in the pure metal is small and comparable with the phonon friction forces, therefore growth of the latter makes an essential contribution to the drag of the dislocations. Note that the test temperature was far below the melting temperature of titanium. As a result, a possible temperature-dependent contribution of defect concentration to the drag forces should be insignificant. In contrast to pure metals, alloys contain numerous obstacles such as inter-phase boundaries, inclusions, etc., that have been created specifically to increase the

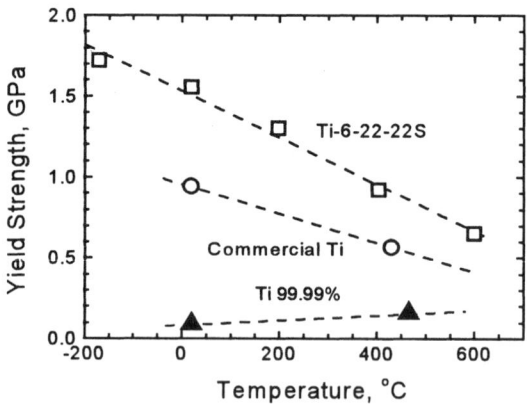

Figure 3.22. Temperature dependencies of dynamic yield strengths of Ti-6-22-22S alloy, commercial grade titanium, and titanium of 99.99% purity. Data by Kanel et al., 2003.

yield strength. The stress needed to overcome these obstacles far exceeds the forces of phonon drag which are, therefore, unable to make a significant contribution into the resistance of the alloys to plastic flow.

Figure 3.23 compares the temperature dependencies of spall strength data for the titanium alloy, commercial titanium, and pure titanium. Despite the multifold difference in the yield strength, the dynamic strength values do not differ very much for these materials. In the case of pure titanium, the temperature dependence of the dynamic yield strength and tensile strength does not agree even in its sign.

It may be assumed that both the yield stress and the tensile strength of high-strength alloys decrease monotonically with heating and increase with increasing strain rate. However, the response of complicated materials may not necessarily follow this common trend. Figures 3.24 and 3.25 show the temperature dependence of the dynamic yield strength and the spall strength of Inconel IN 738 LC alloy. Inconel is a nickel-based superalloy used as turbine-blade material. This two-phase alloy consists of an fcc γ-phase matrix with a high volume fraction of embedded γ' particles. The strengthening precipitates have a long-range-ordered structure with suitable γ/γ' lattice mismatch. The alloy maintains high strength properties at temperatures up to ~800°C due to structural rearrangements that occur during heating. In particular, the structural transformations manifest themselves in the heat capacity. The increase of heat absorption was attributed by Brooks et al., 1978, to equilibration of the short-range order in the γ-phase matrix in the temperature interval 500–700°C. The heat capacity anomaly at 670°C is accompanied by a small decrease of the shear modulus whereas the bulk modulus varies insignificantly over the temperature range 20–1000°C (Fukuhara and Sanpei, 1993).

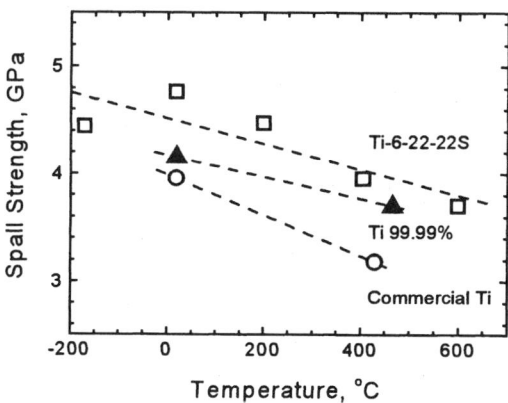

Figure 3.23. Temperature dependence of the spall strength of Ti-6-22-22S alloy, commercial grade titanium, and titanium of 99.99% purity.

The data in Fig. 3.24 show non-monotonic variation of the flow stress with increasing temperature. The general trend is that the flow stress decreases with increasing temperature up to ~500 °C, then sharply recovers with further heating to ~550 °C, and after that decreases again. The response is highly unstable at temperatures around 550 °C, which is associated with the short-range disordering. The spall strength, again, is less sensitive to the temperature than is the yield stress.

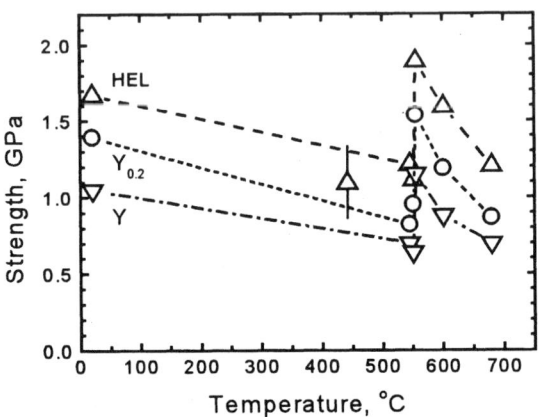

Figure 3.24. Hugoniot elastic limit, initial yield stress, Y, and the flow stress at 0.2% plastic strain, $Y_{0.2}$, for IN 738 LC as a function of temperature.

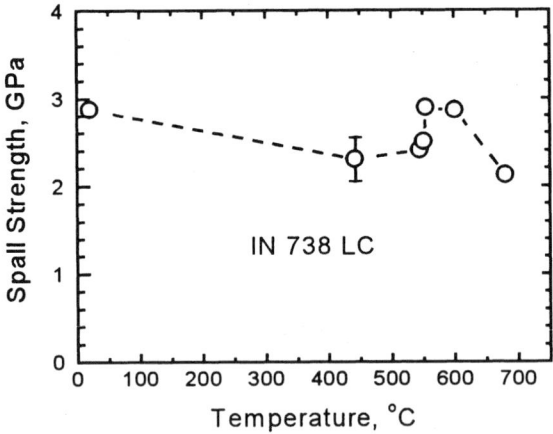

Figure 3.25. Temperature dependence of spall strength for polycrystalline IN 738 LC alloy.

The experimental data for aluminum single crystals shown in Fig. 3.10 demonstrate a decrease of temperature sensitivity of the spall strength with increasing strain rate. Figure 3.26 presents similar data for cobalt. The temperature sensitivity of the dynamic tensile strength of cobalt also disappears with increasing the strain rate. Obviously this is a quite natural trend: If the temperature dependence of the dynamic strength is less than that of the quasi-static strength, it should decrease with a further increase of the strain rate.

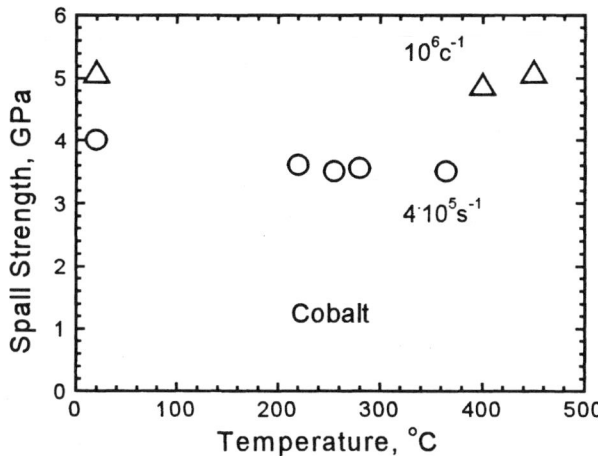

Figure 3.26. Spall strength of cobalt as a function of the temperature and the rarefaction rate. Data by Razorenov et al., 2002.

3.4. Discussion

Even the first studies of the yield and strength properties of metals and alloys at elevated temperatures and very high strain rates of shock-wave tests disclosed new interesting phenomena which could be expected but had not been discussed previously. The data show that the influence of temperature on the yield stress at high strain rates may be opposite to that at low and moderate strain rates. The dynamic tensile strength of metal single crystals and polycrystalline metals exhibit different temperature dependencies near the melting point. Different response of single crystals and polycrystalline metals is treated in terms of superheated states and pre-melting. Thus, a new way to study superheated solid states and melting at negative pressures is opened.

It may be expected that materials with thermal hardening behavior should have an increasing resistance to high-rate deformation under adiabatic conditions. As a result, these materials should not be able to form adiabatic shear bands. In the process of high-speed cutting, thermal hardening may produce the same effect as blunting of the cutter does.

However, the known observations are not systematic and do not allow establishing a correlation between the material structure and its high-strain-rate properties or predicting its response at elevated temperatures. A more systematic study is needed in order to establish a criterion for determining the transition from thermal softening to thermal hardening of materials of different structure.

Thermal hardening and temperature-independent response was observed in pure metals, but high-strength alloys are much more important for applications. In the context of the simplified hypothesis of thermal hardening discussed above, alloys contain higher barriers, such as inclusions, inter-phase boundaries, etc., for movement of dislocations. As a result, one may suppose that the transition to thermal hardening may be shifted toward higher strain rates. On the other hand, the hypothesis does not account for other structure-sensitive components of the plastic deformation mechanisms, such as multiplication of dislocations, their splitting, and twinning, that may have contributed even more to the yield stress.

The problem of real mechanisms and kinetics of melting is still unsolved and attractive. Whereas superheated solid states are allowed by theory and were observed in experiments, nothing is known about mechanical and thermophysical properties of matter in these exotic metastable states, and about real temperature and time limits of their existence. The presence of inter-granular surfaces in polycrystalline material introduces local distortion of the crystal lattice. The excess free energy of the grain boundary can be reduced by means of pre-melting at temperatures below the bulk melting temperature. We may suppose a correlation exists between the pre-melting and the stored energy of grain boundaries. Actually, the link between micro-level and meso-level behavior of solids occurs through the grain boundaries and so the grain-boundary effects are essential to the overall mechanical response of the material. In particular, pre-melting under tension near crack tips may stimulate nucleation of damage and growth of the crack. However, whereas pre-melting phenomena were observed in experiments, their influence on mechanical properties is not yet entirely clear.

The results of experiments stimulate an interest in studying phase transitions and polymorphic transformations in the negative pressure region and promise verification of the existing models of melting and the limit criteria for stability of superheated solid states, to prolong the melting curve $T_m(p)$ into the negative pressure region, to evaluate kinetics of the melting process, to provide measurements of elastic constants and strength properties of crystals in metastable states, to relate pre-melting and strength properties of polycrystalline solids, and to establish a link between multiscale levels of melting phenomena.

References

Antoun, T., L. Seaman, D.R. Curran, G.I. Kanel, S.V. Razorenov, A.V. Utkin (2003). *Spall Fracture*. Springer, New York

Asay, J.R. (1974). "Shock-Induced Melting in Bismuth," *J. Appl. Phys.* **45**, p. 4441.

Asay, J.R., and D.B. Hayes (1975). "Shock compression and release behavior near melt states in aluminum," *J. Appl. Phys.* **46(11)**, pp. 4789–4800.

Berner, R., and H. Kronmuller (1965). *Plastische Verformung von Einkristallen*, Springer-Verlag, Berlin.

Besold, G., and O.G. Mouritsen (1994). "Grain-boundary melting: a Monte Carlo study," *Physical Review B* **50(10)**, pp. 6573–6576.

Bhate, N., R.J. Clifton, and R. Phillips (2002). "Atomistic simulations of the motion of an edge dislocation in aluminum using embedded atom method," in: *Shock Compression of Condensed Matter — 2001* (eds. M.D. Furnish, N.N. Thadhani, and Y. Horie) American Institute of Physics, New York, pp. 339–342.

Bogach, A.A., G.I. Kanel, S.V. Razorenov, A.V. Utkin, S.G. Protasova, and V.G. Sursaeva (1998). "Resistance of zinc crystals to shock deformation and fracture at elevated temperatures," *Phys. Solid State* **40(10)**, pp. 1676–1680 [trans. from *Fiz. Tverd. Tela* **40**, pp. 1849–1854 (1998)].

Boyer, L.L. (1985). "Theory of melting based on lattice instability" *Phase Transitions* **5(1)**, pp. 1–48.

Brooks, C.R., M. Cash, and A. Garcia (1978). "The heat capacity of Inconel 718 from 313 to 1053K," *J. Nuclear Materials* **78(2)**, pp. 419–421.

Cahn, R.W. (1986). "Melting and the surface," *Nature* **323**, pp. 668–669.

Cheremskoy, P. G., V.V. Slezov, and V.I. Betehtin (1990). "Pores in Solids," *Energoatomizdat*, 376 p. (in Russian).

Clifton, R.J. (1971). "Plastic waves: theory and experiment," in: *Shock Waves and the Mechanical Properties of Solids* (eds. J.J. Burke and V. Weiss), Syracuse University Press, pp. 73–116.

Cotterill, R.M. (1980). *J. Cryst. Growth* **48**, p. 582.

Curran, D.R., L. Seaman, and D.A. Shockey (1987). "Dynamic failure of solids," *Physics Reports* **147(5&6)**, pp. 254–388.

Dash, J.D. (1999). "History of the search of continuous melting," *Rev. Mod. Phys.* **71(5)**, pp. 1737–1743.

Fecht, H.J. and W.L. Johnson (1988). *Nature* **334**, p. 50.

Follansbee, P.S. and J. Weertman (1982). "On the question of flow stress at high strain rates controlled by dislocation viscous flow," *Mech. Mater.* **1**, pp. 345–350.

Fukuhara, M. and A. Sanpei (1993). "Elastic moduli and internal frictions of Inconel 718 and Ti-6Al-4V as a function of temperature," *J. Mater. Sci. Lett.* **12(14)**, pp. 1122–1124.

Gu, Zhuowei, and Xiaogang Jin (1998). "Temperature dependence on shock response of stainless steel," In: *Shock compression of condensed matter — 1997*, in: *Shock Compression of Condensed Matter—1997* (eds S.C. Schmidt, D.D. Dandekar, and J.W. Forbes) American Institute of Physics, New York, pp. 467–470.

Guinan, M.W., and D.J. Steinberg (1974). "Pressure and temperature derivatives of the isotropic polycrystalline shear modulus for 65 elements," *J. Phys. Chem. Solids* **35**, pp. 1501–1512.

Iwamatsu, M. (1999). "Homogeneous nucleation for superheated crystal," *J. Phys.–Condensed Matter* **11(1)**, pp. L1–5.

Johnson, J.N. (1981). "Dynamic fracture and spallation in ductile solids," *J. Appl. Phys.* **52(4)**, pp. 2812–2825.

Johnson, J.N., and D.L. Tonks (1992). "Dynamic plasticity in transition from thermal activation to viscous drag," in: *Shock Compression of Condensed Matter—1991* (eds. S.C. Schmidt, R.D. Dick, J.W. Forbes and D.G. Tasker) North-Holland, Amsterdam, pp. 371–378.

Kanel, G.I. (2000). "The temperature limit of the dynamic strength of metals," *High Temp.* **38(3)**, pp. 481–491 [trans. from *Teplofiz. Vys. Temp.* **38(3)**, pp. 512–514 (2000)].

Kanel, G.I., S.V. Razorenov, K. Baumung, and J. Singer (2001). "Dynamic yield and tensile strength of aluminum single crystals at temperatures up to the melting point," *J. Appl. Phys.* **90(1)**, pp. 136–143.

Kanel, G.I., S.V. Razorenov, A.A. Bogatch, A.V. Utkin, V.E. Fortov, and D.E. Grady (1996). "Spall Fracture Properties of Aluminum and Magnesium at High Temperatures," *J. Appl. Phys.* **79(11)**, pp. 8310–8317.

Krüger, L., G.I. Kanel, S.V. Razorenov, L. Meyer, and G.S. Bezrouchko (2002). "Yield and strength properties of the Ti-6-22-22S alloy over a wide strain rate and temperature range," in: *Shock Compression of Condensed Matter—2001* (eds. M.D. Furnish, N.N. Thadhani, and Y. Horie) American Institute of Physics, New York, pp. 1327–1330.

Kumar, A., and R.G. Kumble (1969). "Viscous drag on dislocations at high strain rates in copper" *J. Appl. Phys.* **40(9)**, p. 3475.

Lu, K., and Y. Li (1998). "Homogeneous nucleation catastrophe as a kinetic stability limit for superheated crystal," *Phys. Rev. Lett.* **80(20)**, pp. 4474–4477.

Lu, J., and J.A. Szpunar (1995). "Molecular dynamics simulation of the melting of a twist grain boundary," *Interface Sci.* **3(2)**, pp. 143–150.

Lynden-Bell, R.M. (1995). "A simulation study of induced disorder, failure and fracture of perfect metal crystals under uniaxial tension," *J. Phys.–Condensed Matter* **7**, pp. 4603–4624.

Ma, D., and Y. Lip (2000). "Heterogeneous nucleation catastrophe on dislocations in superheated crystals," *J. Phys.–Condensed Matter* **12(43)**, pp. 9123–9128.

Mabire, C., and P.L. Hereil (2000). "Shock induced polymorphic transition and melting of tin" in: *Shock Compression of Condensed Matter—1999* (eds. M.D. Furnish, L.C. Chhabildas, and R.S. Hixon) American Institute of Physics, New York, pp. 93–96

Ninomura, T. (1974). *J. Phys. Soc. Jpn.* **36**, p. 399.

Razorenov, S.V., G.I. Kanel, K. Baumung, and H. Bluhm (2002). "Hugoniot elastic limit and spall strength of aluminum and copper single crystals over a wide range of strain rates and temperatures," in: *Shock Compression of Condensed Matter—2001* (eds. M.D. Furnish, N.N. Thadhani, and Y. Horie) American Institute of Physics, New York, pp. 503–506.

Razorenov, S.V., G.I. Kanel, E. Kramshonkov, and K. Baumung (2002). "Shock compression and spalling of cobalt at normal and elevated temperatures," *Comb. Expl. Shock Waves* **38(5)**, pp. 598–601 [trans. from *Fiz. Goreniya Vzryva* **38(5)**, pp. 119–123 (2002)].

Rohde, R.W. (1969). "Dynamic Yield Behavior of Shock-Loaded Iron from 76 to 573°K," *Acta Met.* **17**, pp. 353–363.

Sakino, K. (2000). "Transition in the rate controlling mechanism of FCC metals at very high strain rates and high temperatures," *J. Phys. IV France* **10**, Pr 9-57–62.

Tallon, J. L. (1989). *Nature* **342**, p. 658.

Tallon, J.L., and A. Wolfeden (1979). "Temperature dependence of the elastic constants of aluminum," *J. Phys. Chem. Solids* **40**, pp. 831–837.

Ubbelohde, A.R. (1965). *Melting and Crystal Structure*, Clarendon Press, Oxford.

Zwicker, U. (1974). *Titan und Titanlegierungen*, Springer-Verlag, Heidelberg.

CHAPTER 4

Behavior of Brittle Materials under Shock-Wave Loading

4.1. Introduction

Interest in the response of brittle materials to dynamic loading is related to many applications including explosive excavation of rocks, design of hard ceramic armor, meteorite impacts on spacecraft windows, impact of condensed particles on turbine blades, etc. When a brittle material is subjected to an impact or explosive attack, inelastic deformation, fracture, and fragmentation occur under conditions of three-dimensional stress, where at least one stress component is compressive. It is known that the behavior of brittle materials under quasi-static compression is characterized by such features as compressive fracture, dilatancy, and pressure-sensitive yielding which do not permit the use of classical elastic–plastic constitutive models. Recent review papers (Bombolakis, 1973, Kranz, 1983, and Wang and Shrive, 1995) show that, although the fracture and fragmentation of brittle materials under tension is more or less clear, the governing mechanisms of compressive fracture are not quite clear even for quasi-static conditions. The rapid application of load can also introduce the effect of rate dependencies which make the problem even more complicated.

There are many definitions of ductile and brittle behavior of solid materials. Failure refers to the loss of most or all of the ability of the material to support loads. Phenomenologically, brittle response is associated with failure without significant plastic deformation, whereas ductile response is associated with large plastic deformation which does not necessarily result in failure. Microscopically, plastic deformation occurs through microshears produced by dislocation motion and is not associated with the appearance of cracks or other discontinuities. The plastic deformation does not influence the density of the material. Brittle behavior, in contrast, is accompanied by nucleation and growth of a crack or distribution of cracks which increase the volume of the material. The tensile strength of brittle materials is usually much less than the compressive strength.

Griffith originally explained the low tensile strength of elastic solids in terms of an inequality between the rate at which elastic energy is released and the rate at which surface energy is absorbed as a flaw or crack extends within the material. This approach works reasonably well for brittle materials. For ductile materials, the work of plastic deformation in the vicinity of growing flaws must be added to the surface energy if agreement between the theory and the observations is to be achieved. When the work that is absorbed in producing an incremental increase of the flaw surface area increases, the resistance to crack growth also increases. For brittle materials this work is two orders of magnitude less than for ductile materials.

One should say that no absolutely brittle or absolutely ductile materials exist. Depending on the stress state and the loading conditions, the response of almost any solid can be either brittle or ductile. However, although cracks can form in most ductile materials in states of stress approaching 3D tension, most brittle materials can be deformed in a ductile manner only under conditions of high pressure. In fact, any solid material should exhibit ductility if the loading conditions suppress the cracking and create sufficiently high shear stresses in the body.

In this chapter we summarize and compare the behavior of various kinds of brittle materials, including single crystals, glasses, and ceramics, under shock-wave loading. Of course, impact and penetration problems are associated with divergent flows but the simple uniaxial strain conditions of a planar shock-wave test are a limiting case. If it is found that the response of a material under shock compression is brittle in nature, this means its behavior will also be brittle when it is subjected to more general impact conditions. Other dynamic tests, such as rod impact and Hopkinson pressure bar tests, are performed under uniaxial stress conditions.

4.2. General Behavior of Brittle Materials Under Compression

The shape of a body should change in response to application of a non-hydrostatic stress field. Above the elastic limit, inelastic deformation changes the form of a body by shearing along inclined slip planes. In isotropic media the slip planes are inclined at 45° to the principal stress directions. In ductile crystalline solids, the slip occurs as a result of the movement of dislocations. For crystals with high lattice symmetry and for polycrystalline ductile solids, many slip planes with different orientations are activated more or less simultaneously. As a result, the global shear is accommodated by slip along many planes and in many directions, and plastic deformation is macroscopically homogeneous. Hard solids with covalent inter-atomic bonds have a lower crystallographic symmetry, higher Peierls energy barriers to dislocation motion, and a smaller number of crystallographic planes on which dislocations can move. As a consequence, slip on optimally oriented planes starts when the applied force becomes high enough, but it is not accompanied by slip on other planes. If dislocation motion is stopped inside the body, for example, by a grain boundary, this should result in opening of voids or cracks. It is hardly possible that such cracks can appear on planes that are perpendicular to the axis of compression, but cracks can appear on planes that are parallel to the compression direction. It is natural to suppose also that a lateral stress or confining pressure will hinder the opening of an axial crack. Cracking leads to an irreversible loss in the resistance of the material to deformation.

Contrary to the response of ductile materials, the compressive fracture of brittle materials under one-dimensional stress conditions or at relatively low confining pressure often occurs by axial splitting. For greater confining pressures, failure occurs by shear faulting at an angle less than 45° to the loading axis. Extensive compressive fracture is preceded by microcracking. The orientation of the microcracks is predominantly within 10° of the direction of compression (Wawersik and Brace, 1971). Crack density increases as the macroscopic deviatoric stress increases above a distinct threshold level. Faults and other macroscopic fractures appear to form after attainment of the ultimate compressive stress, which is called the failure stress. Beyond the point of peak load, the failure becomes unstable. In the post-failure region of compression, the load-carrying capacity of the material drops rapidly to a low value.

Open cracks and other voids can appear in solids only under tensile stress. It is known that, even when the applied stress is wholly compressive, the local stress can become tensile at certain points of non-uniformity (McClintock and Argon, 1966) and, hence, cracks may appear and grow in response to the tensile stress in this localized region. Based on Inglis's solution of the stress distribution around an elliptical hole in a plate, and assuming that the isotropic material contains cracks of all directions and that the maximum crack length, as well as the radius of curvature at the tip of the crack, is the same for cracks of all orientations, Griffith, 1924, derived a fracture criterion for biaxial stress conditions. He assumed that fracture occurs when the highest local tensile stress at the longest crack of the most vulnerable orientation reaches a fixed critical value. Localized regions of tensile stress may be created by stress concentrations at grain boundary contacts, around microcracks and cavities in the undeformed materials, at points where twin lamellae interact with grain boundaries and other twins, at kink bands, or, at a larger scale, as a result of grain rotations and translations.

Complete macroscopic fracture under compression is a result of the coalescence of many microcracks. As deviatoric stress increases and material failure is approached, the microcrack population changes spatially from random to locally intense zones of cracking. Scholz, 1968, observed (for a Westerly granite) rapid acceleration in the micro-seismic activity within a few per cent of the failure stress and distinct clustering of microfractures in a single inclined plane coincident with the eventual fault plane. Post-failure examination of microcracks associated with the fracture indicate the density is very high in the fault region and dies off rapidly to the background level a few grains away (the references have been collected by Kranz, 1983).

Since cracks occupy volume, their formation is accompanied by a decrease in the average density of the material. This nonlinear inelastic volume change is commonly referred to as dilatancy or bulking (Brace et al., 1966). Typically, the onset of dilatation occurs at a stress between one third and two thirds of the failure stress. The bulking effect increases with increasing deviatoric stress and decreases under confining pressure. Unloading from this stress region yields a permanent residual volume increment. For rocks, the magnitude of the dilatancy ranges from 20 to 200% of the elastic volume change that would have occurred were the rock simply elastic; the maximum porosity below the failure stress is 2%.

The dilatancy is accompanied by a hysteresis of the physical properties between loading and unloading which is manifested mostly in the lateral strains. The axial strain is nearly elastic and almost completely recoverable. The lateral dilatational strains are attributed to opening of axial cracks. Formation of open axial cracks at a fraction of the maximum stress that can be applied is also suggested by variations of sound velocity in axial and transverse directions: The sound velocity in the axial direction is hardly changed by stress whereas the sound velocity in a transverse direction begins to decrease at about half the failure stress and may drop 10–20% (Brace et al., 1966).

A confining pressure strongly affects the strength and inelastic behavior of brittle materials. The deviator stress at which microcracking starts or failure occurs increases as the confining pressure increases. At some high pressure a transition from brittle to ductile response usually occurs. Heard and Cline, 1980, observed failure followed immediately after an essentially elastic deformation of Al_2O_3, AlN, and BeO when the confining pressure was low, but there was a transition to more ductile response at high pressures. As an example, Fig. 4.1 presents data by Heard and Cline for a BeO ceramic. The ultimate compressive

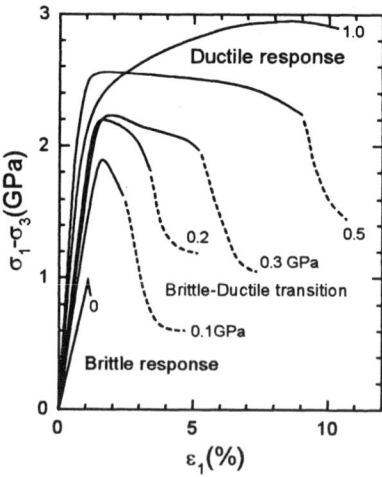

Figure 4.1. Uniaxial compressive loading of BeO ceramic. Data by Heard and Cline, 1980.

strength of ceramics increases rapidly with pressure below the brittle–ductile transition; above this threshold the ultimate strength is nearly constant (Fig. 4.1). At low confining pressure, the failure occurred by macroscopic shear on fault planes oriented at about 30° to the compression axis, although all the rest of the material contained numerous axial microcracks. At a confining pressure exceeding 0.5 GPa, the dominant deformation mechanism of BeO and AlN became intracrystalline slip by dislocation motion, with attendant pile-up at grain boundaries. These ceramics also exhibit increasing ductility when the confining pressure rises above the brittle–ductile transition pressure.

The transition pressure from brittle to ductile response is different for different materials. Alumina, for example, remains brittle at confining pressures at least up to 1.25 GPa. The transitions between damage accumulation, structural failure, and ductile response of a test specimen are often not sharp. It is better to speak of dominant mechanisms: Below some pressure, brittle fracture dominates whereas above the transition ductile flow dominates. For example, Hockey, 1971, using transmission electron microscopy, found extensive evidence for both slip and twinning when alumina was indented at room temperature. Chan and Lawn, 1988, also observed basal twinning and pyramidal slip from indentation deformation of sapphire. These shear processes operated close to the cohesive limit. The factor which permits plastic flow under indentation is the intense confining pressure induced within the elastic field surrounding the indentation.

4.3. Possible Mechanisms of Microcracking Under Compression

Fracture does not occur under hydrostatic compression. Some stress difference is needed to open cracks when the body is being squeezed. A non-hydrostatic stress field implies non-zero shear stresses and shear strains along some planes inside the body. Let us try first to image qualitatively the reasons for nucleation of cracks by a shear. Even intuitively, it appears that, if easy shear is allowed within a limited band with fixed tips inside a body, it should create rarefaction and compression regions near the tips, as illustrated schematically in Fig. 4.2. There should also be concentration of shear stresses in the crack plane ahead of the tip. The rarefaction may initiate a tensile crack that can grow out of the crack plane in the direction perpendicular to that of maximum tension, whereas localized shear may propagate further in the crack plane. For different materials and various loading rates, Kalthoff, 2000, observed both these modes of shear failure initiated at the crack tip depending, in general, on a ratio of yield strength to tensile strength. In his experiments, a relatively brittle epoxy resin showed failure by tensile cracks, ductile aluminum alloy samples failed due to shear band processes, and high strength steel showed a transition of the failure mode from tensile cracking at low loading rates to shear banding at higher rates of loading. The data indicate that shear bands require and absorb more energy for propagation than tensile cracks.

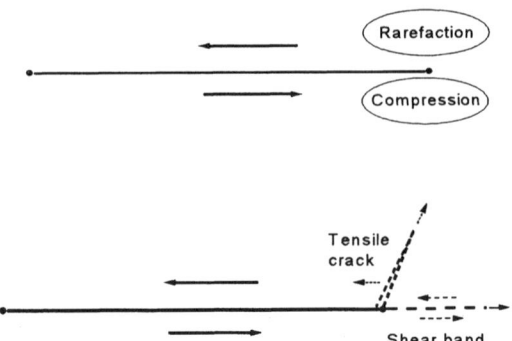

Figure 4.2. Schematic of failure initiation at a mode-II crack tip.

Brace and Bombolakis, 1963, observed the growth of cracks in glass and polymer plates under compression. They found that the most severely stressed cracks were inclined at about 30° to the axis of compression. The cracks, when either isolated or placed in an array, grow along a curved path that becomes parallel to the direction of compression as shown in Fig. 4.3. When this direction is attained, growth stops. The resultant kinked crack, called a "wing crack," consisted of a central crack with sliding surfaces, which is inclined to the direction of compression, and two cracks emanating from its ends. Modern theories of brittle fracture and dilatancy under compression are mostly based on development of the wing crack model.

More recently, a series of similar experiments was performed by Nemat-Nasser and Horii, 1982, and Horii and Nemat-Nasser, 1985. They have also shown that the relative sliding of the faces of one or even an array of pre-exist-

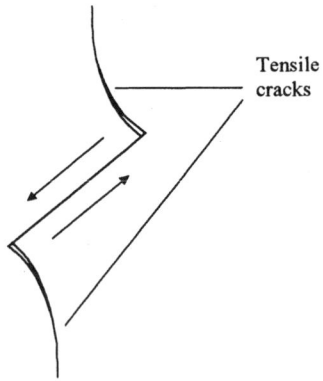

Figure 4.3. Wing crack.

ing cracks leads to the formation of tensile cracks which grow in the direction of maximum compression. Lateral compression reduces the final crack length whereas even a small amount of lateral tension increases it. In experiments with specimens containing a number of randomly oriented cracks or rows of inclined coplanar cracks, axial splitting rather than localized shear failure was observed. When models that initially contained flaws of various sizes were tested, cracks were nucleated at the larger flaws first. In the presence of confinement, smaller flaws soon become activated as the cracks emanating from the larger flaws are arrested because of the confining pressure. Then the smaller flaws interact with each other and nucleate cracks that grow in an unstable manner, leading to final failure.

Griffith, 1924, postulated that isotropic materials contain randomly oriented flaws or cracks in all directions which significantly alter the stress field within the material. The basic hypothesis of Griffith's model is that fracture occurs when the most vulnerably oriented crack begins to extend under applied stress. The extension of the crack is assumed to occur when the maximum tensile stress component at any point around the crack reaches the critical value needed to overcome the inter-atomic cohesion of the material. The Griffith theory, or at least its basic premise that fracture starts from flaws, is fundamental to all investigations of brittle fracture.

Thus, crack growth is directed preferentially along the compression axis and does not immediately produce mechanical instability as it does in tension. The axial tensile cracks, which can open and grow at stress concentrators under overall compression, cannot themselves be considered as a mechanism of inelastic shear strain, but they may facilitate shear and rotation of blocks of material relative to each other and in this way to contribute to deformation. The stress required to provide subsequent crack growth increases after some crack growth has occurred. Macroscopic faults are formed out of a system of cracks.

Modern theories of brittle fracture and dilatancy under compression are based mostly on development of wing crack models. The wing crack system is really two or three interacting cracks: a shear crack that slides under non-hydrostatic stress to pull open one or two tensile cracks at the tips of the shear crack. However, such idealized wing cracks are very rarely observed in rocks after compression tests. Even with a resolution of 0.01 μm, Tapponier and Brace, 1976, were able to find only a few of them. Sets of grain boundary, low-aspect cavities, as well as suitably oriented interfaces of two different minerals, produced most of the microcracks, which were long (a few hundreds microns) and were preferentially oriented with their long axes approximately parallel to the axis of maximum compressive stress.

On the other hand, it seems reasonable that localized sliding would take place along a plane of high shear, especially if weak areas such as sets of grain boundary cavities or soft inclusions occur at these orientations. In this way, the wing crack model still presents a reasonable idealization. A sliding element of this kind may have frictional resistance, or cohesive shear resistance, or both. The shear resistance along the sliding surfaces acts in opposite directions depending on whether shear stress on the sliding surfaces is increasing or decreasing. This produces hysteresis in the loading–unloading cycle.

Within the framework of linear fracture mechanics, Nemat-Nasser and Horii, 1982, analyzed the out-of-plane extension of a pre-existing straight crack, induced by overall far-field compression, and quantified various parameters that characterize the growth process. For this, the stress intensity factors at the tip of the kinked extension were obtained, and the kink angle and the corresponding kink length were calculated. For a wide range of pre-existing crack orientations, it was shown that the out-of-plane crack extension initiates at an angle close to 70° from the direction of the pre-existing crack. It was shown that under 1D stress conditions, the stress intensity factor, K_I, decreases monotonically and approaches zero as the ratio of the kink length to the incipient crack length, l/c, increases. Thus, to increase the kink length, the axial compression must be increased, and therefore the crack growth process is stable. For biaxial loading with lateral compression the K_I value becomes zero at some kink length. This implies that the crack extension is stable and the kink stops at some finite length. On the other hand, when even a small lateral overall tension is applied normal to the compressive force, the value of K_I first decreases until a critical kink length is attained, and then begins to increase as l/c increases. This implies that the compressive axial force required to attain the corresponding kink length must first be increased with increasing l/c until the critical l/c is reached, and then it must be decreased with further kink extension. Thus, kink growth under compression with a lateral tension is first stable and then becomes unstable.

These results explain the axial splitting of brittle bodies under compression, although it is evident that interaction between microcracks, which can lead to unstable growth of a tension crack, must dominate the intermediate stages of the failure process. At certain stages of crack growth the interaction effects become so dominant that only certain cracks continue to grow in an unstable fashion while others are arrested or even closed. It is important also that even when the initial flaws are randomly distributed, their preferential activation, and the nucleation and growth of tension cracks at preferred flaws, render the overall response of the solid highly anisotropic. Mechanistic models of flaw nucleation and growth are helpful in understanding mechanisms but they are too complicated to be used in practical calculations.

The dilatancy effect produced by wing cracks was calculated by Holcomb, 1978, Stevens and Holcomb, 1980, Moss and Gupta, 1982, and Nemat-Nasser and Obata, 1988. The various wing crack models provide excellent qualitative (and reasonable quantitative) agreement with experimental observations of dilatancy. It is also found that including sliding friction gives larger hysteresis than observed, so the friction coefficient during unloading must approach zero (Moss and Gupta, 1982).

4.4. Dynamic Strength Properties of Brittle Single Crystals

4.4.1. Sapphire

Figure 4.4 presents particle velocity histories measured at the interface between a sapphire or ruby sample and a water window (Kanel et al., 1993, 1999). These data exhibit most of the peculiarities of the response of single crystals of hard brittle materials to shock-wave loading. In the experiment with ruby, the peak stress did not exceed its HEL. The interface velocity history is smooth and mimics the shape of the stress pulse inside the sample. The large velocity pullback indicates that the dynamic tensile strength (the spall strength) is as high as 10 GPa in this case. In the other shot, the peak stress exceeded the HEL. The spall strength drops practically to zero, and irregular oscillations appear in the wave profile. Vanishing resistance to tension after shock compression above the HEL was also observed for quartz single crystals (Kanel et al., 1992). Presumably the absence of fracture nucleation sites enables high spall strength at peak shock stresses below the HEL. However, fracture nucleation sites obviously appeared as a result of shock compression above the elastic limit.

The high-frequency particle velocity jitter is evidence for heterogeneity of the inelastic deformation process. Similar records have been obtained for quartz (Kanel et al., 1992) and olivine (Furnish et al., 1986). Another characteristic feature is the significant stress relaxation behind the elastic precursor front that

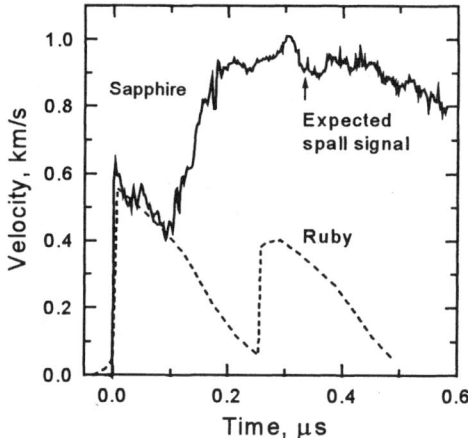

Figure 4.4. Shock-wave profiles at the interface between alumina single crystals and a water window. The shock amplitude exceeded the HEL for the sapphire sample and was less than the HEL for the ruby sample.

is caused by intense multiplication of the deformation carriers. This also is typical for brittle crystals (Furnish et al., 1986, Grady, 1977).

Graham and Brooks, 1971, compared the Hugoniot of sapphire to hydrostatic data. Shock-wave experiments were performed with sapphire samples of three different crystallographic orientations to the load direction: 0° (crystallographic orientation [0001]), 90° (crystallographic orientation [$\bar{1}2\bar{1}0$]), and 60° (crystallographic orientation [11$\bar{2}$3]). The results are shown in Fig. 4.5. Even though sapphire has trigonal symmetry, the elastic stiffness does not vary significantly with orientation: The longitudinal wave speeds in the 0° and 90° orientations differ by only 0.5 per cent. The observed HEL varied from 12 to 21 GPa around the average, 15 GPa, for sample thicknesses of 6–13 mm. Barker and Hollenbach, 1970, detected erratic optical behavior of sapphire at a shock stress of about 15 GPa. They interpreted this as an indication of the onset of plastic yielding. Over the elastic range, sapphire exhibits slightly nonlinear longitudinal shock compressibility which is described as $U_s = 11.19 + u_p$ km/s (Barker and Hollenbach, 1970) or $U_s = 11.17 + 0.95 u_p$ (Mashimo et al., 1988). Above the HEL, all the states of shock-compressed sapphire fall on a common compression curve, independent of the orientation, the sample thickness, and the HEL value. The Hugoniot shows a collapse toward the isotropic compression curve: The stress offset for states above the HEL is 3.8–4.3 GPa whereas, at the HEL, it ranges from 5.5–11 GPa. The collapse is confirmed by a direct comparison of the second-wave shock velocity for stresses just above the HEL with the bulk sound speed: Subsonic waves were observed in this stress region. On the other

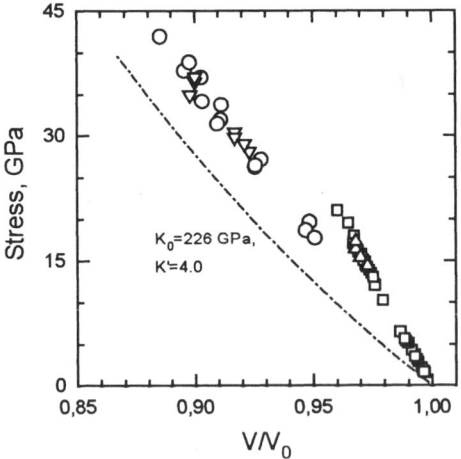

Figure 4.5. Stress–volume relations for sapphire under shock-wave compression. Points present the data by Graham and Brooks, 1971, and Mashimo et al., 1988. The dot-dashed line shows the isotropic compression curve used by Mashimo et al., 1988.

hand, Mashimo et al., 1988, using a different hydrostat from that of Graham and Brooks, found a stress offset of 6.5–8 GPa above the HEL, values that do not differ so much from these values at the HEL. One should say also that a similar collapse was observed for some ductile metals (see, for example, data by Arnold, 1992, for Armco iron). Strength collapse cannot be considered as an unique feature of brittle crystals.

Wang and Mikkola, 1992, examined recovered sapphire samples with transmission electron microscopy after shock compression up to 23 GPa. They observed a significant number of slip bands in different crystallographic directions and suggested that a large amount of plastic deformation had occurred at shock stresses of 12 GPa and more. Changing the crystallographic orientation with respect to the shock-wave direction had a significant influence on the substructure developed.

4.4.2. Quartz

According to Wackerle, 1962, the Hugoniot elastic limit for quartz single crystals is between 5.5 and 8.5 GPa for shocks in the x-axis direction and 10–15 GPa along the z-axis. Above the HEL, a catastrophic loss of shear strength was observed, so the shock-compressed states were no longer related to the sample orientation. Careful measurements performed by Graham, 1974, have shown a definite offset between the isotropic hydrostatic equation of state and the shock-compression data. Nevertheless, this offset is considerably less than that corresponding to the shear strength at the 6 GPa HEL in x-cut quartz. From an analysis of piezoelectric output of shock-compressed quartz it was concluded also that between 4 and 6 GPa the stress at the impact surface relaxed to lower values in times of about 10^{-8} s. Photographs of shock-compressed x-cut quartz (Brannon et al., 1984, Schmitt et al., 1986) display a spotty, heterogeneous radiation field that indicates heterogeneous shock deformation. The onset of luminosity between 4 and 6 GPa coincides with dynamic yielding. The emission occurs along discrete lines that lie along the intersections of known fracture planes with the yz plane. As the stress is increased moderately beyond the HEL, the emission pattern becomes homogeneous. The emission spectra represent nonthermal radiation combined with gray body radiation and are similar to the photoluminescence from structural defects.

Anan'in et al., 1974, have revealed glass-like interlayers between quartz blocks in recovered single crystals. This lamellar structure indicates the heterogeneous nature of shock deformation of quartz, accompanied by melting. Grady, 1980, has developed a model of localized dissipation of elastic strain energy in strong, low thermal conductivity solids which results in an adiabatic shear band. It is supposed that the local dissipation of the elastic shear strain energy leads to a local temperature increase that causes the shear strain and the energy release to be localized within narrow bands in which the temperature may reach the melting point.

Direct high-resolution measurements of shock wave profiles for quartz are complicated by piezoelectric effects and, it seems, by a nonuniformity of the shock-wave process. Figure 4.6 shows VISAR free surface velocity profiles measured below the HEL in experiments with impact loads of long duration. At a 3.5 GPa peak stress the velocity history corresponds to the purely elastic response of the material. The negative velocity pullback is a result of the wave re-reflection from a copper base plate used in the experiment. Reproducibility of the measurements decreased when the peak stress was increased to ~5.5 GPa and above. Chaotic oscillations appearing in the velocity history near the HEL are probably evidence of localized inelastic deformation or cracking.

Figure 4.7 shows free surface velocity profiles for x-cut quartz samples for very short load durations (Kanel et al., 1992). The free surface velocity profile corresponding to a shock pulse amplitude of 2.8 GPa replicates the form of the compression pulse inside the sample and does not display any symptoms of spallation. Increasing the pulse amplitude to 4.6 GPa causes spall damage as is evident from the spall pulse in the free surface velocity profile. A further increase in pulse amplitude to 5 GPa modifies the free surface velocity profile dramatically. The unloading part of the pulse is not manifest at the surface, which means that the tensile strength is practically zero in this case. Cracking in the brittle single crystal under compression is a plausible explanation for the diminishing tensile strength near the HEL.

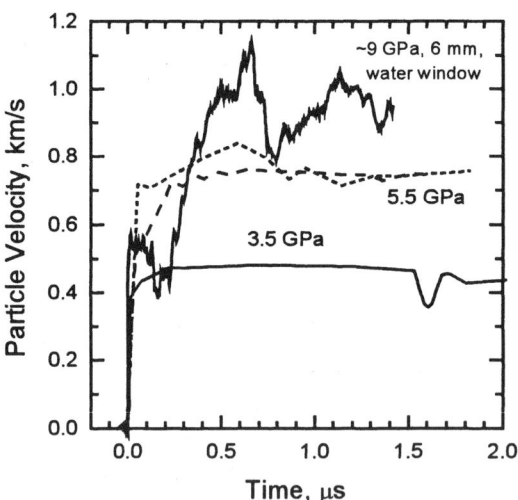

Figure 4.6. Free surface velocity profiles for 4-mm-thick x-cut quartz samples. The shot at the highest stress, ~9 GPa, was done with a 6-mm-thick sample and a water window.

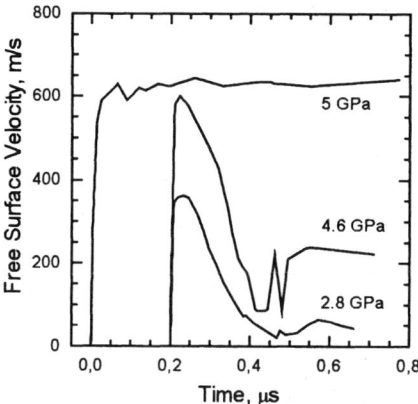

Figure 4.7. Free surface velocity profiles for x-cut quartz at different peak stresses generated by 0.2- or 0.4-mm-thick aluminum impactor plates.

4.4.3. Magnesium Oxide (Periclase)

Using the inclined-mirror technique, Ahrens, 1966, has obtained Hugoniot data for magnesium oxide single crystals shocked in the [001] direction at pressure levels of 17, 42, and 66 GPa. He found that the HEL is markedly dependent on both the final shock state and, possibly to a lesser extent, on the specimen thickness for a fixed final shock state. The elastic-shock amplitude increased from 3.5 to 8.9 GPa with increasing final shock stress from 16.6 to 42.3 GPa. The elastic-shock amplitude decreased from 4 to 3.5 GPa with increasing shock propagation distance from 4.7 to 9.7 mm at a final shock stress of 16.6 GPa. Over the pressure range of experimental data above the HEL, the difference between the calculated and experimental 20°C isotherms and the Hugoniot is negligible. This result implies a fluid-like behavior of MgO in the shock-deformed state.

Grady, 1977, measured stress and particle velocity wave profiles in [100]-oriented MgO shock loaded to stress levels between 4.8 and 11.2 GPa using more precise quartz gauges and VISAR techniques. The character of the observed shock loading process exhibits a significantly more complex behavior than simple elastic–plastic response would predict. The initial elastic loading path consisted of a sharp rise to a stress level of 2.5–3.1 GPa followed by significant stress relaxation. The observed width of the second (deformation) wave decreased with increasing peak stress. The Hugoniot stress–volume data indicate total collapse to the hydrostat. The structure of the release wave was similar to the elastic–plastic response observed in release waves for metals. This means that the material strength persists in periclase at the Hugoniot state. However, no leading purely elastic release wave of a finite amplitude was recorded. The initial elastic release velocities are below the expected [100] longitudinal velocities but are substantially above the expected bulk sound velocities, whereas the subsequent plastic release velocities are very close to the bulk sound velocity. The release paths deviate significantly from the hydrostat as the axial stress drops by 1.4–1.5 GPa at which point the subsequent unloading path appears to lie parallel to the hydrostat.

4.4.4. Zirconia

Zirconia (ZrO_2) in its pure form exists in a stable monoclinic crystal structure. Through doping with other metal oxides, zirconia can be stabilized in a cubic or tetragonal structure. Mashimo et al., 1995, have measured the Hugoniot of Y_2O_3-doped cubic zirconia single crystals. The Hugoniot elastic limits for compression parallel to the $\langle 100 \rangle$ and $\langle 110 \rangle$ lattice directions were 13.7 and 24.6 GPa, respectively. These anisotropic HELs can be reasonably well understood in terms of the primary [111] cleavage plane in the cubic structure. However, the yield stresses along these two orientations were estimated by the von Mises yield criterion to have almost the same value of 10.2 and 9.3 GPa, respectively, despite the large difference in HEL structure. Above the HELs the Hugoniot data for impacts parallel to the $\langle 100 \rangle$ and $\langle 100 \rangle$ axes converged on each other, and showed large stress relief to a state of isotropic compression. A phase transformation began at approximately 53 GPa independent of the shock direction and was completed by about 70 GPa. These values are much higher than those of phase transitions under static compression. The volume decrement at the shock induced phase transition was estimated to be ~20%.

4.4.5. Iron-Silicate Almandine-Garnet

Iron-silicate almandine-garnet single crystals were tested by Graham and Ahrens, 1973 who were trying to detect a polymorphic transformation which occurs when static pressure exceeds 19.5 ± 2 GPa. The HEL values observed for the individual garnet samples covered an extended range of stress levels around the average value of 8.1 ± 1.7 GPa for shocks in the [100] direction. The corresponding stress offset between the HEL and the hydrostatic Hugoniot is 2.7 ± 0.7 GPa. Since the final shock-compressed states beyond the elastic deformation region correspond well to the hydrostatic data, it was concluded that garnet behaves as a fluid above the HEL.

4.4.6. Olivine

Olivine single crystals of the approximate composition $(Mg_{0.9}Fe_{0.1})_2SiO_4$ were tested by Furnish et al., 1986. The measured particle velocity profiles are similar to that for sapphire shown in Fig. 4.4. An elastic precursor wave with a rise time <10 ns and a peak stress of 6.5 GPa for shocking along the a axis and 7.8 GPa for shocking along the bisector of the [100] and [013] directions is accompanied with stress relaxation and small high-frequency excursions. The stress relaxation is much more pronounced for loading in the bisector direction. The shock-compressed state behind the second shock wave corresponds well to a calculated "bulk" isentrope. The release wave gradually decreases the particle velocity and the stress. The release curves in the stress–density coordinates are close to the isentrope, indicating much less shear stress than at the Hugoniot elastic limit.

Thus, hard single crystals show a more-or-less substantial reduction in shear strength following shock compression beyond their Hugoniot elastic limits. Within the elastic strain range they demonstrate very high dynamic tensile strength which is attributed to lack of flaws and heterogeneities. For shock compression above the HEL they show vanishing dynamic tensile strength.

4.5. Shock Wave Properties of Silicate Glasses

In the early days, silicate glass was in some ways the archetypal material for studies of brittle fracture. It was natural to test it under impact conditions also. Silicate glasses exhibit a high yield strength that is associated with their amorphous state because dislocation slip is impossible in irregular structures, and low fracture toughness as a result of their high homogeneity. The fracture of glasses under compression occurs by axial splitting. Schardin, 1959, investigated velocity effects in fracture of glasses. Growing tensile cracks usually start with a low velocity which increases up to an ultimate value as the crack propagates. Depending on the glass composition, the ultimate crack speed varies from $0.38 c_s$ (lead flint glass) to $0.61 c_s$ (fused quartz), where c_s is the shear wave speed. A surprising observation was that the rapidly growing cracks never decelerate. If the driving stress decreases, rapid cracks either maintain speed or stop.

At high pressures, brittle glasses become ductile. Ductility of glass is caused by a loose microstructure with a large concentration of molecular-size voids. It is known that glasses show gradual structural changes resulting in increased density (Arndt and Stöffer, 1969). A microscopic examination of the deformed zone in glass under pyramidal indentations (Hagan, 1980) shows that inelastic deformation is concentrated in shear faults of negligible thickness produced by genuine shear displacements without cracking. Once plastic flow starts, the stress relaxation reduces the stress concentration at the crack tip and thus stops the propagation of cracks. Since densification occurs under Vickers indentation, it is supposed (Ernsberger, 1968) that the irreversible densification of the silicate structure is responsible for the plastic flow properties of glasses under high pressure. The degree of densification can be varied to some extent by variation of pressure, temperature, and shear strain, and remains irreversible under normal conditions. Irreversible densification of some glasses also occurs under shock compression above the HEL (Gibbons and Ahrens, 1971).

Figure 4.8 shows a typical stress history for shock compression of a crown glass impacted by a 6-mm-thick PMMA flyer plate at an impact velocity of 3.15 km/s (Kanel and Molodets, 1976). The stresses and particle velocities were recorded with manganin pressure gauges and magneto-electric gauges, respectively. Strains were calculated from particle velocity histories. The dispersion of the elastic precursor wave is a result of anomalous compressibility at moderate pressures, which is a property of many silicate glasses. The data demonstrate the irreversible densification of glass which exhibits itself in the residual strains after unloading. As discussed above, the inelastic deformation of glass occurs by means of the densification mechanism. Onset of densification and inelastic deformation are accompanied by a sharp increase of the propagation velocities in unloading waves, as shown in Fig. 4.9.

Figures 4.10 and 4.11 present free surface velocity profiles for K8 crown glass and soda lime glass (Kanel et al., 1998, 2002). Spallation was not observed in these shots, which means that the spall strength of the glass exceeds 6.8 GPa

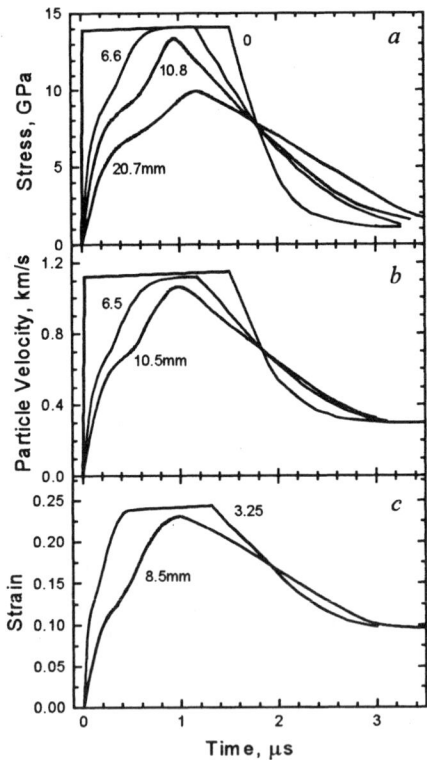

Figure 4.8. Stress, particle velocity, and strain histories at shock compression of K8 crown glass. Numbers at wave profiles show the distance inside the glass sample at which the measurements were made.

4. Behavior of Brittle Materials Under Shock-Wave Loading   127

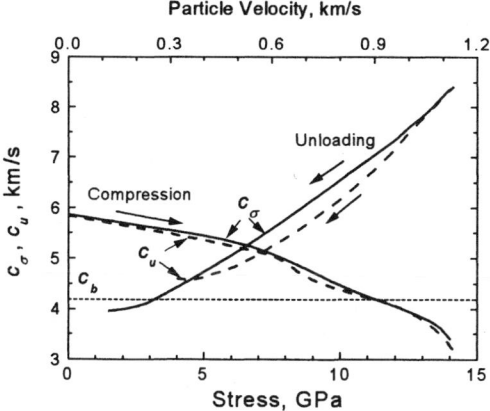

Figure 4.9. Phase speeds c_σ and c_u determined from the stress histories and particle velocity histories, respectively, measured in K8 crown glass at the input surface and at a distance of 6.5 mm. The constant c_b is the bulk sound speed at zero pressure.

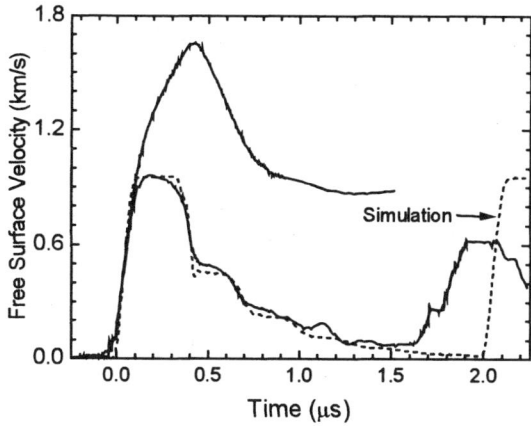

Figure 4.10. Experimental results for 6.1-mm-thick K8 crown glass samples impacted at 670 ± 30 m/s by a 0.9-mm-thick steel flyer plate and at 1900 ± 50 m/s by a 2-mm-thick aluminum flyer plate backed by paraffin. The dashed line shows results of computer simulations assuming no failure to occur.

below the HEL and remains very high above the HEL. For comparison, the static tensile strength of glasses is around 0.1 GPa. The reason for such a large discrepancy is that the fracture nucleation sites in homogeneous glass are concentrated on the surface. These incipient microcracks are activated and determine the strength magnitude in the static measurements, whereas spall strength is an intrinsic property of matter.

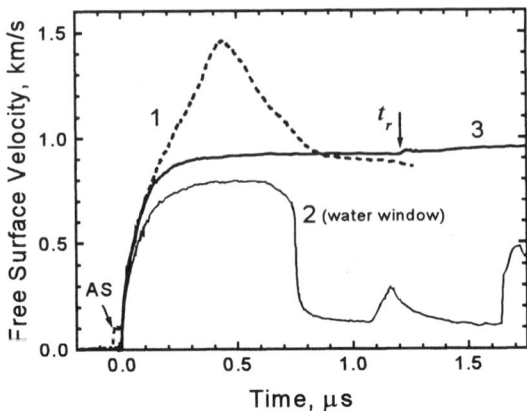

Figure 4.11. Free surface velocity histories of 5.9-mm-thick soda lime plate glass samples. The wave profile 1 corresponds to impact by a 2-mm-thick aluminum flyer plate backed by paraffin, with the impact velocity being 1900 ± 50 m/s. The weak velocity step AS before the main front is obviously the result of an air shock propagating ahead of the flyer plate. The wave profile 2 corresponds to impact by a 2.1-mm-thick aluminum flyer plate at the impact velocity 970 m/s, measured through a water window. The wave profile 3 corresponds to impact by a 7-mm-thick aluminum flyer plate through a copper base plate, with the impact velocity being 1.17 km/s. The arrow at $t_r = 1.21$ μs points to a recompression pulse which is a result of wave re-reflection at the failure front.

Thus, in contrast to the behavior of quartz single crystals, the spall strength of glasses does not drop when prior inelastic compression occurs. Single crystals and glasses are initially homogeneous in bulk, so the fracture nucleation sites can only be formed in the course of plastic deformation. In this sense, the difference between single crystals and glasses is in that hard crystals have only a limited number of crystallographic planes and directions in which the usual mechanisms of ductility can operate, whereas the ductility of amorphous glasses is completely isotropic. The impossibility of plastic shear along arbitrary directions in crystals results in stress concentration at points of intersections of slip bands or twins that, in turn, may result in cracking during compression or unloading. Isotropic glasses are free of this limitation. The high spall strength revealed in the stress range above the HEL means that the ductility is preserved even under subsequent tensile stressing of the material.

Glasses do not exhibit a distinct Hugoniot elastic limit (HEL) in the profiles of shock compression waves. Therefore, we need to use indirect arguments in order to determine the transition from elastic to plastic deformation. An upper estimate may be based on the approach of the wave phase speed to the bulk sound speed. At this point the shear stresses do not increase with further compression and that should result in shear instability of the glass. As a result of an anomalous decrease of longitudinal sound speed with increasing stress, a rarefaction shock wave should be formed in glass at unloading from a shock-compressed state. Obviously, this may occur only in the case that the compression is completely reversible. Since the reversibility of stress–strain processes is a main attribute of elastic deformations, observation of the rarefaction shock (demonstrated by the waveform 2 in Fig. 4.11) may be considered as evidence of an elastic regime of deformation. Finally, above the HEL the unloading wave speed becomes greater than the compression wave speed.

On the other hand, the rounded transition from the compression wave to the plateau indicates that the response of glass is not purely elastic even in the stress range where rarefaction shocks are observed. Figure 4.12 shows free surface velocity histories measured for soda lime glass plates of different thicknesses at peak stresses of 4 GPa and 7 GPa. At both stresses the evolution of wave profiles is, in general, everywhere close to that of a centered simple wave. Such behavior is not typical for elastic–viscous response. Probably we have to account for internal friction or other effects in glass in order to understand this peculiarity.

To complete discussion of the structure of compression waves in glasses, Figs. 4.13 and 4.14 present examples of the free surface velocity histories of shocked fused quartz (Razorenov et al., 1991) and TF1 heavy flint glass (Kanel et al., 1998). Fused quartz exhibits anomalous compressibility in the low stress region whereas the flint glass forms an elastic shock that gradually turns into a ramped compression wave. The smoothed shape of the wave profiles complicates unambiguous direct determination of the Hugoniot elastic limits of glasses. This complication stimulates the use of indirect methods. Table 4.1 summarizes results of measurements of shock-wave response of silicate glasses.

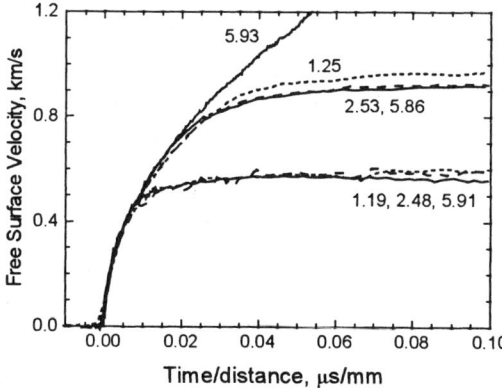

Figure 4.12. Profiles of compression waves of 4.1 GPa, 6.7 GPa and 9 GPa peak stress in soda lime glass plates of different thicknesses (indicated by numbers at the waveforms).

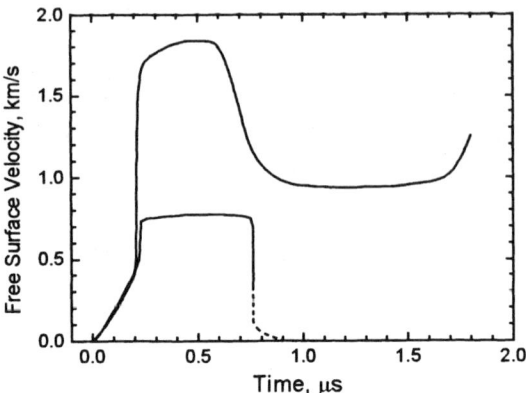

Figure 4.13. Free surface velocity histories of 5.4-mm-thick fused quartz plates impacted by 2-mm-thick aluminum flyer plates having velocities of 0.7 km/s and 1.9 km/s. The high-speed impactor plate was backed by paraffin.

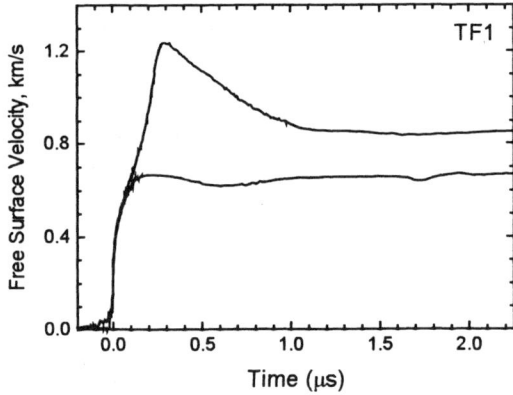

Figure 4.14. Free surface velocity histories of 6.1-mm-thick heavy flint glass plates impacted by a 10-mm-thick copper plate at 0.48 km/s and a 2-mm-thick aluminum plate backed by paraffin at 1.9 km/s.

Kanel and Molodets, 1976, observed localization of inelastic deformation following shock compression of K8 crown glass above the HEL. In this study, distortion of internal interfaces of the two-piece glass targets resulted in elongation of the manganin piezoresistive foil gauges used for recording the stress profiles. Figure 4.15 presents both the stress history and the component of gauge resistance increase as a result of its elongation. The foil gauges sense the surface distortion if the size of the non-uniformities is comparable to, or larger than, the thickness of the gauge foil. The distortion is higher in the gauge next to the receiving block. An important observation is that the maximum of the surface distortion is delayed relative to the stress maximum. Obviously, relative shifting of the blocks and, correspondingly, the shear stress relaxation continues until these stresses are equilibrated.

Table 4.1. Shock-wave response of silicate glasses

Glass	ρ, g/cm^3	c_l, km/s	ν	HEL, GPa	Reference
Fused quartz	2.205	5.95	0.18	8.7	Chhabildas and Grady, 1984
K8 crown glass	2.52	5.85–6.05	0.209	8 ± 1	Kanel and Molodets, 1976; Kanel et al., 1977
Borosilicate glass	2.23	6.05	0.2	8.0	Bourne et al., 1999
TF1 heavy flint glass	3.86	4.04	0.227	3.1–4.7	Kanel et al., 1998
DEDF heavy flint glass	5.18	3.49	0.25	4.3 ± 0.2	Bourne et al, 1996
Soda lime glass	2.45–2.53	5.7–5.85	0.23	3–8	Dandekar, 1998; Bourne et al., 1999; Kanel et al., 2002

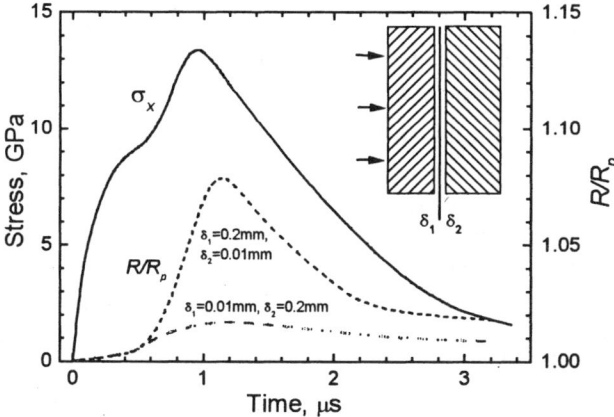

Figure 4.15. Stress history and distortions of plate surfaces following shock compression of K8 crown glass. Measurements were carried out at a distance of 10 mm from the impact surface with manganin and constantan gauges insulated by Teflon. Dashed lines show the component of the gauge resistance increase due to elongation. The quantity δ_1 is the thickness of film placed toward the impact side, and δ_2 is the film thickness toward the sample rear surface.

4.6. Failure Waves in Glasses

The impact loading of a glass and, probably, other brittle materials can result in the appearance of a failure wave. The failure wave is a network of cracks that are nucleated on the surface and propagate into the stressed body. There are many observations of fracture front propagation in glasses under high tensile stresses. Schardin, 1959, recorded expansion of circular fractured areas with a sharp front formed by bifurcated cracks. He also observed detonation-like propagation of fracture in glass pipes filled with compressed gas. Chandrasekar and Chaudhri, 1994, recorded the explosive disintegration of thermally tempered glasses in which the internal central region was under tensile stress. Galin et al., 1966, reported an explosion-like fracture under bending of high-strength glass from which the surface defects had been removed. The explosive fracture resulted in formation of micrometer-size glass particles and was treated by Galin and Cherepanov, 1966, in terms of a self-propagating failure wave. A similar fracture mode under compression was revealed in shock-wave experiments. However, failure waves present a mode of catastrophic fracture in elastically compressed media that is not limited to impact events. One may hope that the investigations of failure waves in shock-compressed glasses provide information about the mechanisms and general rules of nucleation, growth, and interaction of the multiple cracks, and will help us to better interpret experiments with other hard brittle materials such as ceramics and rocks.

In modeling the behavior of materials subjected to intense impulsive loading, it is usually implied that local material response can depend only on characteristic material properties and the local state. It seems that the failure wave is an example of non-local behavior. The response of each elementary volume in the body depends not only on its local state, but also on whether or not the failure wave has approached the point of interest.

A failure wave was originally observed as a response to planar shock-wave compression below the Hugoniot elastic limit. In experiments described by Razorenov et al., 1991, and Kanel et al., 1992, a failure wave was observed in the following experimental configuration. A triangular, long-duration 4.5 GPa shock pulse was introduced into a plane sample of glass or fused quartz through a thick copper baseplate. In fused quartz samples, the free surface velocity profiles contained short negative velocity pullbacks (Fig. 4.16a). This is a result of re-reflection of the rarefaction wave at the sample–baseplate interface. The baseplate had a higher shock impedance, so the rarefaction wave reflected with conservation of a sign and produced a short tensile pulse in the unloaded sample, which, in turn, produced a negative velocity pullback when it reached the rear surface. No re-reflected tensile pulses were observed in experiments with K19 crown glass (Fig. 4.16b). Instead, a small velocity rise was noted on the free surface and the moment of the second velocity jump was earlier than the elastic wave reverberation time for the sample. A similar modification of the wave

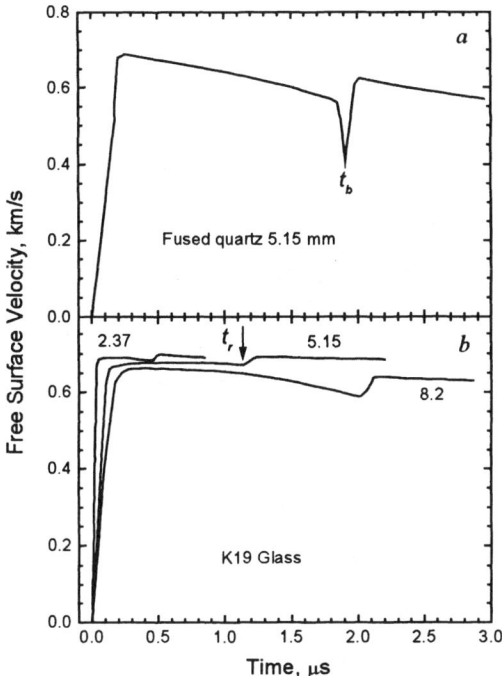

Figure 4.16. Free surface velocity profiles in cases of failure wave formation (a) and without it (b).

process was explained in terms of formation of a failed layer near the interface. This layer should have lowered impedance and zero resistance to tension. Measurements with variations of the sample thickness have shown that the thickness of the failed layer increases with time. This process was interpreted as propagation of a failure wave.

It follows from consideration of the time–distance diagram shown in Fig. 4.17 that the failure wave meets the unloading wave reflected from the free surface of the glass at the distance x and time t_x, as determined by the relationships

$$t_r = 2\frac{\delta - x}{c_\ell}, \text{ and } t_x = \frac{\delta}{c_\ell} + \frac{\delta - x}{c_\ell},$$

where δ is the thickness of the glass plate. Measurements for glass plates of different thicknesses provide information about steadiness of the failure wave. In particular, Fig. 4.18 shows that, at a stress of 4.5 GPa, the propagation velocity of the failure wave decreases with distance.

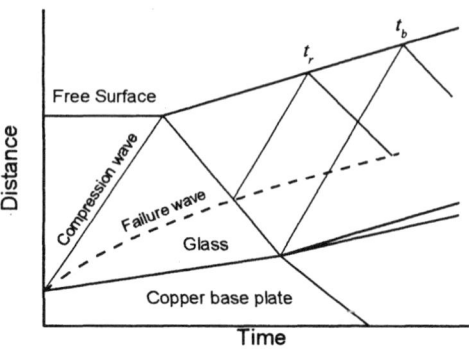

Figure 4.17. Distance-time diagram of experiments shown in Fig. 4.16.

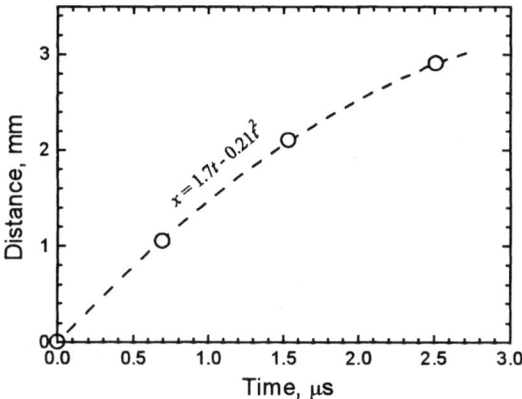

Figure 4.18. Propagation of a failure wave in K19 crown glass at a decreasing stress of 4.5 GPa. peak value

Brar et al., 1991a, 1991b, Brar and Bless, 1992, and Bless, et al., 1992, have shown by direct measurements on a soda lime glass that behind the failure wave the tensile strength drops to zero, or almost to zero, and the transverse stress increases, indicating a decrease in shear strength. Figure 4.19 shows an example of earlier measurements of longitudinal and lateral stresses at shock compression of glass near the elastic limit made by Kanel et al., 1977, which were first treated as a delayed collapse. Figure 4.20 summarizes the results of stress difference measurements by Brar et al., 1991a, and Kanel et al., 1977. The failure waves were recorded in the longitudinal stress range 4–10 GPa; for this stress range the diagram shows the stress difference ahead of, and behind, the failure front. It looks quite reasonable that reduction of the final stress difference with an increase of the shock amplitude, and, respectively, an increased degree of comminution occurs. At peak stresses exceeding 10 GPa, the densification processes start in glass. This produces shear stress relaxation without cracking. A second rise in the stress difference at ~15 GPa may be evidence that maximum densification has been achieved.

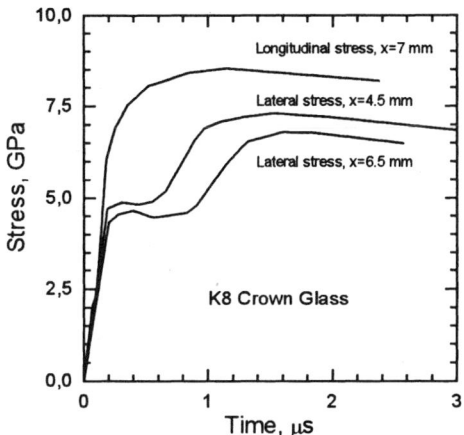

Figure 4.19. Profiles of longitudinal and lateral stress in K8 crown glass measured with manganin gauges for shock amplitudes near the HEL. The distance of wave propagation is noted on the curves.

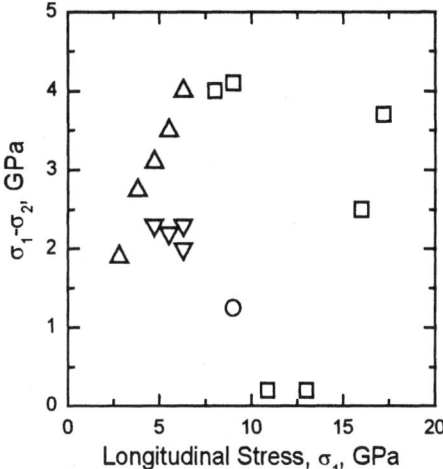

Figure 4.20. Results of measurements by Brar et al., 1991a (triangles) and Kanel et al., 1977 (squares and circles) of the stress difference in shock compressed glass as a function of the peak stress.

Raiser and Clifton, 1994, and Raiser et al., 1994, have found that a surface roughness of aluminosilicate glass between 0.04 and 0.52 μm does not appear to play a significant role in the formation of a failure wave. Independently of the surface roughness, they observed a high spall strength of the glass when the compressive stress was around 3.5 GPa whereas, at peak stresses of 7.5–8.4 GPa, the spall strength was high ahead of the failure front and was low behind it. They also obtained evidence of increasing failure wave velocity with increasing impact stress. Bourne et al., 1997, have found that deliberately introducing flaws by roughening the surface speeds the fracture of a glass, increasing the average failure-wave velocity. Dandekar and Beaulieu, 1995, have found that a failure wave is initiated in soda lime glass at an impact stress between 4.7 and 5.2 GPa. The propagation velocity of the failure wave is determined to be 1.56 km/s and it was independent of the thickness of the glass sample between 3.1 and 9.4 mm at the impact stress of 5.2 GPa. Espinosa et al., 1997, observed a progressive reduction in normal stress behind the failure wave and close to the impact surface. They have concluded that the inelastic process responsible for the reduction in shear strength has well defined kinetics. Although much work has concentrated around relatively open structure, lower density glasses, Bourne et al., 1996, and Kanel et al., 1998, observed a failure wave in a higher density filled lead glass.

Whether the failure wave is steady or it decays and stops at some distance is an important issue for understanding the mechanism and nature of the phenomenon. Figure 4.21 summarizes measurements of failure waves in soda lime glass plates of different thickness at various impact stresses. It is seen that, at stresses exceeding ~4 GPa, the failure wave is steady but its velocity is stress-dependent. Steady failure waves were not recorded for peak stresses below 4 GPa. The stress dependence of the failure wave speed explains its decay for the triangular stress pulse in the experiments illustrated in Fig. 4.16.

Brar and Bless, 1992, Bourne and Rosenberg, 1996, and Senf et al., 1995, have photographed failure waves. Very interesting results were obtained in experiments with large specimens of optical crown glass impacted by blunt steel cylinders (Senf et al., 1995). The experiments demonstrated several modes of fracture nucleation. Around the edges of the projectiles many cracks are initiated by the propagating surface wave. These cracks form a conical fracture zone. In the central part, the damage zone exhibits a planar front that propagates with a velocity equal to the terminal crack velocity of 1.55 km/s in this type of glass. Besides fracture nucleation at the sample surface, separate crack nucleation sites were also activated at microdefects inside the stressed glass target. As a result, several spherical or nearly spherical failure waves were found to form ahead of the main front.

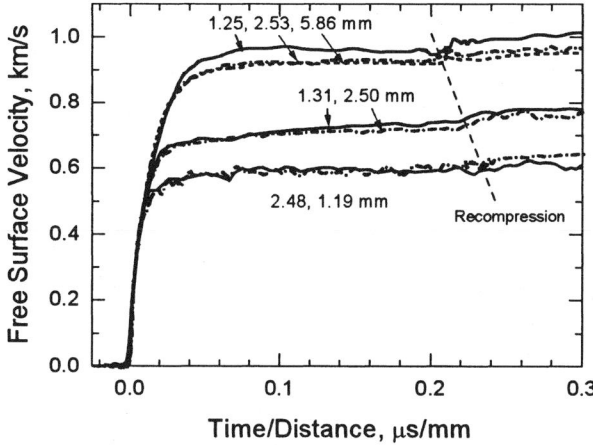

Figure 4.21. Relative positions of the recompression signal as a function of glass plate thickness and impact stress. The experiments were conducted on soda lime glass.

Since the first experimental observations of failure waves in shock-compressed glass it was believed that the failure wave is accompanied by increasing lateral stresses and is not accompanied by any change in longitudinal stresses. However, recent experiments of Dandekar, 1998, and of Millet et al., 1998, revealed a disagreement between the longitudinal stress measured on the impact surface of a shock-loaded glass plate and the stress measured at some distance from the impact surface. Although the recorded wave profiles had a rectangular shape without any signature of a second compression wave, the measured stresses at some distance were less than at the impact surface when the incident shock amplitude exceeded some threshold. These observations may be treated as evidence of formation of an unrecorded second compression wave, and perhaps this should be identified as the failure wave. The kinematics of failure wave phenomena were investigated in the experimental arrangement discussed below.

Figure 4.22 presents stress histories measured by Kanel et al., 2002, on input and output surfaces of glass samples as shown in Fig. 4.23. The wave profiles shown in graph *a* were obtained in an experiment with thick, non-lapped glass plates. The other experiments were conducted with lapped plates. First of all, measurements confirm the difference between peak stresses measured on the impact surface of a shock-loaded glass plate and the stresses measured at some distance from the impact surface, as observed by Dandekar and by Millet et al. The stress measured by the first gauge in Fig. 4.22*a* is 6.6 GPa, whereas the second gauge recorded a 6.1 GPa stress between two glass plates at a distance of 5.85 mm from the input surface of the glass sample.

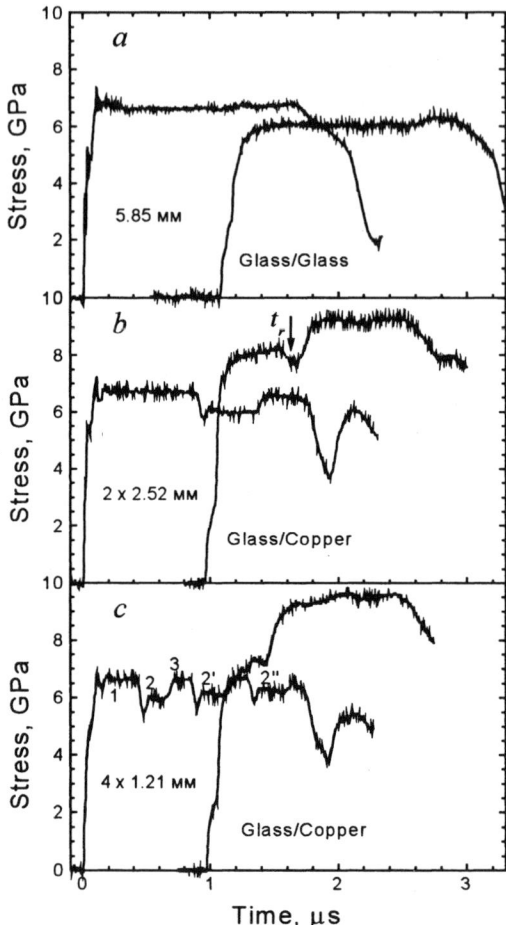

Figure 4.22. Stress histories on the input and output surfaces of glass samples. Graph *a* represents the experimental result with a thick glass plate backed by another glass plate of the same thickness. Graph *b* shows the experimental result with a sandwiched sample consisting of two glass plates of the same thickness, 2.52 mm, backed by a 5.5-mm-thick copper plate. The time marked t_r is the expected arrival time of the wave reflected from the failure-wave front. Graph *c* presents the experimental result for a layered sample composed of four glass plates of the same thickness, 1.2 mm, backed by a 5.5-mm-thick copper plate.

The response of layered glass samples demonstrates some specific features. In order to discuss these features, let us consider the time–distance ($t-x$) diagram of the wave interaction shown in Fig. 4.24, and the stress–particle velocity ($\sigma_x - u_p$) diagram shown in Fig. 4.25. The conjunctions between lapped plates are porous interlayers with lowered impedances so that short reflected rarefac-

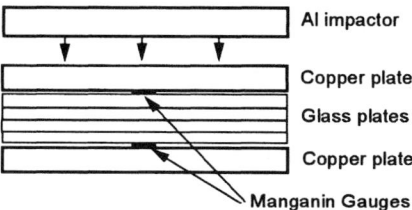

Figure 4.23. Schematic of shock compression of layered assemblies of glass plates and gauges for recording the stress histories.

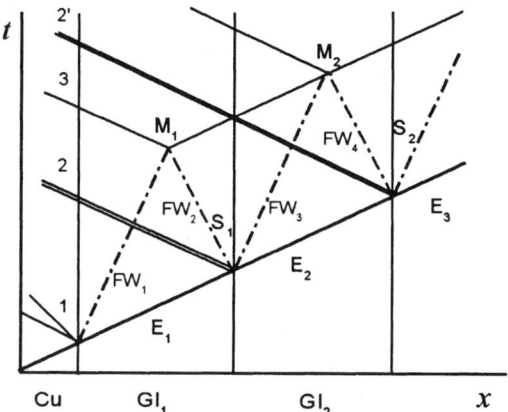

Figure 4.24. The assumed scheme of wave interactions during shock compression of glass samples composed of several lapped plates indicated as Gl_i, with E_i denoting the compression wave trajectory and FW_i being trajectories of the failure waves.

tion pulses are generated when the compression wave passes through them. These reflected rarefaction pulses propagate back to the first stress gauge and generate short negative spikes in the stress histories recorded by it. The measured time intervals between these spikes correspond approximately to the time of reverberation of an elastic wave in the glass plates.

The measurements show that, after the rarefaction pulse, the stress on the first gauge returns to its initial magnitude in a step-like way. From the comparison of data in Fig. 4.22b and c one can see that the time intervals between the steps are proportional to the thickness of glass plates. We may conclude that the step-like recovery of stress is associated with a two-wave recompression process. In other words, after the initial compressive wave passes through the con junction between plates, two reflected compressive waves are created which propagate back with different velocities. Since the stresses correspond to the

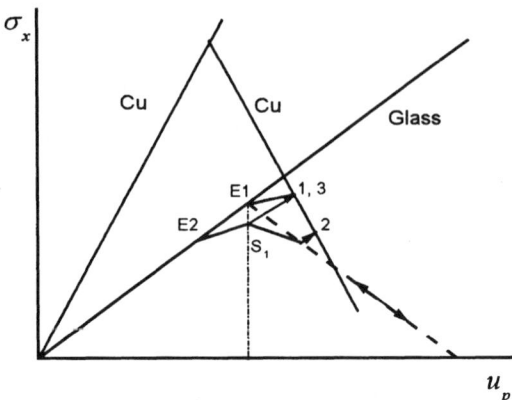

Figure 4.25. The assumed schematic stress–particle velocity diagram of wave interactions during shock compression of glass samples composed of two lapped plates, with the marks being the same as those in Figs. 4.22c and 4.24.

elastic deformation region, the second wave is the failure wave. However, the failure wave can pass through a given layer of a sample only one time because the fracture can occur only once. It seems that the second steps correspond to compression waves that are created as a result of interactions between two opposing failure waves. Assuming that the velocity of all waves, except for the failure waves, is equal to the longitudinal sound speed, c_ℓ, based on the consideration of the time–distance diagram, we can obtain the following two equations to calculate the failure wave speed, c_f, using the measured time intervals and the thickness of the glass plate:

$$\frac{x}{c_f} = \frac{\delta}{c_\ell} + \frac{\delta - x}{c_f}, \quad \text{and} \quad t_{2-3} = \frac{\delta - x}{c_f} - \frac{\delta - x}{c_\ell},$$

where δ is the plate thickness, x is the coordinate of the meeting point M_i of two failure waves (see Fig. 4.24), and t_{2-3} is the time interval between the reflected rarefaction pulse and the reflected compression wave indicated by points 2 and 3 in Figs. 4.22 and 4.24. By solving this set of equations we obtain

$$c_f = c_\ell \frac{c_\ell\, t_{2-3} + \delta - \sqrt{c_\ell\, t_{2-3}(c_\ell\, t_{2-3} + 2\delta)}}{\delta}.$$

There is some uncertainty with the sound speed value. The shape of the compression wave profile indicates a decrease in the longitudinal sound speed with increasing stress. On the other hand, the sound speed in comminuted material behind the failure waves may differ from that ahead of it as a result of the changed state of stress. Using $c_\ell = 5.58$ km/s for the longitudinal sound speed as estimated from the time intervals between reflected rarefaction pulses 2 and 2' in Fig. 4.22c, we find the failure wave speed presented in Table 4.2. All these estimates are close to the value 1.55 km/s of the failure wave speed evaluated from the recompression wave in the free surface velocity history.

Table 4.2. Evaluated parameters of failure waves

Data	C_f, km/s	σ_f, GPa	u_f, m/s	ε_f
Fig. 4.22b	1.57	0.15	41	0.0261
Fig. 4.22c, 1st plate	1.61	0.19	52	0.0323
Fig. 4.22c, 2nd plate	1.48	0.16	44	0.0297

Thus, we could come to the conclusion that, when a failure wave is formed, shock compression of glass leads to a two-wave structure. Since the failure waves nucleate at each surface, this conclusion means that, on each interface, the magnitude of the leading elastic wave should decrease as a result of its decomposition into two waves. The proposed process is illustrated in Fig. 4.25. After entering the first plate of a layered glass assembly, the shock wave is split into an elastic compression wave with the stress and particle velocity corresponding to the point E_1, and a failure wave which is described by the point 1. When the elastic wave passes through the interface between the first and the second of the glass plates, a partial release and recompression occurs in the porous interlayer. After recompression, two new elastic waves and two failure waves are formed and propagate both forward into the sample and back to the base plate. The resulting stress and particle velocity at the interface correspond to the point S_1 in Fig. 4.25, whereas the image point of the elastic compression wave propagating forward shifts from E_1 down to E_2. The decreasing magnitude of the leading elastic wave is visible from a comparison of the second stress profiles in Figs. 4.22b and 4.22c.

A more thorough consideration of the first stress histories shows that the stress at point 2' is larger than that at point 2, and that next reverberation makes the stress at point 2" larger than that at 2' whereas the maximum stress remains approximately constant. In other words, the magnitude of the second stress step decreases with each new interface being passed by the compression wave.

Having the stress increments recorded by the first gauge after the meeting of two failure waves, we can estimate the stress increments, σ_f, in the failure waves. The appropriate relationship,

$$-\frac{\sigma_f}{\rho_g c_f} + \frac{\Delta\sigma - \sigma_f}{\rho_g c_g} = -\frac{\Delta\sigma}{\rho_{Cu} c_{Cu}},$$

is derived from the stress–particle velocity diagram shown in Fig. 4.25. In this equation ρ_g and c_g are the density and sound speed for the glass, ρ_{Cu} and c_{Cu} are the density and sound speed for copper, ρc is the dynamic impedance, and $\Delta\sigma$ is the stress increment from point 2 to point 3. Using $c_g = 5.58$ km/s, $c_f = 1.5$ km/s, and $c_{Cu} = 4.5$ km/s at a pressure of 6 GPa, we obtained $\sigma_f \approx 0.28 \cdot \Delta\sigma$ and evaluated the stress increments to be ~0.17 GPa, as are presented in Table 4.2. Corresponding increments of particle velocity,

$$u_f = \sigma_f / (\rho_g c_f),$$

and compressive deformations in the failure waves

$$\varepsilon_f = u_f / c_f$$

are also presented in the table. The estimated σ_f values are less than 0.5 GPa of the difference between stresses measured by the first and second gauges in the shot with thick glass plates. However, we have to account for the stress decrease as a result of formation of failure waves in the plane of the second gauge in this shot.

The decrease of elastic wave amplitude should repeat at each interface until the failure threshold is reached. As a result, for a sufficiently large number of plates in the assembly, an elastic precursor wave having an amplitude equal to the failure threshold should be formed. In such an assembly, fracture will occur upon further compression and should form a second compression wave.

Figure 4.26 presents results of two shots where free surface velocity histories were recorded for layered assemblies of 8 glass plates of average thickness 1.21 mm, subjected to the same impact loading. The results with good reproducibility show the waveform that is typical for elastic–plastic solids. The magnitude of the elastic precursor wave is 4.0 GPa, its rise time (from 10 to 90% of its amplit-

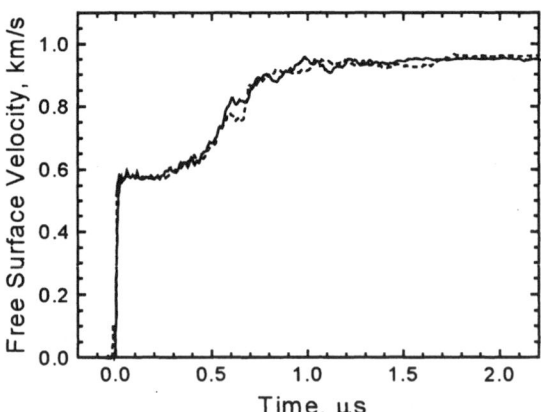

Figure 4.26 The free surface velocity histories recorded in two shots with layered assemblies of 8 soda lime glass plates of 1.21 mm average thickness.

ude) is ~15 ns, which is only about one-fourth of the 63 ns rise time of the same velocity increment in an elastic wave propagating through a thick plate. The reason for this disagreement is obviously associated with the porous interlayers between the lapped plates of the glass assembly. The rise time of the second wave corresponds to approximately one-half of the time needed for the failure wave to pass through one plate. The final free surface velocity is practically equal to that of a single glass plate.

The failure wave is really a wave process, but its kinematics differ from those of elastic–plastic waves. The shock compression wave in an elastic–plastic body becomes unstable as a result of the sudden decrease of longitudinal compressibility that occurs when yielding begins. As a result, the wave splits into an elastic precursor wave and a plastic shock wave. The peak stress behind the elastic precursor front is the HEL, which is determined by the yield stress. A similar wave structure should be seen in a polycrystalline brittle solid where the HEL corresponds to the failure threshold stress and fracture occurs locally in each grain (or around grains) immediately when an applied stress exceeds the failure threshold. Note that, in both these cases, the propagation velocities of the elastic precursor wave front and the second compression wave are determined by the longitudinal and bulk compressibility, respectively.

However, the propagation velocity of the failure wave is determined by the crack growth speed, which is not directly related to the compressibility. On the other hand, the final longitudinal stress in the comminuted glass behind the failure wave is determined by the impact conditions, whereas the deviator stress component is controlled by the post-failure material properties. Thus, since the propagation velocity of a failure wave and the final stress are fixed, the stress in the leading elastic wave should be governed by these values and should not necessarily be equal to the failure threshold. The glass surface plays an important role in the failure-wave process because the surface is a source of cracks. In this sense the process is similar to diffusion. When the stressed state is maintained, the subsonic failure wave may evidently propagate in a self-supported mode like a combustion wave.

Having the stress and deformation increments in the failure wave, we may estimate the final state behind it. Figure 4.27 presents the stress–strain diagram of shock compression of soda lime glass. The slope of the $\sigma_x(\varepsilon_x)$ curve decreases with increasing stress as a result of anomalous compressibility of the glass. The bulk compressibility of glass is assumed to be constant: $p = -\rho_0 c_b^2 (\Delta V/V_0)$, where c_b = 4.24 km/s is the bulk sound speed at zero pressure and V is the specific volume. In Fig. 4.27 the Rayleigh line for the failure wave that is shown has been calculated using average values for the failure wave speed, c_f = 1.55 km/s, the stress increment, σ_f = 0.17 GPa, deformation, ε_f = 0.02925, and the final stress, σ_x = 6.78 GPa at the impact surface. The estimated final state is above the pressure curve by the amount $\sigma_x - p$ = 1.4 GPa. Since the hydrostatic pressure is the average stress $p = (\sigma_x + 2\sigma_y)/3$, we may estimate the principal

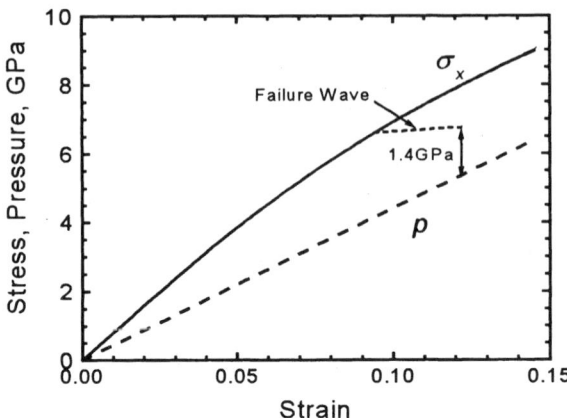

Figure 4.27. The stress–strain diagram of glass under shock compression. The dashed line shows the assumed linear bulk compressibility, and short-dashed line illustrates the failure wave process.

stress difference to be $\sigma_x - \sigma_y = 3(\sigma_x - p)/2 = 2.1$ GPa. For comparison, Brar et al., 1991a, carried out direct measurements of the principal stress difference and got 2.0–2.3 GPa behind the failure wave in soda lime glass (see Fig. 4.20).

The response of a layered assembly of thin brittle plates as compared to that of one thick plate is a simple way to diagnose nucleation of the failure process on the plate surfaces and determine the failure threshold. As an illustration, Figs. 4.28 and 4.29 show the results of such experiments with fused quartz and K8 crown glass. The failure wave phenomenon is obviously a common one for different glasses but the threshold stress differs depending on the glass properties.

The mechanism of failure wave propagation under compression is not yet completely clear. Note that the compressive fracture of glasses under quasistatic conditions occurs by axial splitting (Bridgman, 1964). Since the failure wave speed reaches the ultimate speed of cracks in glasses, it is natural to conclude that the failure wave consists of cracks propagating in the direction of compression. Probably, some mechanism of self-supporting propagation of rapid cracks exists. However, even with self-propagating axial cracks it is difficult to understand the observed shear stress relaxation. Branching or transverse cracking must also take place in the failure wave.

In modeling the behavior of materials subjected to intense impulsive loading, it is usually assumed that local material response can depend only on characteristic material properties and the local state. In the case of cracking, the material ahead of the fracture zone should not "know" that cracks even exist. The principal features of fracture of glass are related to different structures of the homogeneous bulk material and the damaged surface with microfissure nuclei. The experimental data reported here imply an urgent need to develop a nonlocal constitutive model for the failure-wave phenomenon.

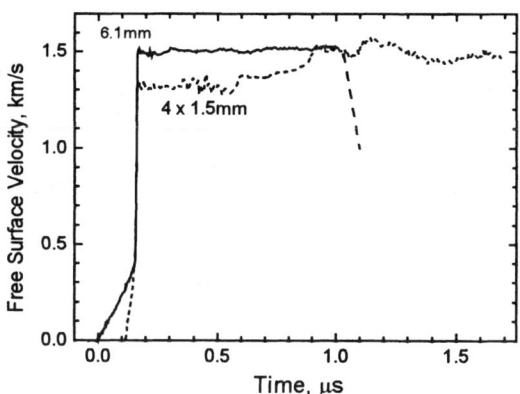

Figure 4.28. The free surface velocity histories of one thick plate and a layered assembly of four thin plates of fused quartz under the same impact conditions.

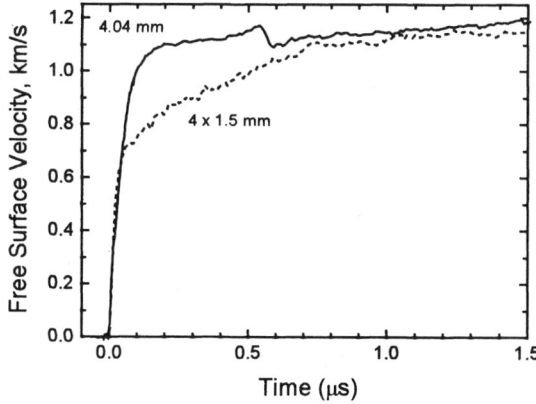

Figure 4.29. The free surface velocity histories of one thick plate and a layered assembly of four thin plates of K8 crown glass under the same impact conditions.

4.7. Dynamic Strength Properties of Polycrystalline Ceramics

The response of ceramics has been studied by many authors since their low densities and high Hugoniot elastic limits have made them attractive potential armor materials. Extensive studies have been conducted to obtain Hugoniot data, longitudinal wave profiles, and spall strength data for various ceramics under shock-wave loading. The most widely tested and applied ceramics are Al_2O_3, B_4C, SiC, TiB_2, and AlN. They are densified by liquid-phase sintering, solid-state sintering, or hot pressing. The dynamic mechanical properties of these ceramics are discussed below.

Modern shock-wave tests of ceramics include measurements of the Hugoniot over a wide stress range, shock front rise time, Hugoniot elastic limit, stress state immediately after shock compression, high pressure stress–strain path, tensile (spall) strength after shock compression below and above the elastic limit, and post-test examination of recovered samples. Figure 4.30 shows typical particle velocity histories recorded by Kipp and Grady, 1990, for SiC and B_4C ceramics. The wave profiles exhibit two extreme examples of behavior of ceramics in plane shock waves. The response of silicon carbide is very similar to that of ductile materials. The waveform consists of an initial elastic wave whose amplitude (the HEL) is the limit stress for elastic behavior. This is followed by a second shock corresponding to bulk compression, and this is followed, in turn, by elastic and bulk unloading waves that originate from the rear surface of the flyer plate. The post-yield strength of silicon carbide, determined by comparison of Hugoniot uniaxial strain and calculated hydrodynamic response, increases considerably beyond the initial dynamic yield stress. The release trajectories for silicon carbide indicate reverse yielding and continued elastic–plastic bulk behavior, probably with a Bauschinger effect at higher peak stresses. The shock response of B_4C is quite different, at least beyond the HEL. The bulk compression wave is much slower. According to Grady, 1992, the Hugoniot collapses to the hydrostat at stresses approaching twice the HEL. The dispersed character of the unloading wave indicates that the amplitude of elastic unloading is very small.

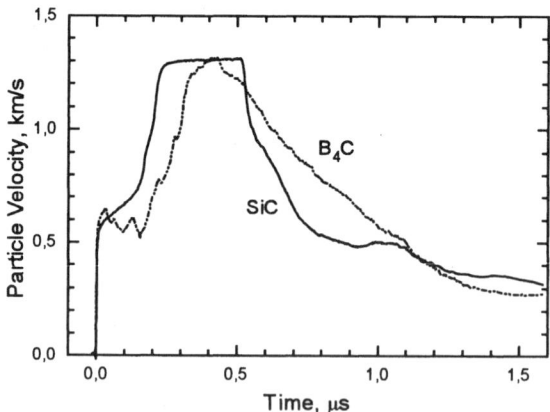

Figure 4.30. Particle velocity profiles for SiC and B_4C ceramics measured at the interface with a LiF window (Data by Kipp and Grady, 1990).

The inelastic strain starts almost immediately behind the rarefaction-wave front. The stress–strain trajectory for the B_4C ceramic shows evidence of dilatancy when the compressive stress approaches zero on unloading.

The post-yield characteristics of these materials are qualitatively different, as is apparent from shape of the wave profile behind the elastic precursor front. For silicon carbide, positive slope of the wave demonstrates strain hardening. The stress drop after the HEL in boron carbide, in contrast, indicates post-yield softening. Spall strength is sustained for shocks above the HEL in SiC, but not in B_4C.

The shock-wave properties of hard ceramic materials are reviewed in more detail below.

4.7.1. Aluminum Oxide

Knowledge of the dynamic yield behavior of polycrystalline aluminum oxide (alumina) is important because of technical uses of this material. Table 4.3 summarizes the HEL data for Al_2O_3 ceramics in different states. Ahrens et al., 1968, reported Hugoniot data for the high density ceramic, Lucalox (0.2% porosity), and a lower density ceramic, Wesgo Al-995 (3.5–4.3% porosity). The measured HEL values are 11.2 ± 1.3 GPa for Lucalox and 8.3 ± 0.5 GPa for Al-995 ceramics. Using the values of Poisson's ratio obtained from ultrasonic data, the yield strength was determined to be 7.8 GPa and 6 GPa respectively. Gust and Royce, 1971, performed Hugoniot measurements for five aluminas. The general trend is that the HEL decreases with decreasing ceramic density, but they found that the impurity content and material processing also influence the HEL value. Compaction of more porous ceramics occurs within the stress range from the yield point to about 30 GPa. At the higher stresses the states of all alumina ceramics are described by one curve in stress–volume coordinates. Beyond the compaction region, the yield strength estimated from the stress offset between the Hugoniot and isotropic compression curve is comparable to, but somewhat smaller than, the yield strength at the HEL.

Figure 4.31 presents a set of free surface velocity histories measured for alumina ceramic of submicron grain size and 3.46 g/cm³ density. The wave profiles in general are similar to that for silicon carbide. They exhibit an elastic jump and a subsequent dispersed rise to the bulk wave which compresses the material to a final state. This gradual transition from elastic to inelastic portions of the compression wave is typical for strain hardening materials. There is no sharp distinction between the elastic and inelastic parts. The rounded transition from the elastic to the inelastic portion of the compression wave makes unambiguous determination of the HEL value difficult. Perhaps this is a reason for the existing controversy over whether aluminas exhibit precursor decay.

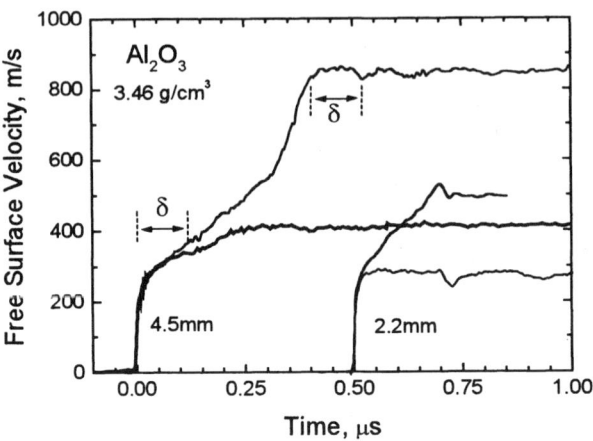

Figure 4.31. Free surface velocity histories of alumina ceramic samples of different thickness at various impact stresses. In this figure, the time interval δ designates the spall thickness

In the case of rate-dependent response of ceramics, precursor decay should be observed. The results of Grady's VISAR experiments on AD995 plates 2.5–10 mm in thickness show that, although it appears as if the precursor is decaying, a self-similar scaling plot shows that, within a natural scatter, there is no decay. In contrast, Yaziv et al., 1988, observed precursor decay from 6.1 to 4.7 GPa in AD-85 ceramic when the propagation distance increased from 6.15 to 36.8 mm. Murray et al., 1996, also measured decay of 20–25% for lower density aluminas. It is necessary to note, however, that the interpretation may have been incorrect because of the limited time resolution of the manganin gauges used in these studies. Cagnoux and Longy, 1988, measured free surface velocity profiles for alumina at various rise times of the compression wave entering the sample. The HEL was found to be independent of the wave propagation distance, the peak shock stress, and the stress gradient of the entering compression wave. This means that there is no influence of strain rate on the yield strength of alumina in a range of 5×10^4 to 6×10^5 s^{-1}. On the other hand, Furnish and Chhabildas, 1998, found evidence of rate-dependent behavior of AD995 ceramic subjected to step-like compression. According to many measurements, the unloading wave front in shock-compressed alumina is elastic. However, there is no sharp distinction between the elastic and inelastic parts of the unloading process.

Table 4.3 and Figs. 4.32 and 4.33 present the shock data [longitudinal sound speed, c_ℓ, the Hugoniot elastic limit, HEL, and the von Mises yield stress,

$Y = \text{HEL} \cdot (1 - 2\nu)/(1 - \nu)]$ as functions of the initial density, ρ_0, for different Al_2O_3 ceramics. The lateral stress, $\sigma_t = \text{HEL} - Y$, in these uniaxial strain shock tests is within a range of 2–3.5 GPa, which exceeds maximum confining pressure of 1.25 GPa in the quasi-static experiments of Heard and Cline, 1980.

Besides the porosity, the yield strength of ceramics depends on the grain size. Longy and Cagnoux, 1989, found that alumina ceramics with 2% porosity but with grain size of 5 or 60 μm exhibit an HEL of 8.5 GPa and 5 GPa, respectively. Microscopic examination of impure alumina showed microcracks in the inter-granular glassy phase after shock stress ≥0.9 HEL without any correlation between the HEL and microcracking. According to the results of TEM observations, only part of grains were strained at 2 HEL.

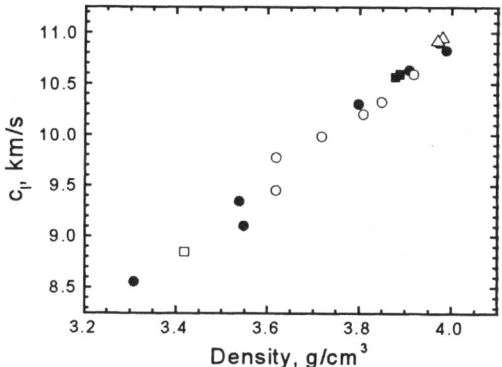

Figure 4.32. Longitudinal sound speed, c_ℓ, in different Al_2O_3 ceramics as a function of their density.

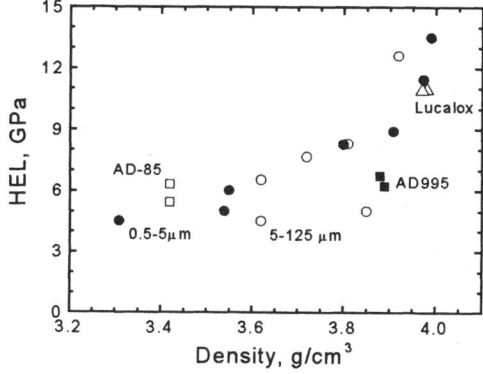

Figure 4.33. Hugoniot elastic limits of Al_2O_3 ceramics of different density and grain size.

Table 4.3. Hugoniot elastic limits of aluminum oxide ceramics

Material, (wt. % Al_2O_3) grain size	Density, g/cm^3, (Void %)	Longitudinal Sound Velocity, c_ℓ, km/s	Poisson's ratio	Hugoniot Elastic Limit, GPa	Dynamic Yield Strength, Y, GPa	Reference
Lucalox (99.8%)	3.98 (<0.2)	10.95	0.2363	11.2±1.3	7.8	Ahrens, Gust, and Royce, 1968
Lucalox (99.9%) 25–40 µm	3.969	10.92	—	9.1±0.4	6.0	Munson and Lawrence, 1979
MTU JS-I (99.99%), 1.5 µm	3.974	10.9	0.237	11–11.9	7.6–8.2	Staehler et al., 1994
D999 (99.9%), 4 µm	3.99	10.82	0.232	13–14	9–9.8	Murray et al., 1996
Carborundum, hot pressed	3.92 (0.8)	10.59	0.243	9.2–16	—	Gust and Royce, 1971
Wesgo Al-995 (99.5%)	3.81 (3.5–4.3)	10.2	0.218	8.3±0.5	6.0	Ahrens, Gust, and Royce, 1968
D975 (97.5%), 4 µm	3.8	10.3	0.234	7.5–9	5.2–6.2	Murray et al., 1996
Coors AD995, alumino-silicate glass binder	3.88 (2)	10.56	—	6.7±0.1	—	Dandekar and Bartkowski, 1994
Coors AD995	3.89	10.59	0.234	6.2±0.4	4.3	Grady, 1994
Coors AD-85	3.42 (6.6)	8.84	0.256	6.1–6.5	4.1	Gust and Royce, 1971
Coors AD-85 (84%)	3.42 (6.6)			4.7–6.1	—	Yaziv et al., 1988
H880 (88%), 2 µm	3.55	9.1	0.226	5.5–6.5	3.9–4.6	Murray et al., 1996
Diamonite P-3142-1	3.72 (5.5)	9.98	0.234	7.2–8.1	—	Gust and Royce, 1971
Desmarquest alumina	3.62 (5.3)	9.45	—	4.5	—	Cagnoux and Longy, 1988
ENSCI, 4.7 µm	3.91 (2)	10.63	—	8.7±0.4	—	Longy and Cagnoux, 1989
ENSCI, 1 µm	3.54 (11)	9.34	—	5	—	Longy and Cagnoux, 1989
ENSCI, 0.6 µm	3.31 (17)	8.55	—	4–5	—	Longy and Cagnoux, 1989
ENSCI T60 (99.7%), 5–125µm	3.85 (3.5)	10.32	—	5	—	Longy and Cagnoux, 1989
UL500 (93.8%), 11 µm	3.62 (0.2)	9.77	—	6.5	—	Longy and Cagnoux, 1989

Cagnoux, 1990, carried out microscopic examination of alumina samples of two different grain sizes (4.7 μm and 10–20 μm) with 99.7% Al_2O_3 content and 3.91 g/cm^3 density. The samples were recovered after compression above their HEL by spherical shock waves. In the region of maximum peak stress, the fine-grain alumina remained uncracked whereas the coarse-grain sample was microfragmented. The SEM photographs of the fine-grain samples show a reduction in porosity with no slip-nucleated microcracks. In the coarse-grained sample numerous twins were observed. It was concluded that twinning is favored by large grain size whereas slipping is favored by small grain size. Activation of twinning as an additional mechanism of inelastic deformation led to a reduction of the yield stress. Wang and Mikkola, 1992, observed twins in fragments of recovered shock-compressed samples of sapphire and high-purity alumina with 25 μm average grain size. The defect densities in polycrystalline samples were lower than in single crystals, especially at the highest shock stress of ~23 GPa. In polycrystals, the grain boundaries serve as crack initiation sites and crack propagation paths, thus contributing to fragmentation.

The uniaxial strain conditions of the shock-wave tests do not permit varying the relationship between longitudinal and transverse stresses. Table 4.4 presents the results of rod-impact experiments that provide the dynamic failure threshold under uniaxial stress conditions. For unconfined rods, the maximum axial stress increases up to some ultimate magnitude with increasing impact velocity. This stress magnitude is considered as the failure threshold. Comparison of data of Tables 4.3 and 4.4 shows that the failure threshold for rod impacts is lower than the dynamic stress, Y, at the HEL for the same material. For the ceramic rods confined within a close-fitting high-impedance sleeve, the maximum axial stress may approach the HEL value. Chhabildas et al., 1998, performed experiments with confined and unconfined AD995 alumina rods shock loaded by gradient-density impactors. The time-dependent stress pulse generated by such an impactor allows an efficient transition from the initial uniaxial strain loading to a uniaxial stress state as the stress pulse propagates along the rod.

Table 4.4. Failure thresholds for alumina ceramics in rod impact experiments

Material, impact conditions	Confinement	Failure, Threshold, GPa	Reference
AD94, direct impact	—	2.7	Bless et al., 1990
AD99, direct impact	—	4	Bless et al., 1990
AD995, direct impact	—	3.15	Wise and Grady, 1994
AD995, direct impact	Ta	5.8–6.3	Wise and Grady, 1994
AD995, dispersed impact	—	3.5–4.2	Chhabildas et al., 1998
AD995, dispersed impact	Steel	4.6	Chhabildas et al., 1998
AD995, direct impact	—	3.6–3.7	Simha et al., 2000
AD995, direct impact	Steel	4.2	Simha et al., 2000
AD995, direct impact	—	3.8	Cuzamias et al., 2002

Figure 4.34 presents results of spall strength measurements as a function of normalized peak stress for alumina ceramics (Rosenberg, 1992; Dandekar and Bartkowski, 1994; Staehler et al., 1994; and Song et al., 1994). It seems that the spall strength undergoes a transition, first decreasing near the HEL then increasing with increasing pressure above the HEL. The reduction of spall strength near the HEL is especially significant for aluminas with a large glassy phase content. This observation correlates with observations by Longy and Cagnoux, 1989, of microcracks in the inter-granular glassy phase at shock stresses exceeding 90% of the HEL. Within the elastic region, the spall strength decreases with increasing porosity and grain size.

It is necessary to mention that the high spall strength at peak stresses much above the HEL may be an artifact. The free surface velocity histories shown in Fig. 4.31 demonstrate a remarkable velocity pullback at peak stresses below and much above the HEL whereas the pullback is not recorded in the experiment at a peak stress just slightly exceeding the HEL. It is seen also that δ, the period of wave reverberation in the spall plate, is less than the time interval between elastic and plastic waves at the highest peak stress. This means that, as a result of unloading after reflection of the elastic precursor wave from the sample surface, the spallation actually occurs within a layer near the impact surface where the peak stress did not significantly exceed the HEL.

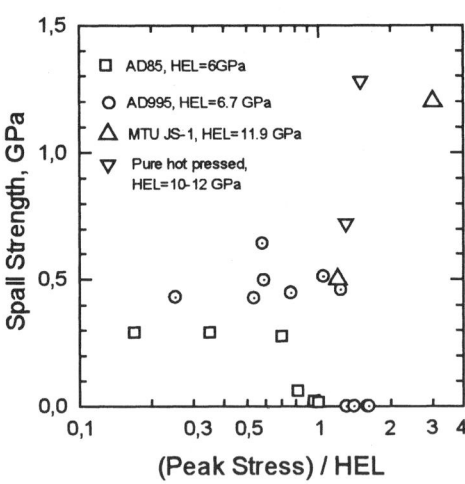

Figure 4.34. Spall strength of alumina ceramics as a function of peak stress. Data for AD85 from Rosenberg, 1992; for AD995 from Dandekar and Bartkowski, 1994; for hot pressed pure alumina from Staehler et al., 1994, and Song et al., 1994.

4.7.2. Silicon Carbide

Lightweight intermetallic compounds are characterized by large dynamic impedances and large yield strength. Gust et al., 1973, measured the Hugoniot of hot-pressed silicon carbide ceramic. They found that SiC exhibits evidence of a very sluggish phase transition or a phase transition with a small volume change. However, subsequent measurements of shock-wave profiles at 27.6 GPa and 36.5 GPa peak stress made by Kipp and Grady, 1990, as well as measurements of hydrostatic compressibility in a diamond anvil cell by Bassett et al., 1993, have not confirmed this finding. The hydrostatic data have been described up to 68.4 GPa by the Birch–Murnagan equation of state with a bulk modulus $K_0 = 230.2 \pm 4.0$ GPa and pressure derivative $K_0' = 4$.

Table 4.5 presents results of measurements of the Hugoniot elastic limits of silicon carbide ceramics. The HEL values obtained from the particle velocity profile measurements exceed the value reported by Gust et al., 1973. It is speculated that this is due to the substantially lower silica content of modern silicon carbides. Post-yield strength of silicon carbide, determined by comparison of Hugoniot uniaxial strain and calculated hydrodynamic response, reveals significant strengthening with subsequent deformation beyond the initial dynamic yield. According to Grady, 1992, the maximum shear stress in Eagle–Picher SiC ceramic increases from 7 GPa at the HEL to 10 GPa at ~30–50 GPa shock stress. The release trajectories for silicon carbide indicate reverse yielding and continued strength characteristics of elastic–plastic material behavior, probably with a Bauschinger effect at higher peak stresses. According to Feng et al., 1998, the maximum shear stress in the Cercom SiC–B hot-pressed ceramic increases from 4.5 GPa at the HEL to 7.0 GPa at a shock stress of approximately twice the HEL.

Figure 4.35 summarizes the results of measurements of the spall strength as a function of the peak stress. As in the case of alumina, the results are very scattered and show a tendency to decrease when the peak stress is in the vicinity of the HEL. However, there is no evidence that the spall strength falls to zero at the HEL. Bartkowski and Dandekar, 1996, compared the spall properties of sintered and hot-pressed silicon carbide under the same test conditions. The results demonstrate that the hot-pressed ceramic has better cohesion between grains than the sintered material.

Pickup and Barker, 1998, made split Hopkinson pressure bar (SHPB) experiments at a strain rate of $\sim 10^3 \, \text{s}^{-1}$ with three SiC ceramics that were manufactured using a reaction bonding technique, by pressureless sintering, and using a pressure-assisted densification method. The measured dynamic compressive strength is 6.7 GPa, 7.5 GPa, and 8.2 GPa, respectively, whereas the quasi-static

Table 4.5. Hugoniot elastic limits of silicon carbide ceramics

Material	Density, g/cm³, (Void fraction, %)	Longitudinal Sound Velocity, c_l, km/s	Poisson's ratio	Hugoniot Elastic Limit, GPa	Dynamic Yield Strength, Y, GPa	Reference
Carborundum Co., type KT, 99.2% SiC	3.09 (4)	11.4	0.157	8 ± 3	6.5	Gust, Holt, and Royce, 1973
Eagle-Picher α-SiC, 7 μm grain size	3.177 (1)	12.06	0.16	15–16	—	Kipp and Grady, 1990
Cercom type B hot-pressed, 99.3% SiC, 2 μm grain size	3.221	12.22	0.164	11.7	—	Feng et al., 1998
Reaction bonded, 1.2 μm grain size, 4.5 GPa static comp. strength	3.21	11.89	0.18	13.2 ± 0.3	10.4	Bourne and Millett, 1997
Sintered, 4.5 μm grain size, 5.2 GPa static comp. strength	3.16	11.94	0.16	13.5 ± 0.3	11	Bourne and Millett, 1997
Hot pressed, 2.9 μm grain size, 5.2 GPa static comp. strength	3.24	12.34	0.17	15.7 ± 0.3	12.6	Bourne and Millett, 1997

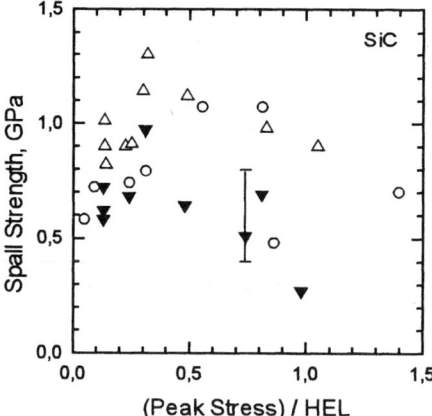

Figure 4.35. Spall strength of silicon carbide ceramics vs. impact stress. The bar shows data by Grady, 1992, and Kipp and Grady, 1992 for Eagle–Picher SiC; circles are data by Winkler and Stilp, 1992a, triangles show data by Bartkowski and Dandekar, 1996, for hot-pressed Cercom SiC–B of 2 μm grain size (open triangles) and sintered Sohio SiC of 15 μm grain size (solid triangles).

compressive strength of these materials is 4.5 GPa, 5.2 GPa, and 5.2 GPa, respectively. The first material failed in an apparently brittle manner whereas the others demonstrated a significant amount of apparently inelastic flow. For the second (sintered) material, they also observed ejection of failed microparticles from the sample surface, beginning at a stress approximately equal to the quasistatic failure strength. For the third (hot-pressed) material, there was no evidence of any ejecta being formed from the point at which the quasi-static compressive strength was exceeded and up to the highest stress attained. The ranking of relative resistance to damage in the SHPB experiments correlates with the results of shear strength measurements under shock compression made for the same three ceramics. Bourne and Millett, 1997, found hardening of hot pressed SiC over an applied longitudinal stress range of ~1–1.4 times the HEL, whereas the shear strength of two other SiC ceramics decreased over the same stress range.

The lateral stress profiles measured by Bourne and Millett, 1997, indicate that, above the HEL, the failure of SiC ceramics was delayed for some time after the maximum stress had been achieved. This finding was interpreted in terms of the failure wave phenomenon.

4.7.3. Boron Carbide

The inelastic deformation of boron carbide behind the precursor front is accompanied by a dramatic loss in shear strength. This is exhibited by the precursor shape and by the subsonic velocity of the second compression wave. According to Grady, 1992, Hugoniot and hydrodynamic response for boron carbide converge at stresses approaching about twice the HEL. The dispersive character of the unloading wave indicates that the amplitude of elastic unloading is very small. The stress–strain trajectory for the B_4C ceramic shows evidence of dilatancy when the compressive stress approaches zero on unloading.

The velocity profiles for boron carbide, as well as stress histories measured by Brar et al., 1992b, show instabilities behind the elastic precursor front. In experiments with a gapped flat mirror, Mashimo and Uchino, 1997, observed jagged profiles of the moving free surface in both the elastic and the plastic regions. The average period of the nonuniformities formed on the B_4C sample surface was of order of 1 mm and the maximum local displacement was a few tens of µm. Such a heterogeneous free surface motion is evidence of operating macroscopic slip systems, such as cracks, cleavages, and melting zones.

Brar et al., 1992b, measured the HEL values for B_4C ceramics of different porosity and the same grain size. They have found the HEL decreases with porosity according to the relationship

$$\sigma_{HEL}^p = (Z_p / Z_s)^2 \sigma_{HEL}^s$$

where σ_{HEL}^p and σ_{HEL}^s are the Hugoniot elastic limit of the porous and solid matter, and Z_p and Z_s are impedances of the porous and solid matter, respectively. When the porosity exceeded ~10% the elastic precursor wave shape changed: A ramped rise appeared behind the precursor front in porous ceramics instead of the spike observed at higher density. The experimental data on the Hugoniot elastic limits of boron carbide ceramics are summarized in Table 4.6. According to Winkler and Stilp, 1992a, the spall strength of boron carbide is within a range of 0.6–0.8 GPa at the peak stresses up to 11 GPa, although at a higher strain rate Kipp and Grady, 1992, observed a much larger spall strength of 1.8 GPa within this peak stress range. Above the HEL the spall strength of B_4C is presumed to be zero.

4.7.4. Titanium Diboride

Figure 4.36 shows the shock wave profile in a TiB_2 ceramic plate, as measured by Kipp and Grady, 1990. The material exhibits significant dispersion of the compression wave to at least 50 GPa. The structure of the compressive wave caused difficulty in unambiguously selecting a particle velocity corresponding to the HEL. There are two smoothed breaks at ~160 m/s and ~430 m/s which correspond to ~5 GPa and ~13.5 GPa, respectively. Actually, this wave structure in TiB_2 ceramic is not well reproduced. At the same impact parameters Winkler and Stilp, 1992b, recorded free surface velocity profiles without the second break and, in one shot, with the second break at 9 GPa stress level. In the shot

Table 4.6. Hugoniot elastic limits of boron carbide ceramics

Material	Density, g/cm^3, (Void fraction, %)	Longitudinal Sound Velocity, c_ℓ, km/s	Poisson's ratio	Hugoniot Elastic Limit, GPa	Dynamic Yield Strength, Y, GPa	Reference
Norton Co., Hot pressed, 99.7% B$_4$C	2.5 (0.8)	13.78	0.188	15.4±1	11.9±0.5	Gust and Royce, 1971
B$_4$C, 10μm grain size	2.516	14.04	0.164	14–14.8	13.7	Kipp and Grady, 1990
Dow Chemical, 3 μm grain size	2.506 (1)	14.03	0.17	18–20	15.1	Grady, 1994
Dow Chemical, 2.5 μm grain size	2.52	13.42	0.17	19.4	—	Brar et al., 1992b
Dow Chemical, 2.5 μm grain size	2.43	13.02	0.18	17.1	—	Brar et al., 1992b
Dow Chemical, 2.5 μm grain size	2.33	12.8	0.18	16.3	—	Brar et al., 1992b
Dow Chemical, 2.5 μm grain size	2.25	12.52	0.18	13.5	—	Brar et al., 1992b
Dow Chemical, 2.5 μm grain size	2.13	11.85	0.17	9.6	—	Brar et al., 199b

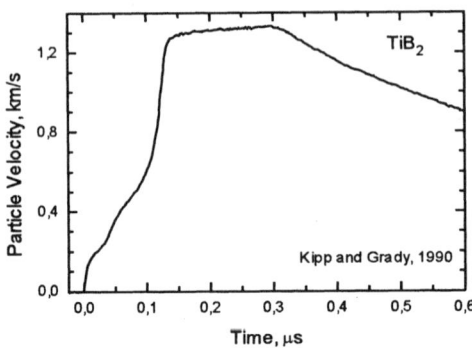

Figure 4.36. Particle velocity history measured by Kipp and Grady, 1990, at the interface between titanium diboride sample plate and LiF window.

where the second HEL was observed the grain size of the ceramic was much less than in others (5–10 μm and 30–50 μm, respectively). Also, the samples had different kinds of initial porosity: intergrain pores localized mainly in triple grain junctions in the case of fine-grain ceramic and mainly intragrain pores in other ceramics. The appearance of plastic defects in recovered samples has been found to be connected with stresses exceeding the second HEL. Microscopic examination of recovered TiB_2 samples by Ewart and Dandekar, 1994, indicated that the average number density of cracks and the average crack length increase with increasing peak compressive stress. The largest generation of new cracks occurs when the peak stress exceeds the lower Hugoniot elastic limit. Damage in the form of cracking was distributed homogeneously throughout the body.

Table 4.7 summarizes the results of measurements of the Hugoniot elastic limits for the titanium diboride ceramics. Rosenberg et al., 1992, measured the transverse stresses in shock-compressed states. The results presented in Fig. 4.37 show that the shear strength of TiB_2 at a shock stress of 24 GPa increases to more than twice its value at the lower HEL. Dandekar, 1994, calculated the shear sustained by TiB_2 from the stress offset between the normal shock stress and calculated hydrostat curve and found good agreement with the data shown in Fig. 4.37. On the other hand, the stress–strain load-release trajectories calculated by Kipp and Grady, 1990, do not demonstrate as large a value of hysteresis as could be expected with so much hardening. Figure 4.38 presents the results of spall strength measurements. The dramatic rise in crack lengths and densities at HEL1 observed by Ewart and Dandekar, 1994, indicates that the first HEL is a threshold above which there is increased generation of irreversible damage. The result is a falling spall threshold.

4. Behavior of Brittle Materials Under Shock-Wave Loading   159

Table 4.7. Hugoniot elastic limits of titanium diboride ceramics

Material	Density, g/cm^3, (Void fraction, %)	Longitudinal Sound Velocity, c_ℓ, km/s	Poisson's ratio	Lower and upper Hugoniot Elastic Limit, GPa	Dynamic Yield Strength, Y, GPa	Reference
Union Carbide, Hot pressed, 98.5% TiB$_2$	4.51 (0)	11.21	0.141	8.6±3	7.2	Gust, Holt, and Royce, 1973
TiB$_2$, .12 μm grain size	4.452	10.93	0.1	4.7–5.2 (13.1–13.7)	4.2–4.6 (11.7–12.2)	Kipp and Grady, 1990
Hot pressed, 5–10 and 30–50 μm grain size	4.36	10.79	—	4.2–4.9 (9)	—	Winkler and Stilp, 1992b
Ceradyne, hot pressed, 98.9% TiB2, 10 μm grain size	4.49 (1)	11.23±0.21	0.1144	5.8±0.2 (13±0.4)	—	Dandekar and Gaeta, 1992

4.7.5. Aluminum Nitride

The published shock-wave data for aluminum nitride ceramics are presented in Table 4.8 and Fig. 4.39. For shock compression above 20–22 GPa peak stress, a phase transition, presumably wurtzite to rocksalt, occurs. As a result, the plastic shock wave is split in two, or perhaps three, steps. The 20% volume strain associated with the transition leads to a rapid attenuation of the peak wave amplitude (Grady, 1998). Measurements by Brar et al., 1992a, of the longitudinal and transverse stresses in shock-compressed AlN ceramic show some moderate strain hardening which was observed also in quasistatic compressive tests under confining pressure (Heard and Cline, 1980).

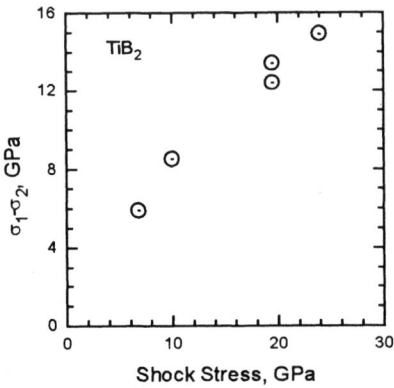

Figure 4.37. Difference in principal stresses, $\sigma_1 - \sigma_2$, as a function of peak shock stress in TiB$_2$ ceramic (Rosenberg et al., 1992).

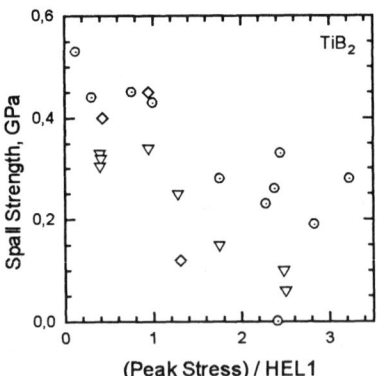

Figure 4.38. Spall strength of titanium diboride ceramics vs. impact stress. Diamonds show data by Grady, 1992; circles are data by Winkler and Stilp, 1992a; triangles show data by Dandekar, 1994. The peak stresses are normalized by the lower Hugoniot elastic limit of 4.2 to 7 GPa.

Table 4.8. Hugoniot elastic limits of aluminum nitride ceramics

Material	Density, g/cm^3, (Void fraction, %)	Longitudinal Sound Velocity, c_ℓ, km/s	Poisson's ratio	Hugoniot Elastic Limit, GPa	Dynamic Yield Strength, Y, GPa	Reference
Dow Chemical Co., hot pressed 2 μm grain size	3.226 (1)	10.72	0.238	9.4±0.2	6.4	Brar et al., 1992a
Dow Chemical Co., sintered 4–6 μm grain size	3.20 (1–2)	10.45	0.237	7.0±0.4	4.8	Brar et al., 1992a
AlN	3.254	10.73	0.234	7.9	5.5	Grady, 1994

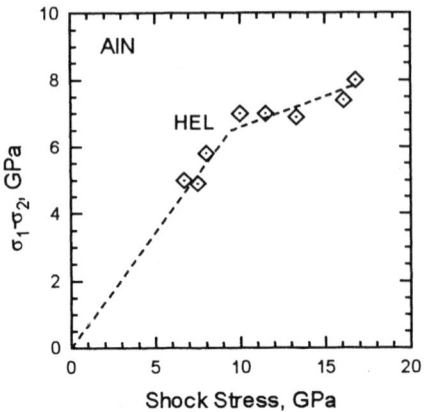

Figure 4.39. Difference in principal stresses ($\sigma_1 - \sigma_2$) as a function of peak shock stress in a hot-pressed aluminum nitride ceramic (Brar et al., 1992a).

According to Heard and Cline, 1980, a transition from brittle to ductile inelastic compression in the hot-pressed AlN ceramic occurs at 0.5 GPa confining pressure, which is much less than the transition pressure for alumina. At a confining pressure of 0.8 GPa the quasi-static yield strength for AlN reached 4.7 GPa whereas without confinement it was 3.2 GPa. Microscopic examination confirmed the change in mechanical response of AlN at high pressure due to intracrystalline slip by dislocation generation and motion. Chen and Ravchandran, 1996, observed an increase in the compressive strength of sintered AlN by approximately 1.5 GPa as the strain rate was increased from 4×10^{-4} to 5×10^2 s^{-1} so, at the confining pressure of 230 Mpa and high strain rate, the strength exceeded 5 GPa. A comparison of these data with dynamic yield strength data in Table 4.5 suggests a ductile behavior of the ceramic under shock-wave compression.

4.7.6. Zirconia and Other Ceramic Materials

Table 4.9 presents yield strength data for ZrO_2 ceramics and some other ceramics that have not been widely studied. Mashimo et al., 1995, compared Hugoniots of polycrystalline and monocrystalline cubic yttria-dopped zirconia and found the polycrystal preserves a considerably larger shear stress above its HEL than do the single crystals of various orientations. It is thought that, in polycrystals, the growth, multiplication, and motion of cracks or dislocations are disturbed by grain boundaries and, as a result, the material may become toughened. A shock-induced polymorphic transition of cubic zirconia starts at approximately 53 GPa and is completed by about 70 GPa. The volume decrement between the initial cubic phase and the final phase at zero pressure was estimated to be 20%. In the case of tetragonal yttria-dopped zirconia Mashimo et al.,

Table 4.9. Hugoniot elastic limits of zirconia and other ceramics

Material	Density, g/cm^3, (Void fraction, %)	Longitudinal Sound Velocity, c_ℓ, km/s	Poisson's ratio	Hugoniot Elastic Limit, GPa	Dynamic Yield Strength, Y, GPa	Reference
ZrO$_2$	5.602	6.61	0.299	5.0–5.4	3.2	Kipp and Grady, 1990
ZrO$_2$	5.954	6.87	0.306	13.2	7.4	Grady, 1994
ZrO$_2$	6.028	7.11	0.311	16.2	8.9	Grady, 1994
ZrO$_2$ + 8 mol% Y$_2$O$_3$, cubic	5.879 (1.5)	7.45 (calculated)	0.2973	13±2	7–8	Mashimo et al., 1995
ZrO$_2$ + 3 mol% Y$_2$O$_3$, tetragonal	6.03 (<0.6)	7.07	0.3124	15–17	8.2–9.3	Mashimo et al., 1995
BeO	2.84 (5.6)	11.54	0.198	8.2±1	6.2±0.5	Gust and Royce, 1971
Be$_4$B	1.98 (1.5)	12.42	0.094	7.4±1	6.6	Gust, Holt, and Royce, 1973
Si$_3$N$_4$	3.15	10.66	—	12.1	—	Nahme et al., 1994
Si$_3$N$_4$	2.28	8.6	—	1.9	—	Nahme et al., 1994

1995, observed a three-wave structure with transition points of 15–17 and 33–35 GPa. The velocity of the second wave was so high that the first transition could not be identified in terms of the elastoplastic response. The second transition point differed from that of the cubic material, but was higher than that of the tetragonal-to-orthorhombic II phase transition point under static compression.

4.8. Brittle Failure Criteria and Models

Models and criteria for brittle fracture that occurs during quasi-static compression have been exhaustively reviewed by, for example, Wang and Shrive, 1995, and Nikolaevskii and Rice, 1979. Most models consider the consequences of three sequential events: crack initiation, crack propagation, and crack coalescence. According to Griffith's criterion, crack initiation occurs when the highest local tensile stress at the longest crack of the most vulnerable orientation reaches a fixed critical value. For a biaxial stress state, the corresponding relationship is

$$(\sigma_1 - \sigma_2)^2 - 8\sigma_f(\sigma_1 + \sigma_2) = 0, \qquad (4.1)$$

where σ_1, and σ_2 are principal stresses and σ_f is a material constant which is assuming to be the ordinary tensile stress for uniaxial stressing. Thus, Griffith's criterion predicts the value of the uniaxial compressive strength to be eight times the value of the uniaxial tensile strength. This ratio is smaller than the ratio commonly measured for rocks and other brittle materials.

Chen and Ravichandran, 1997, 2000, have found that the Mohr–Coulomb criterion, generally used as a yield criterion for granular materials, fits the experimental data on failure of ceramics under lateral confinement rather well. The criterion is expressed by a relationship

$$|\tau| + \alpha p = \tau_0, \qquad (4.2)$$

where $|\tau|$ is the absolute value of shear stress, p is the pressure or the mean stress, τ_0 is the shear strength of the material at zero pressure, and α is the internal friction coefficient. It is expected that the value of α is between -1.5 and 0. For AlN ceramic $\alpha = -1$ and for Macor glass ceramic $\alpha = -0.74$. Actually, in Coulomb's interpretation, there should be a normal stress component perpendicular to the shear direction used in Eq. (4.2) instead of the mean stress (Paul, 1968). In this case, the competition between the resolved shear force τ and the friction forces αp explains the failure angle $<45°$ observed in rocks and granular materials under compression.

In general, theoretical criteria, such as that of Griffith, often give an inadequate fit to the data. Because of that, empirical criteria are developed to meet the practical requirements of accurate strength prediction and simplicity of use. Since different mechanisms of inelastic deformation and failure may operate depending on the region of stress space, different strength criteria should be used at different levels of stress. For computations of impact problems, continuum damage mechanics models are needed. Such models operate with defined damage parameters and include an evolutionary equation for the damage and a constitutive equation that relates the stress and strain to the damage.

In this sense, the model of Ashby and Sammis, 1990, is very illustrative. They constructed failure surfaces in stress space which combine information about brittle failure with data describing onset of plastic yielding. The criterion for crack initiation under axisymmetric loading has been chosen to be linear,

$$\sigma_1 = c_1 \sigma_3 + \sigma_0, \tag{4.3}$$

where c_1 and σ_0 are material properties that are determined by the coefficient of friction, crack size, and the fracture toughness of the material. The quantity σ_1 is the axial stress and $\sigma_2 = \sigma_3$ is the radial stress component. Once initiated, the cracks grow longer. A damage parameter D is introduced as a state variable given by the equation

$$D = \pi(a - a_0 + \alpha a)^2 N_A, \tag{4.4}$$

where α is a geometric constant, N_A is a number of cracks per unit area, and a and a_0 are the current and the initial sizes of microcracks, respectively. The surfaces of constant damage in stress space are assumed linear. The surface corresponding to final macroscopic fracture is also approximated by a linear relationship,

$$\sigma_1 = c_2 \sigma_3 + \sigma_c, \tag{4.5}$$

where σ_c is the compressive strength for states of uniaxial stress. As the confining pressure is increased, brittle fracture is made increasingly difficult. A critical pressure may be reached at which true plasticity replaces crack extension. The yield surface is described by the von Mises criterion

$$\sigma_y^2 = \tfrac{1}{2}[(\sigma_1 - \sigma_2)^2 + (\sigma_2 - \sigma_3)^2 + (\sigma_3 - \sigma_1)^2], \tag{4.6}$$

where σ_y is the yield strength. As a result, the failure surface takes the shape shown in Fig. 4.40. This model has been fit to data for a number of rocks.

4.9. Comparison of 1D Stress and 1D Strain Data for Ceramics at Various Strain Rates

A critical issue for impact problems is how much the strength properties of ceramics depend on the strain rate. It is known that crystalline solids exhibit rate-dependent stress–strain behavior. The strain rate effect is more pronounced for relatively soft crystals and less so for high-strength materials. For brittle materials cracking accompanies the inelastic deformation, so the kinetics of crack nucleation and growth may contribute to the total strain-rate dependence of the resistance to inelastic deformation.

166 Shock-Wave Phenomena and the Properties of Condensed Matter

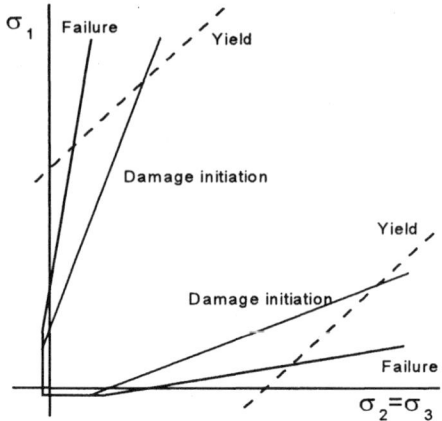

Figure 4.40. Failure and yield surface according to Ashby and Sammis, 1990.

Conclusions about rate sensitivity of ceramics should be based on comparison of compressive strength properties over a wide strain rate range. However, this procedure is not so trivial. In Fig. 4.41, the yield strength, Y, determined from the HEL is compared with the failure stress values measured in uniaxial stress conditions (quasi-static tests, Hopkinson bars, rod impact) for alumina. The data indicate weak rate dependence at strain rates less than 10^3 s^{-1} and greater than 10^5 s^{-1} with a sharp transition between these ranges. Grady, 1998,

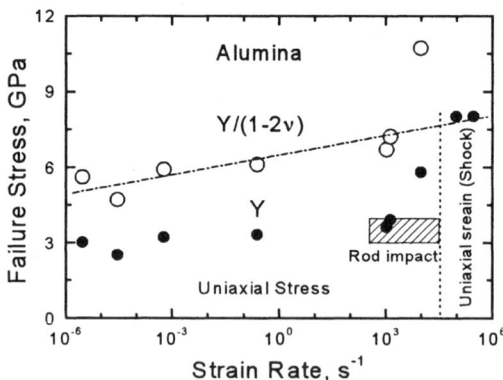

Figure 4.41. Dynamic failure properties of alumina ceramic under quasi-static and dynamic compressive loading. Solid points reproduce data by Grady, 1998, for failure strength measured under uniaxial stress conditions and the yield strength at the HEL (1D strain) calculated with the von Mises yielding criterion. The hatched rectangle shows the region of rod impact data. Open points are the failure strength values multiplied by a factor of $1/(1-2v)$ according to Rosenberg, 1993, 1994.

associated the transition with the kinetics of fracture damage. On the other hand, in order to compare 1D stress and 1D strain data we have first to choose the parameter for comparison. In fact, in Fig. 4.41 we compare effective stresses of the yielding criteria of von Mises or Tresca which are commonly used for ductile metals. Both those criteria relate the HEL to the equivalent dynamic compressive yield stress, Y, corresponding to the 1D-stress conditions:

$$\sigma_{HEL} = Y \frac{(1-\nu)}{(1-2\nu)}. \qquad (4.7)$$

However, for the brittle failure criteria discussed in the previous section we obviously should use some other effective stress. Rosenberg (1993, 1994) has found that the use of Griffith's failure criterion instead of the yield criteria provides better reconciliation of 1D-stress and 1D-strain data for ceramics. In that case the HEL is given by the equation

$$\sigma_{HEL} = Y \frac{(1-\nu)}{(1-2\nu)^2}. \qquad (4.8)$$

Thus, Griffith's failure criterion predicts the HEL to be higher by a factor of $1/(1-2\nu)$ than that predicted by the yield criterion (4.7) used for ductile materials, and this reduces the discrepancy between the 1D stress and 1D strain data. Figure 4.41 shows that Griffith's failure criterion provides very good agreement between 1D stress and 1D strain dynamic data.

Although the rod impact and Hopkinson bar experiments permit a fairly confident judgment as to whether failure is brittle or ductile, interpretation of the shock-wave experiments under 1D strain conditions is more ambiguous. The Hugoniot data and the shock-wave profiles are not sufficient to make definite conclusions about ductile or brittle response. The spall strength measurements confirm that the mode of deformation is mainly ductile as long as the observed spall strength is not zero.

4.10. Evidence of Ductile Response of Alumina Ceramic Under Shock-Wave Compression

Planar impact experiments provide a way to obtain quantitative information about the resistance of materials to inelastic deformation during high-rate compression and subsequent unloading. However, when ceramics or rocks are tested, one needs to know with certainly whether the observed response is brittle or ductile. To diagnose the kind of response, different sensitivities of the elastic limits of brittle and ductile materials to the confining pressure can be used. In the range below the brittle-to-ductile transition, the compressive strength of ceramics increases rapidly with the confining pressure, whereas above the transition the compressive strength is much less sensitive to further increase of the confining pressure.

Under one-dimensional strain conditions of shock-wave loading the transverse stress component σ_2 is non-zero. Below the Hugoniot elastic limit (HEL) it is equal to $\sigma_2 = \sigma_1 v/(1-v)$, where σ_1 is the longitudinal stress and v is Poisson's ratio. At the HEL the transverse stress component is $\sigma_2 = \sigma_1 - Y$. In particular, for alumina ceramics the transverse stress at the HEL varies from 1.5 to 3.5 GPa depending on the material porosity, composition, and grain size.

The transition from brittle to ductile response should be accompanied by a decrease of sensitivity of the HEL value to additional transverse (confining) stress. With additional transverse stress, π, the criteria (4.7) and (4.8), respectively, give

$$\sigma_{HEL}^{duct} = Y\left(1 + \frac{\pi}{Y}\right)\frac{(1-v)}{(1-2v)} \qquad (4.9)$$

and

$$\sigma_{HEL}^{brit} = Y\frac{1-v}{(1-2v)^2}\frac{\left[2(1-2v)(\pi/Y)+1+\sqrt{8(1-2v)(1-v)(\pi/Y)+1}\right]}{2}. \qquad (4.10)$$

In the case that $\pi/Y \ll 1$, differentiating Eqs. (4.9) and (4.10) yields

$$\frac{d\sigma_{HEL}^{duct}}{d\pi} = \gamma^{duct} = \frac{(1-v)}{(1-2v)} \qquad (4.11)$$

and

$$\frac{d\sigma_{HEL}^{brit}}{d\pi} = \gamma^{brit} = \frac{(1-v)(3-2v)}{(1-2v)}. \qquad (4.12)$$

Measuring the HEL for an increased value of π allows one to verify whether the criterion of brittle failure or ductile yielding is applicable and what is the mode of the inelastic deformation that takes place in the investigated material under shock-wave loading. In particular, for aluminas having $v \approx 0.23$, the ratio $\gamma^{brit}/\gamma^{duct}$ is about 2.5.

Chen and Ravichandran (1996, 1997), in experiments conducted using the split Hopkinson pressure bar technique, provided a controlled confining pressure on the specimen by installing a shrink-fit metal sleeve on the lateral surface of the cylindrical ceramic specimen. This technique of varying the transverse stress was used in planar impact experiments by Zaretsky and Kanel, 2002.

The experiments were performed with hot-pressed alumina of 96% purity. The confining pressure on the alumina samples was produced by shrink-fit steel rings machined from rods of normalized 4340 steel. The inside of each ring was machined to a diameter smaller than the diameter of the samples by the amount $\delta = 0.1 \pm 0.005$ mm. Prior to insertion of the ceramic discs, the rings were heated to 600 °C together with the witness samples machined from the same rod. The witness samples were used for further measurements of the yield stress of the heat-treated ring. The confining stress, π, produced by the ring can be estimated using the known solution of an axisymmetric boundary value problem (Cook and Young, 1985). For purely elastic deformation of the ring, the solution is

$$\pi = \frac{\delta}{2R_c \left[\frac{1-v_c}{E_c} + \frac{R_s^2 + R_c^2 + v_s(R_s^2 - R_c^2)}{E_s(R_s^2 - R_c^2)} \right]}, \quad (4.13)$$

where R_c is the sample radius, R_s is the ring outer radius, and E_c, v_c, and E_s, and v_s are the Young's moduli and Poisson's ratios of the sample and the ring materials, respectively. For the case of elastic–perfectly plastic behavior of the ring material, the confining stress can be found from the expression

$$\pi = \sigma_Y \left[\ln \frac{R}{R_c} + \frac{R_s^2 - R^2}{2R_s^2} \right], \quad (4.14)$$

where R is the radius of the boundary between the elastically and the plastically deformed regions of the ring and may be found from the equation

$$\left[(1-2v_s)(1+v_s) \frac{\sigma_Y}{E_s} - (1-v_c) \frac{\sigma_Y}{E_c} \right] \left[\ln \frac{R_c}{R} - \frac{R_s^2 - R^2}{2R_s^2} \right] + \frac{(1-v_s^2)}{E_s} \sigma_Y \frac{R^2}{R_c^2} = \frac{\delta}{2R_c}.$$
(4.15)

For $\delta = 0.1$ mm the state of the steel ring falls just a little outside the elastic range and the confining stress is $\pi = 0.3$ GPa.

Figure 4.42 summarizes the results of planar impact experiments with confined and unconfined alumina samples. Each free surface velocity profile shown was obtained by averaging the profiles of two shots, confined and unconfined, respectively. The transition from elastic to inelastic response occurs at the free surface velocity $u_{HEL} = 285$ m/s that corresponds to

$$\sigma_{HEL} = \rho_0 C_\ell u_{HEL} / 2 = 5.35 \text{ GPa}.$$

Assuming that the inelastic response of the alumina is ductile, the confining pressure is $\pi = 0.3$ GPa, which should result in an increase of the HEL by $\Delta\sigma_{HEL} = 0.43$ GPa. The latter corresponds to an increase of the sample surface velocity equal to $\Delta u_{HEL} = 23$ m/s. In the case of brittle inelastic response, Griffith's failure criterion results in HEL stress and particle velocity increments that are 2.5 times greater, so the expected free surface velocity increment should

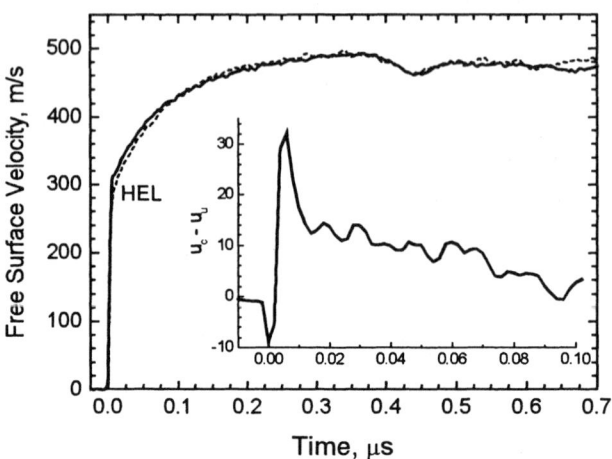

Figure 4.42. The average free surface velocity profiles of confined ceramic samples, $u_c(t)$, at impact velocities of 500 ± 5 m/s (solid line) and unconfined samples, $u_u(t)$, at impact velocity 505 ± 5 m/s (dashed line) and their difference $u_c - u_u$.

be close to 60 m/s. The measured difference between the average $u(t)$ data for confined and unconfined ceramics for the first 50 ns is 10–15 m/s.

These results show unambiguously that the response of alumina under conditions of one-dimensional shock compression is ductile. The lateral stress at the HEL in those experiments was 1.6 GPa, which means that the transition from brittle to ductile response occurs between this stress value and the 1.25 GPa confining pressure at which alumina remained brittle as was shown by Heard and Cline, 1980. The wave profiles exhibit a spall strength of 0.4–0.5 GPa. The applied lateral stress did not decrease the spall strength of alumina under those loading conditions.

4.11. Discussion

It seems that compressive fracture is (or may be) a secondary effect of initiation of plastic deformation in brittle materials. Experiments with single crystals demonstrate that microcracking may nucleate at stress concentrators formed during inelastic compression even if the material does not initially contain any defects. A comparison of the behaviors of single crystals and glasses shows that the choice of ductile or brittle response is controlled by the capability of the material to accommodate local shears in various directions.

Most of the shock-wave tests of ceramics, with the exception, probably, of boron carbide, do not show unambiguous evidence of fracture under uniaxial compression. Only boron carbide demonstrates dilatancy effects on the stress–strain trajectory when the compressive stress approaches zero during unloading. the behavior of other ceramics during shock compression and unloading is very similar to that of ductile metals and alloys which, meanwhile, is also not quite clear in detail. Even the observed decrease in spall strength after shock compression above the HEL may be the result of cracking not under compression but at the end phase of unloading as an effect of transverse stress when the axial stress becomes zero.

Obviously, the effects of cracking and dilatancy become much more significant when at least one normal stress component is less than the failure strength measured in uniaxial stress tests. Additional efforts are needed to answer the question whether or not microcracking of ceramic materials occurs under shock compression. On the other hand, the shock-wave tests provide information about the resistance to inelastic compressive deformation which is useful itself independently of whether or not microcracking occurs. Additional information over a wider range of stress and strain states may be provided by experiments with spherical (Tranchet and Collombet, 1995) and cylindrical (Kanel et al., 1998) divergent shock waves and experiments with granular ceramics and ceramic powders (Meyer and Faber, 1997).

References

Ahrens, T.J. (1966). "High-pressure electrical behavior and equation of state of magnesium oxide from shock wave measurements," *J. Appl. Phys.* **37(7)**, pp. 2532–2541.

Ahrens, T.J., W.H. Gust, and E.B.and Royce (1968). "Material strength effect in the shock compression of alumina," *J. Appl. Phys.* **39(10)**, pp. 4610–4616.

Anan'in, A.V., O.N. Breusov, A.N. Dremin, S.V. Pershin, A.I. Rogacheva, and V.F. Tatsii, (1974). "The effect of shock waves on silicon dioxide: I. Quartz," *Comb. Expl. Shock Waves* **10(4)**, pp. 372–376 [trans. from *Fiz. Goreniya Vzryva* **10(4)**, pp. 426–436 (1974)].

Arndt, J., and D. Stöffler, (1969). "Anomalous changes in some properties of silica glass densified at very high pressures," *Phys. Chem. Glasses* **10(3)**, pp. 117–134.

Arnold, W. (1992). "Influence of twinning on the elasto-plastic behavior of Armco iron," in: *Shock Compression of Condensed Matter—1991*, (eds. S.C. Schmidt, R.D. Dick, J.W. Forbes, and D.G. Tasker) North-Holland, Amsterdam, pp. 539–542.

Ashby, M.F., and C.G. Sammis (1990). "The damage mechanics of brittle solids in compression," *PAGEOPH* **133(3)**, pp. 490–521.

Barker, L.M., and R.E. Hollenbach (1970). "Shock-wave studies of PMMA, fused silica, and sapphire," *J. Appl. Phys.* **41**, pp. 4208–4226.

Bartkowski, P., and D.P. Dandekar (1996). "Spall strength of sintered and hot pressed silicon carbide," in: *Shock Compression of Condensed Matter—1995* (eds: S.C. Schmidt and W.C. Tao) American Institute of Physics, New York, pp. 535–538.

Bassett, W.A., M.S. Weather, and T.C. Wu (1993). "Compressibility of SiC up to 68.4 GPa," *J. Appl. Phys.* **74(6)**, pp. 3824–3826.

Bless, S.J., N.S. Brar, G.I. Kanel, and Z. Rosenberg (1992). "Failure Waves in Glass," *J. Amer. Ceram. Soc.* **75(4)**, pp. 1002–1004.

Bless, S.J., N.S. Brar, and Z. Rosenberg (1990). "Failure of ceramic and glass rods under dynamic compression," in: *Shock Compression of Condensed Matter—1989* (eds. S.C. Schmidt, J.N. Johnson, and L.W. Davison) North-Holland, Amsterdam, pp. 939–942.

Bombolakis, E.G. (1973). "Study of the brittle fracture process under uniaxial compression," *Tectonophysics* **18**, pp. 231–248.

Bourne, N.K., and J.C.F. Millett (1997). "Delayed failure in shocked silicon carbide," *J. Appl. Phys.* **81(9)**, pp. 6019–6023.

Bourne, N.K., J.C.F. Millett, and Z. Rosenberg (1996). "Failure in a shocked high-density glass," *J. Appl. Phys.* **80(8)**, pp. 4328–4331.

Bourne, N.K., J.C.F. Millett, and Z. Rosenberg (1997). "On the origin of failure waves in glass," *J. Appl. Phys.* **81(10)**, pp. 6670–6674.

Bourne, N.K., and J.C.F. Millett (1997). "Delayed failure in shocked silicon carbide," *J. Appl. Phys.* **81(9)**, pp. 6019–6023.

Bourne, N.K., J.C.F. Millett, and J.E. Field (1999). "On the strength of shocked glasses," *Proc. Roy. Soc. Lond.* A **455**, pp. 1275–1282.

Bourne, N.K., and Z. Rosenberg (1996). "The dynamic response of soda-lime glass," in: *Shock Compression of Condensed Matter—1995* (eds. S.C. Schmidt and W.C. Tao) American Institute of Physics, New York, pp. 567–572.

Brace W.F., and E.G. Bombolakis (1963). "A note on brittle crack growth in compression," *J. Geophys. Res.* **68**, pp. 3709–3713.

Brace, W.F., B.V.Paulding, and C. Scholz (1966). "Dilatancy in the fracture of crystalline rocks," *J. Geophys. Res.* **71**, pp. 3939–3953.

Brannon, P.J., R.W. Morris, C.H. Konrad, and J.R. Asay (1984). "Shock-induced luminescence from X-cut quartz and Z-cut lithium niobate," in: *Shock Waves in Condensed Matter—1983* (eds. J.R. Asay, R.A. Graham, and G.K. Straub) North-Holland, Amsterdam, pp. 303–306.

Brar, N.S., and S.J. Bless (1992). "Failure waves in glass under dynamic compression," *High Press. Res.* **10**, pp. 773–784.

Brar, N.S., S.J. Bless, and Z. Rozenberg (1991a). "Impact-induced failure waves in glass bars and plates," *Appl. Phys. Lett.* **59(26)**, pp. 3396–3398.

Brar, N.S., S.J. Bless, and Z. Rozenberg (1992a). "Response of shock-loaded AlN ceramics determined with in-material manganin gauges," in: *Shock-Wave and High-Strain-Rate Phenomena in Materials* (eds. M.A. Meyers, L.E. Murr, and K.P. Staudhammer) Marcel Dekker, New York, pp. 1023–1030.

Brar, N.S., Z. Rosenberg, and S.J. Bless (1991b). "Spall strength and failure wave in glass," *J. de Physique IV, Coll. C3, Suppl. au. J. de Physique III* **1**, pp. C3-639–644.

Brar, N.S., Z. Rosenberg, and S.J. Bless (1992b). "Applying Steinberg model to the Hugoniot elastic limit of porous boron carbide specimens," in: *Shock Compression of Condensed Matter—1991* (eds. S.C. Schmidt, R.D. Dick, J.W. Forbes, and D.G. Tasker) North-Holland, Amsterdam, pp. 467–470.

Bridgman, P., (1964). *Studies in Large Plastic Flow and Fracture*, Harvard University Press, Cambridge, MA.

Cagnoux, J. (1990). "Spherical waves in pure alumina. Effect of grain size on flow and fracture," in: *Shock Compression of Condensed Matter—1989* (eds. S.C. Schmidt, J.N. Johnson, and L.W. Davison) North-Holland, Amsterdam, pp. 445–448.

Cagnoux, J., and F. Longy (1988). "Is the dynamic strength of alumina rate-dependent?" in: *Shock Waves in Condensed Matter—1987* (eds. S.C. Schmidt and N.C. Holmes) North-Holland, Amsterdam, pp. 293–296.

Chen, W., and G. Ravichandran (1997). "Dynamic compressive failure of a glass ceramic under lateral confinement," *J. Mech. Phys. Solids* **45(8)**, pp. 1303–1328.

Cazamias, J.U., B. Reinhart, C. Konrad, L.C. Chhabildas, and S. Bless (2002). "Bar impact tests on alumina (AD995)," in: *Shock Compression of Condensed Matter—2001* (eds. M.D. Furnish, N.N. Thadhani, and Y. Horie) American Institute of Physics, New York, pp. 787–790.

Chhabildas, L.C., M.D. Furnish, W.D. Reinhart, and D.E. Grady (1998). "Impact of AD995 alumina rods," in: *Shock Compression of Condensed Matter—1997*, (eds. S.C. Schmidt, D.P. Dandekar, J.W. Forbes) American Institute of Physics, New York, pp. 505–508.

Chhabildas, L.C., and D.E. Grady (1984). "Shock loading behavior of fused quartz," in: *Shock Waves in Condensed Matter—1983* (eds. J.R. Asay, R.A. Graham, and G.K. Straub) North-Holland, Amsterdam, pp. 175–178.

Chan, H.M., and B.R. Lawn (1988). "Indentation deformation and fracture of sapphire," *J. Ceram. Soc.* **71(1)**, pp. 29–35.

Chandrasekar, S., and M.M. Chaudhri (1994). "The explosive desintegration of Prince Rupert's drops," *Phil. Mag.* **70(6)**, pp. 1195–1218.

Chen, W., and G. Ravichandran (1996). "Static and dynamic compressive behavior of aluminum nitride under moderate confinement," *J. Am. Ceram. Soc.* **79(3)**, pp. 579–584.

Chen, W., and G. Ravichandran (1997). "Dynamic compressive failure of a glass ceramic under lateral confinement," *J. Mech. Phys. Solids* **45(8)**, pp. 1303–1328.

Chen, W., and G. Ravichandran (2000). "Failure mode transition in ceramics under dynamic multiaxial compression," *Int. J. Fracture* **101(1)**, pp. 141–159.

Cook, R.D., and W.C. Young (1985). *Advanced Mechanics of Materials*, Macmillan, New York.

Dandekar, D.P. (1994). "Shear strengths of aluminum nitride and titanium diboride under plane shock wave compression," *Journal de Physique IV* **4**, pp. C8-379–384.

Dandekar, D.P. (1998). "Index of refraction and mechanical behavior of soda lime glass under shock and release wave propagation," *J. Appl. Phys.* **84(12)**, pp. 6614–6622.

Dandekar, D.P. and P. Bartkowski (1994). "Shock response of AD995 alumina," in: *High-Pressure Science and Technology—1993* (eds. S.C. Schmidt, J.W. Shaner, G.A. Samara, and M. Ross) American Institute of Physics, New York, pp. 733–736.

Dandekar, D.P., and P.A. Beaulieu (1995). "Failure wave under shock wave compression in soda lime glass," in: *Metallurgical and Material Applications of Shock-Wave and High-Strain-Rate Phenomena* (eds. L.E. Murr, K.P. Staudhammer, and M.A. Meyers) Elsevier, Amsterdam, pp. 211–218.

Dandekar, D.P., and P.J. Gaeta (1992). "Extent of damage induced in titanium diboride under shock loading," in: *Shock-Wave and High-Strain-Rate Phenomena in Materials* (eds. M.A. Meyers, L.E. Murr, and K.P. Staudhammer) Marcel Dekker, New York, pp. 1059–1068.

Ernsberger F.M., (1968). "Role of densification in deformation of glasses under point loading," *J. Amer. Ceramic Soc.* **51(10)**, pp. 545–547.

Espinosa, H.D., Y. Xu, and N.S. Brar (1977). "Micromechanics of failure waves in glass. Part I: experiments," *J. Amer. Ceramic Soc.* **80(8)**, pp. 2061–2073.

Ewart, L., and D.P. Dandekar (1994). "Relationship between the shock response and microstructural features of titanium diboride (TiB_2)," in: *High-Pressure Science and Technology—1993* (eds. S.C. Schmidt, J. W. Shaner, G.A. Samara, and M. Ross) American Institute of Physics, New York, pp. 1201–1204.

Feng, R., Y.M. Gupta, and G. Yuan (1998). "Dynamic strength and inelastic deformation of ceramics under shock wave loading," in: *Shock Compression of Condensed Matter—1997*, (eds. S.C. Schmidt, D.P. Dandekar, J.W. Forbes) American Institute of Physics, New York, pp. 483–488.

Furnish, M.D., and L.C. Chhabildas (1998). "Alumina strength degradation in the elastic regime," in: *Shock Compression of Condensed Matter—1997*, (eds. S.C. Schmidt, D.P. Dandekar, J.W. Forbes) American Institute of Physics, New York, pp. 501–504.

Furnish, M.D., D.E. Grady, and J.M. Brown (1986). "Analysis of shock wave structure in single crystal olivine using VISAR," in: *Shock Waves in Condensed Matter* (ed. Y.M. Gupta), Plenum Press, New York, pp. 595–600.

Galin, L.A., and G.I. Cherepanov (1966). "On self-supporting failure of a stressed brittle solid," *Sov. Phys.–Dokl.* **11(3)**, pp. 267–269. [trans. from *Dokl. Akad. Nauk SSSR* **167(3)**, pp. 543–546, (1966).]

Galin, L.A., V.A. Ryabov, D.V. Fedoseev, and G.I. Cherepanov (1967). "On failure of a high-strength glass," *Sov. Phys.–Dokl.* **11(8)**, pp. 743–744. [trans. from *Dokl. Akad. Nauk SSSR* **169(5)**, pp. 1034–1036 (1966).]

Gibbons, R.V., and T.J. Ahrens (1971). "Shock metamorphism of silicate glasses," *J. Geophys. Res.* **76(23)**, pp. 5489–5498.

Grady, D.E. (1977). "Processes occuring in shock wave compression of rocks and minerals," in: *High Pressure Research Application in Geophysics*, Academic Press, New York, pp. 389–437.

Grady, D.E. (1980). "Shock deformation of brittle solids," *J. Geophys. Res.* **85(B2)**, p. 913.

Grady, D.E. (1992). "Shock-wave properties of high-strength ceramics," in: *Shock Compression of Condensed Matter—1991* (eds. S.C. Schmidt, R.D. Dick, J.W. Forbes, and D.G. Tasker) North-Holland, Amsterdam, pp. 455–458.

Grady, D.E. (1994). "Impact strength and indentation hardness of high-strength ceramics," in: *High-Pressure Science and Technology—1993* (eds. S.C. Schmidt, J. W. Shaner, G.A. Samara, and M. Ross) American Institute of Physics, New York, pp. 741–744.

Grady, D.E. (1998). "Shock wave compression of brittle solids," *Mech. Mater.* **29**, pp. 181–203.

Graham, R.A. (1974). "Shock-wave compression of X-cut quartz as determined by electrical response measurements," *J. Phys. Chem. Solids* **35(3)**, pp. 355–372.

Graham, R.A., and T.J. Ahrens (1973). "Shock wave compression of iron-silicate garnet," *J. Geophys. Res.* **78(2)**, pp. 375–392.

Graham, R.A., and W.P. Brooks (1971). "Shock-wave compression of sapphire from 15 to 420 kbar. The effects of large anisotropic compression," *J. Phys. Chem. Solids* **32**, pp. 2311–2330.

Griffith, A.A. (1924). "The theory of rupture," *Proc. 1st Int. Con. Appl. Mech. (Delft)*, pp. 55–63.

Gust, W.H., and E.B. Royce (1971). "Dynamic yield strength of B_4C, BeO, and Al_2O_3 ceramics," *J. Appl. Phys.* **42**, pp. 276–295.

Gust, W.H., A.C. Holt, and E.B. Royce (1973). "Dynamic yield, compressional, and elastic parameters for several lightweight intermetallic compounds," *J. Appl. Phys.* **44(2)**, pp. 550–560.

Hagan, J.T. (1980). "Shear deformation under pyramidal indentations in soda-lime glass," *J. Mat. Sci.* **15**, pp. 1417–1424.

Heard, H.C., and C.F. Cline (1980). "Mechanical behavior of polycrystalline BeO, Al_2O_3, and AlN at high pressure," *J. Mat. Sci.* **15**, pp. 1889–1897.

Hockey, B.J. (1971). "Plastic deformation of aluminum oxide by indentation and abrasion," *J. Amer. Ceramic Soc.* **54**, pp. 223–231.

Holcomb, D.J. (1978). "A quantitative model of dilatancy in dry rock and its application to Westerly granite," *J. Geophys. Res.* **83(B10)**, pp. 4941–4950.

Horii, H., and S. Nemat-Nasser (1985). "Compression-induced macrocrack growth in brittle solids: axial splitting and shear failure," *J. Geophys. Res.* **90(B4)**, pp. 3105–3125.

Kalthoff, J.K. (2000). "Modes of dynamic shear failure in solids," *Int. J. Fracture* **101**, pp. 1–31.

Kanel, G.I., A.A. Bogatch, S.V. Razorenov, and Zhen Chen (2002). "Transformation of shock compression pulses in glass due to the failure wave phenomena," *J. Appl. Phys.* **92(9)**, pp. 5045–5052.

Kanel, G.I., and A.M. Molodets (1976). "Behavior of type K-8 glass under in dynamic compression and subsequent unloading," *Sov. Phys.–Tech. Phys.* **21(2)**, pp. 226–232. [trans. from *Zh. Tekh. Fiz.* **46(2)**, pp. 398–407 (1976).]

Kanel, G.I., A.M. Molodets, and A.N. Dremin (1977). "Investigations of features of deforming a glass in intense compression waves," *Comb. Expl. Shock Waves* **13(6)**, pp. 772–773. [trans. from *Fiz. Goreniya Vzryva* **13(6)**, pp. 906–912 (1977).]

Kanel, G.I., and A.N. Pityulin (1984). "Shock-wave deformation of titanium carbide-based ceramics," *Comb. Expl. Shock Waves* **20(4)**, pp. 436–438. [trans. from *Fiz. Goreniya Vzryva* **20(4)**, pp. 85–88 (1984).]

Kanel, G.I., S.V. Razorenov, and V.E. Fortov (1992). "The failure waves and spallation in homogeneous brittle materials," in: *Shock Compression of Condensed Matter—1991* (eds. S.C. Schmidt, R.D. Dick, J.W. Forbes, and D.G. Tasker) North-Holland, Amsterdam, pp. 451–454.

Kanel, G.I., S.V. Razorenov, A.V. Utkin, S.N. Dudin, V.B. Mintsev, S. Bless, and C.H.M. Simha (1998). "Investigation of mechanical properties of ceramics using axi-symmetric shock waves," in: *Shock Compression of Condensed Matter—1997* (eds. S.C. Schmidt, D.P. Dandekar, J.W. Forbes) American Institute of Physics, New York, pp. 489–492.

Kanel, G.I., S.V. Razorenov, A.V. Utkin, and V.E. Fortov (1999). "Investigations of mechanical properties of materials under a shock wave loading," *Mekh. Tverd. Tela* **(5)**, pp. 173–88. (in Russian)

Kanel, G.I., S.V. Razorenov, A.V. Utkin, Hongliang He, Fuqian Jing, and Xiaogang Jin (1998). "Influence of the load conditions on the failure wave in glasses," *High Press. Res.* **16**, pp. 27–44.

Kanel, G.I., S.V. Razorenov, and T.N. Yalovetz (1993). "Dynamic strength of ruby," *Khim. Fiz.* **12(2)**, pp. 175–177 (1993).

Kipp, M.E., and D.E.Grady (1990). "Shock compression and release in high-strength ceramics," in: *Shock Compression of Condensed Matter—1989* (eds. S.C.Schmidt, J.N. Johnson, and L.W. Davison) North-Holland, Amsterdam, pp. 377–380.

Kipp, M.E., and D.E.Grady (1992). "Elastic wave dispersion in high-strength ceramics," in: *Shock Compression of Condensed Matter—1991* (eds. S.C. Schmidt, R.D. Dick, J.W. Forbes, and D.G. Tasker) North-Holland, Amsterdam, pp. 459–462.

Kranz, R.L. (1983). "Microcracks in rocks: a review," *Tectonophysics* **100**, pp. 449–480.

Longy, F., and J. Cagnoux (1989). "Plasticity and microcracking in shock-loaded alumina," *J. Amer. Ceram. Soc.* **72(6)**, pp. 971–979.

Mashimo, T., Y. Hanaoka, and K. Nagayama (1988). "Elastoplastic properties under shock compression of Al_2O_3 single crystal and polycrystal," *J. Appl. Phys.* **63(2)**, pp. 327–336.

Mashimo, T., A. Nakamura, M. Kodama, K. Kusaba, K. Fukuoka, and Y. Syono (1995). "Yielding and phase transition under shock compression of yttria-doped cubic zirconia single crystal and polycrystal," *J. Appl. Phys.* **77(10)**, pp. 5060–5068.

Mashimo, T., A. Nakamura, M. Nishida, S. Matsuzaki, K. Kusaba, K. Fukuoka, and Y. Syono (1995). "Anomalous shock compression behavior of yttria-doped tetragonal zirconia," *J. Appl. Phys.* **77(10)**, pp. 5069–5076.

Mashimo, T., and M. Uchino (1997). "Heterogeneous free-surface profile of B_4C polycrystal under shock compression," *J. Appl. Phys.* **81(10)**, pp. 7064–7065.

McClintock, F.A., and A.S. Argon (1966). *Mechanical Behavior of Materials.* Addison-Wesley, Reading, MA.

Meyer, L.W., and I. Faber (1997). "Investigations on granular ceramics and ceramic powder," *J. Phys. IV France* **7(C3)**, pp. 565–570.

Millet, J., N. Bourne, and Z. Rozenberg (1998). "Observations of the Hugoniot curves for glasses as measured by embedded stress gauges," *J. Appl. Phys.* **84(2)**, pp. 739–741.

Moss, W.C., and Y.M. Gupta (1982). "A constitutive model describing dilatancy and cracking in brittle materials," *J. Geophys. Res.* **87**(B4), pp. 2985–2998.

Munson, D.E., and R.J. Lawrence (1979). "Dynamic deformation of polycrystalline alumina," *J. Appl. Phys.* **50(10)**, pp. 6272–6282.

Murray, N.H., N.K. Bourne, and Z. Rosenberg (1996). "Precursor decay in several aluminas," in: *Shock Compression of Condensed Matter—1995* (eds: S.C. Schmidt and W.C. Tao) American Institute of Physics, New York, pp. 491–494.

Nahme, H., V. Hohler, and A. Stilp (1994). "Determination of the dynamic material properties of shock loaded silicon nitride," in: *High-Pressure Science and Technology—1993* (eds. S.C. Schmidt, J.W. Shaner, G.A. Samara, and M. Ross) American Institute of Physics, New York, pp. 765–768.

Nemat-Nasser, S., and H. Horii (1982). "Compression-induced nonplanar crack extension with application to splitting, exfolation, and rockburst. *J. Geophys. Res.* **87**(B8), pp. 6805–6821.

Nemat-Nasser, S., and M. Obate (1988). "A microcrack model of dilatancy in brittle materials," *Trans. ASME: J. Appl. Mech.* **55(110)**, pp. 24–35.

Nikolaevskii, V.N., and J.R. Rice (1979). "Current topics in non-elastic deformation of brittle materials," in: *High-Pressure Science and Technology, Proceedings of Sixth AIRAPT Conference* (eds. K.D. Timmerhaus and M.S. Barber) Plenum Press, New York, pp. 455–464.

Paul, B. (1968). "Macroscopic criteria of plastic yielding and brittle fracture," in: *Fracture: an Advanced Treatise, Vol. II Mathematical Fundamentals* (ed. H. Liebowitz) Academic Press, New York.

Pickup, I.M., and A.K. Barker (1998). "Damage kinetics in silicone carbide," in: *Shock Compression of Condensed Matter—1997* (eds. S.C. Schmidt, D.P. Dandekar, J.W. Forbes) American Institute of Physics, New York, pp. 513–516.

Raiser, G., and R.J.Clifton (1994). "Failure Waves in Uniaxial Compression of an Aluminosilicate Glass," in: *High-Pressure Science and Technology—1993* (eds. S.C. Schmidt, J. W. Shaner, G.A. Samara, and M. Ross) American Institute of Physics, New York, pp. 1039–1042.

Raiser, G., J.L. Wise, R.J. Clifton, D.E.Grady, and D.E. Cox (1994). "Plate impact response of ceramics and glasses," *J. Appl. Phys.* **75(8)**, pp. 3862–3869.

Razorenov, S.V., G.I. Kanel, V.E. Fortov, and M.M. Abasehov (1991). "The fracture of glass under high-pressure impulsive loading," *High Press. Res.* **6**, pp. 225–232.

Rosenberg, Z. (1992). "The response of ceramic materials to shock loading," in: *Shock Compression of Condensed Matter—1991* (eds. S.C. Schmidt, R.D. Dick, J.W. Forbes, and D.G. Tasker) North-Holland, Amsterdam, pp. 439–444.

Rosenberg, Z. (1993). "On the relation between the Hugoniot elastic limit and the yield strength of brittle materials," *J. Appl. Phys.* **74(1)**, pp. 752–753.

Rosenberg, Z. (1994). "On the shear strength of shock loaded brittle solids," *J. Appl. Phys.* **76(3)**, pp. 1543–1546.

Rosenberg, Z., N.S. Brar, and S.J. Bless (1992). "Shear strength of titanium diboride under shock loading measured by transverse manganin gauges," in: *Shock Compression of Condensed Matter—1991* (eds. S.C. Schmidt, R.D. Dick, J.W. Forbes, and D.G. Tasker) North-Holland, Amsterdam, pp. 471–473.

Schardin, H. (1959). "Velocity effects in fracture," in: *Fracture.* (ed. B.L. Averbach et al.) MIT Press, Cambridge, MA, and Wiley, New York, pp. 297–330.

Schmitt, D., B. Svendsen, and T.J. Ahrens (1986). "Shock induced radiation from minerals," in: *Shock Waves in Condensed Matter—1985* (ed. Y.M. Gupta) Plenum Press, New York, pp. 261–265.

Scholz, C.H. (1968). "Experimental study of the fracturing process in brittle rock," *J. Geophys. Res.*, **73**, pp. 1447–1454.

Senf, H., E. Strausburger, and H. Rothenhausler (1995). "Visualization of fracture nucleation during impact in glass," in: *Metallurgical and Material Applications of Shock-Wave and High-Strain-Rate Phenomena* (eds. L.E. Murr, K.P. Staudhammer, and M.A. Meyers) Elsevier, Amsterdam, pp. 163–170.

Simha, C.H.M., S.J. Bless, and A. Bedford (2000). "What is the Peak Stress in the Ceramic Bar Imact Experiment?" in: *Shock Compression of Condensed Matter —1999* (eds. M.D. Furnish, L.C. Chhabildas, and R.S. Hixson) American Institute of Physics, New York, pp. 615–618.

Song H., S.J. Bless, N.S. Brar, C.H. Simha, and S.D. Jang (1994). "Shock properties of Al_2O_3 and ZrO_2 ceramics," in: *High-Pressure Science and Technology —1993* (eds. S.C. Schmidt, J. W. Shaner, G.A. Samara, and M. Ross) American Institute of Physics, New York, pp. 737–740.

Staehler, J.M., W.W. Predeborn, and B.J. Pletka (1994). "The response of a high-purity alumina to plate-impact testing," in: *High-Pressure Science and Technology —1993* (eds. S.C. Schmidt, J. W. Shaner, G.A. Samara, and M. Ross) American Institute of Physics, New York, pp. 745–748.

Stevens, J.L., and D.J. Holcomb, (1980). "A theoretical investigation of the sliding crack model of dilatancy," *J. Geophys. Res.* **85**(B12), pp. 7091–7100.

Tapponier, R., and W.F. Brace (1976). "Development of stress-induced microcracks in Westerly granite," *Int. J. Rock Mech. Min. Sci. Geomech. Abstr.* **13**, pp. 103–112.

Tranchet, J.-Y., and F. Collombet (1995). "Behavior of pure alumina submitted to a divergent spherical stress wave," in: *Metallurgical and Material Applications of Shock-Wave and High-Strain-Rate Phenomena* (eds. L.E. Murr, K.P. Staudhammer, and M.A. Meyers) Elsevier, Amsterdam, pp. 535–542.

Wackerle, J. (1962). "Shock wave compression of quartz," *J. Appl. Phys.* **33**, pp. 922–937.

Wang, Y., and D.E. Mikkola (1992). "Response of alpha-aluminum oxide to shock impact," in: *Shock-Wave and High-Strain-Rate Phenomena in Materials* (eds. M.A. Meyers, L.E. Murr, and K.P. Staudhammer) Marcel Dekker, New York, pp. 1031–1040.

Wang, E.Z., and N.G. Shrive (1995). "Brittle fracture in compression: Mechanisms, models and criteria," *Eng. Fracture Mech.* **52**(6), pp. 1107–1126.

Wawersik, W.R., and W.F. Brace (1971). "Post-failure behavior of a granite and diabase," *Rock Mechanics* **3**, pp. 61–85.

Winkler, W.-D., and A.J.Stilp (1992a). "Spallation behavior of the TiB_2, SiC, and B_4C under planar impact tensile stresses," in: *Shock Compression of Condensed Matter — 1991* (eds. S.C. Schmidt, R.D. Dick, J.W. Forbes, and D.G. Tasker), North-Holland, Amsterdam, pp. 475–478.

Winkler, W.-D., and A.J. Stilp (1992b). "Pressure induced macro- and micromechanical phenomena in planar impacted TiB_2," in: *Shock Compression of Condensed Matter — 1991* (eds. S.C. Schmidt, R.D. Dick, J.W. Forbes, and D.G. Tasker) North-Holland, Amsterdam, pp. 555–558.

Wise, J.L., and D.E. Grady (1994). "Dynamic, multiaxial impact response of confined and unconfined ceramic rods," in: *High-Pressure Science and Technology —1993* (eds. S.C. Schmidt, J. W. Shaner, G.A. Samara, and M. Ross) American Institute of Physics, New York, pp. 777–780.

Yaziv, D., Y. Yeshurun, Y. Partom, and Z. Rosenberg (1988). "Shock structure and precursor decay in commercial alumina," in: *Shock Waves in Condensed Matter — 1987* (eds. S.C. Schmidt and N.C. Holmes) North-Holland, Amsterdam, pp. 297–300.

Zaretsky, E.B., and G.I. Kanel (2002). "Evidence of ductile response of alumina ceramic under shock wave compression," *Appl. Phys. Lett.* **81**(7), pp. 1192–1194.

CHAPTER 5

Two Examples of Spatially Resolved Shock-Wave Tests

As a rule, kinematic parameters of shock-wave loads are monitored at one point of a sample or are averaged over a transducer cross section. However, there is a set of problems for which it is important to obtain a spatial picture of the phenomenon. In these cases, spatially resolving instrumentation promises to make an essential contribution. The recently developed line-imaging interferometer technique (Baumung et al., 1996b) provides a capability for simultaneously recording the velocity history at many points along a line on the sample surface. In this chapter we present examples of the application of this capability to the study of spall fracture and of adhesion phenomena.

5.1. Dynamic Strength Variations in Metals

Since fracture is a process of nucleation and growth of cracks or voids, it is often important to obtain a spatial picture of the phenomenon. Figure 5.1 presents the line imaging ORVIS interferogram of an experiment conducted on a magnesium sample (Baumung et al., 1997). The experiment was carried out for a short-duration load obtained using a pulsed ion beam (Baumung et al., 1996a) as a shock-wave generator. By this means, the load can be applied on a spatial scale close to that of non-uniformities of the material.

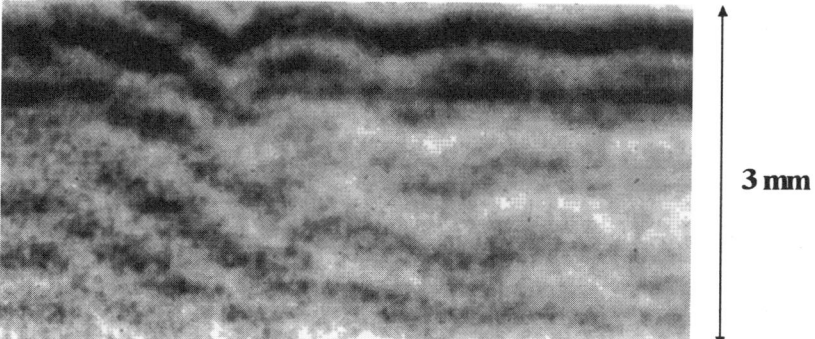

Figure 5.1. Line imaging ORVIS interferogram of an experiment on a 0.82-mm-thick coarse-grain cast magnesium Mg95 sample.

Each point of a horizontal section of the interferogram corresponds to a separate point of the measuring line on the sample surface. The vertical displacement of an interference fringe is proportional to the accrued increment of velocity, so the interferogram provides directly a set of velocity histories along the measuring line. A shift by one fringe period corresponds to a velocity increment of 269 m/s. The amplitude of the initial velocity jump in the shock front is not resolved. However, exact knowledge of the peak velocity is not important for spall strength measurements.

The original magnesium Mg95 casting has a grain size on the order of 1–2 mm. The field of view of the velocimeter included two large grains that are visible on the interferogram due to the different reflectivity. The interferogram demonstrates smooth variations in the velocity pullback from point to point on the sample surface. The distribution of spall strength at points along the measuring line is shown in Fig. 5.2. This diagram indicates that the spall strength is changing within the sample but, despite the grain boundary in the ORVIS field of view, there is no sharp jump in the strength distribution. It is necessary to mention that, under conditions of one-dimensional shock loading of metals with low yield strength, the stress tensor is nearly spherical. In other words, there is no large anisotropy of the loading, and there is no strength jump between grains of different crystallographic orientation. The inter-grain boundary in the ORVIS field of view has to be a site of easier nucleation for fracture, but it seems that such fine details can be observed only at much larger magnifications.

Figure 5.3 shows minimum and maximum spall strength measured in this shot in comparison with the resistance to spall fracture of the same magnesium at lower strain rates. In many previous investigations, such dependencies were well approximated by a power law over a wide range of strain rate. The new data at highest strain rate deviated toward larger values of the spall strength than

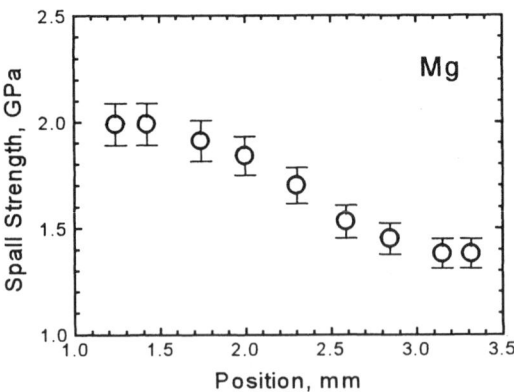

Figure 5.2. The spall strength distribution in magnesium calculated from the interferogram presented in Fig. 5.1.

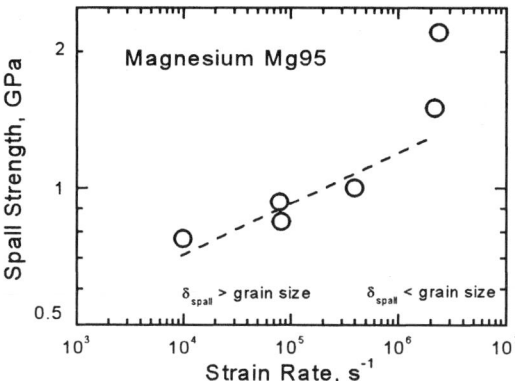

Figure 5.3. Spall strength of magnesium as a function of strain rate during unloading of the incident compression pulse. The two upper points show the range of spall strength covered by the data shown in Fig. 5.2.

that predicted by the power law dependence. The spall thickness in this shot varied between ~60 μm and ~80 μm, a dimension much smaller than the grain size. This means the spall crack had to cross the grains and the measured spall strength should correspond to that of the crystals. It had been found earlier by Kanel et al., 1994, 1996, that the spall strength of metal single crystals exceeds that of polycrystalline samples by a factor of ~2 for molybdenum and ~3 for copper. Since the deviation of the new data from the extrapolated dependence, on average, is not so large, we may suppose that, in this experiment, the main spall crack coincides at least partly with an intergranular boundary below the sample surface. Thus, both intragranular and intergranular strengths were measured, and the intragranular strength is as much as twice as large as the strength of the intergranular boundaries.

The experiment with magnesium shows that a wide spectrum of spall strength values can be realized when the compression wavelength is less than the grain size of the material. In this experiment, the damage did not nucleate as the reflected tensile wave passed through the grains because the tensile stress was not sufficient to activate small intragranular defects. However, the damage was obviously nucleated quite easily when the tensile wave reached a grain boundary below the sample surface where most weak sites in the material are concentrated. When the peak tensile stress reached a sufficient magnitude, initiation of fracture inside the grains started and, owing to that, we observe saturation of the spall strength as shown in Fig. 5.2.

Baumung et al., 1997, have found that even single crystals are not absolutely homogeneous and show variations in the spall strength through their cross sections. Obviously, larger-size fracture nucleation sites are rare and widely separated in crystals, so they may or may not be in the field of view near a future spall plane. When nucleation sites are activated, the stresses relax and, as a result, smaller defects in an adjacent layer are not activated.

5.2. Measurements of Adhesion Strength Using the Spall Technique

The shock-induced spallation technique offers a promising way to measure the tensile strength and adhesion of thin coating layers on substrates of composite systems. For micron-thick coatings, measurements of the interface strength were performed by Gupta et al., 1992, 1993. However, because the coating thickness was much smaller than the length of the stress wave, these experiments did not permit direct recording of the spallation event.

Baumung et al., 2001, applied the spall technique to measurement of the adhesive strength of turbine-blade coatings. The first-stage blades of natural gas turbines for electric power generation are coated with plasma-sprayed protective layers that reduce corrosion at high temperatures. In order to optimize the coating procedure, rapid benchtop-scale tests are required to assess performance. Homogeneity of the coating and its adhesion to the substrate are the criteria for assessing product quality.

Determining the strength at the interface between two plates requires that rupture occur at the interface and that it be recorded in real time rather than being ascertained by post-test examination. The task of producing a rupture at the interface is much simplified if a 1D spatially resolving velocimeter is employed to measure velocity profiles simultaneously at different positions. In this case, it is sufficient that the spall plane and the interface intersect each other within the field of view of the velocimeter. Several means of ensuring that the required intersection occurs are illustrated in Fig. 5.4.

If the pressure amplitude is constant along the measuring line, the spall plane in a plane-parallel sample is oriented parallel to the surface. In this case, the direct approach to ensuring that the spall plane intersects the interface at some point along the measuring line is cutting the test specimen so that the interface is tilted relative to the surface (Fig. 5.4a). Another approach is to use a wedge-shaped sample for which the distance covered by the pressure wave varies considerably along the measuring line. Because of the different amounts by which the rarefaction wave is spread, the pulse length increases with the sample thickness whereas the amplitude decreases. This results in a spall plane that is inclined relative to the surface (Fig. 5.4b).

An inclined position of the spall surface in a plane-parallel specimen can be created by applying a non-uniform load producing a peak stress that varies with position across the surface. To achieve this goal, Baumung et al., 2001, gener-

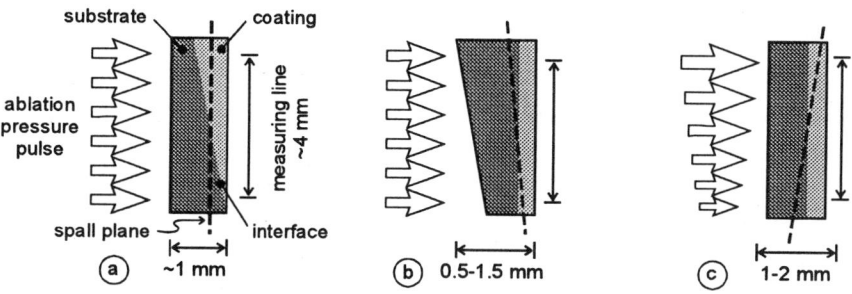

Figure 5.4. Possibilities for ensuring that the spall plane intersects the interface. (a) A plane-parallel specimen in which the interface is tilted. (b) A wedge-shaped specimen for which the longer distance covered by the compression wave at one side results in a longer rarefaction tail and larger distance of the spall from the surface. (c) A plane-parallel specimen subjected to a load producing a peak stress that varies with position across the surface. The higher stress causes steeper stress gradients and rupture closer to surface.

ated shock pulses using the pulsed high-power proton beam of the Karlsruhe Light Ion Facility (KALIF) and took advantage of the bell-shaped power density distribution of the KALIF beam which results in a transversely varying ablation pressure. A variation of the stress gradient during unloading (which determines the spall thickness) was expected because, with the sample thickness and pressure pulse duration given, the gradient depends on the pressure amplitude. The distance of the fracture plane from the free surface depends on the stress gradient in the rarefaction tail of the pulse. Thus, even with plane-parallel sample geometry, the spall plane should be inclined with respect to the interface as shown in Fig. 5.4c. The expected resulting ~1 degree tilt of the spall plane relative to the surface was small enough to approximately meet the assumption of planar geometry used in the data evaluation. Varying the pressure amplitude allows altering the spall-plate thickness.

Three different Inconel 738 LC samples with SICOAT coatings (types A, B, and C) were manufactured using various technological procedures. The surface of the Inconel substrate of the type A and type B samples was cleaned by an electrical arc discharge technique, after which the coating was applied at a 900 °C substrate temperature. After that, the type B samples were annealed for 1 h at 1080 °C, whereas the type A samples where tested without annealing. The type C samples were prepared without the discharge cleaning of the surface, were coated at a temperature lowered to 600 °C, and were annealed for 1 h at the same 1080 °C temperature.

In the experiments, the plane-parallel samples were positioned perpendicularly to the proton beam axis in the scheme shown in Fig. 5.4c. To locate the spall plane at distances $\delta \leq 170$ μm from the surface, a pulse duration of $\tau < \delta/c \approx 30$ ns is required. That is close to the duration of the ablative pressure pulse generated by the direct interaction of the KALIF pulsed proton beam with condensed matter. The spall plane was expected to be inclined with respect to the interface because the pressure amplitude varies whereas the pulse duration is approximately constant. The position of the ~4-mm-long measuring line of the interferometer was varied such that distances between 0–8 mm from the beam axis were covered. This resulted in a range of peak loading pressures of 5–11 GPa for samples of different thickness.

Figures 5.5–5.7 display examples of free surface velocity histories of samples of all three types. The type A samples exhibit the most heterogeneous re-

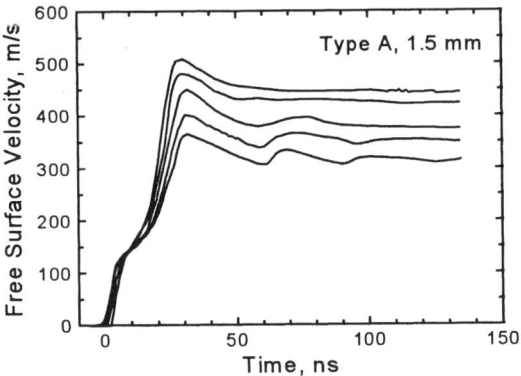

Figure 5.5. A set of free surface velocity histories from a shot with a type A sample. The different curves represent local velocity histories at different places along the measuring line on the sample.

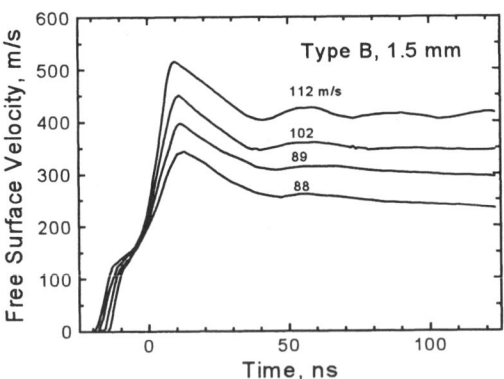

Figure 5.6. A set of free surface velocity histories for a 1.5-mm-thick type B sample. The coating thickness was ~140 μm. The velocity profiles have been shifted along the time axis to align the peak values. The numbers on the curves indicate the velocity pullback values.

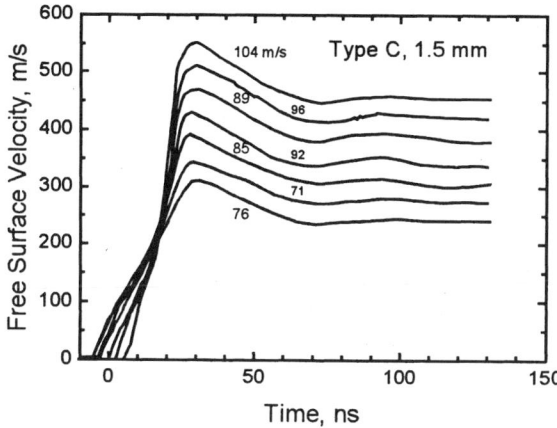

Figure 5.7. A set of free surface velocity histories for type C samples.

sponse of the material. Velocity histories show variations in the spall strength between 0.95 GPa and 1.35 GPa. The velocity pullback did not exceed 70 m/s in this shot which is less than the velocity pullback measured for SICOAT alone. Rupture seems to occur within the coating in all cases. No uniform trend could be observed in the evolution of the wave profiles in a series of experiments with this material.

The type B sample exhibits a much higher spall strength. The free surface velocity profiles show a certain evolution in the material response depending on the radial position of the point relative to the beam axis. This means that the material response depends on the peak stress and the rarefaction gradient. The time interval, Δt, of the pullback in Fig. 5.6 first increases with reduced peak velocity and peak stress, respectively, and then stabilizes. At the same time, the velocity pullback decreases from ~110 m/s to 90 m/s. The maximum measured velocity pullback for this coating material over several shots was 120 m/s. When the spall plane shifts to the substrate–coating interface, a distinct spall signal is replaced by a smooth change in deceleration of the spall plate. The same trend was recorded in a shot with a 2-mm-thick type B sample.

At high stress amplitudes and stress gradients, rupture occurs inside the coating. As the stress gradient decreases, the spall plane moves away from the surface and approaches the interface. If the adhesion strength is markedly smaller than the strength of the coating, a decrease in velocity pullback accompanied by an increase in the spall signal amplitude should be expected when the spall plane reaches the interface. This was not observed in the experiments. If the strength at the interface is equal to the strength of the coating and the strength of the substrate is much higher, spall fracture should stop at the interface and later reappear some distance inside the substrate. Instead of these two expected types of behavior, the observed relatively rapid brittle-like fracture inside the coating is gradually replaced by a relatively slow fracture process near the interface. The observed transition from a distinct spall signal to a smooth deceleration may be interpreted as a decrease in the fracture rate that accompanies the decrease in spall strength when spalling occurs near the coating–substrate interface.

In the experiments with type B samples shown in Figs. 5.6, the fracture stress started to decrease at a distance of about 30–40 μm from the substrate–coating interface. In other words, the first 30–40-μm-thick layer of the coating grains and the remaining coating material have different strengths. In connection with this, one should note that, due to diffusion of the substrate material into the coating, the chemical composition of the near-interface layer differs from that of the bulk material even for fresh coatings. Besides this, the porosity of the coating has its maximum value at the interface. If the tensile strength at the interface were negligible, that should show in the free surface velocity profiles at about ~55 ns in Fig. 5.6.

Figure 5.7 shows the results of an experiment with a type C sample. In this case the coating thickness was ~15 μm greater than that in the previous example involving a type B sample. A decrease in velocity pullback down to ~75 m/s is recorded whereas the maximum velocity pullback reached 120 m/s in the previous case. The largest time interval, Δt, between the velocity peak and the minimum in front of the spall signal increased by ~6–9 ns as compared to the previous case. Nevertheless, the spall plane again failed to approach the interface. Replacement of a distinct spall signal by a smooth deceleration is not observed even at the lowest velocity pullback. This may indicate a lower strength of the near-interface layer of the coating material as compared to the type B sample.

The coating of samples of type B and C shows practically the same internal strength ranging from 2 to 2.2 GPa. The type B samples exhibit larger fracturing stress near the interface than the type C samples. In round numbers, the adhesion strength is 1.3–1.4 GPa for type C samples and ~1.5–1.6 GPa for type B samples.

Comparison of the behavior of three kinds of samples prepared with different technologies demonstrates a great influence of annealing on the homogeneity and strength properties of the coating and a lesser influence of the substrate temperature on the adhesion. This first experience with the use of spatially resolved diagnostics has shown that the materials exhibited more complicated behavior than was expected, obviously as a result of different properties of the bulk coating material and its near-interface layer. No steep spall signal was observed, which should be seen if the adhesion strength is notably smaller than the strength of the coating material. Instead, within the range of load conditions realized, the distinct spall signals turned into smooth plateaus with decreasing peak stress but never disappeared completely. To be sure that the adhesion strength, i.e., the tensile strength of the interface, is measured, one should discern some transition of the spall pulse between spall fracture within the coating or at the interface and fracture inside the substrate.

5.3. Conclusion

The experiments described in this chapter show that spatially resolving diagnostics provide a means for studying phenomena that cannot be investigated with the usual techniques. In this way, Trott et al., 2002, studied the spatial dispersion of a shock wave propagating in a highly porous material. It would be interesting to perform spatially resolved measurements of shock compression of brittle crystals and some ceramics and to discover the source and cause of irregular oscillations seen on wave profiles in these materials. With this technique, we can hope to construct a strength map and to see some new details of the dynamics of shock deformation and initiation and growth of fractures. Undoubtedly, there are other very interesting tasks related to the non-uniformity of shock deformation of some solids that could be effectively investigated using spatially resolving diagnostics.

References

Baumung, K., H.J. Bluhm, B. Goel, P. Hoppe, H.U. Karow, D. Rush, V.E. Fortov, G.I. Kanel, S.V. Razorenov, A.V. Utkin, and O.Yu. Vorobjev (1996a). *Laser and Particle Beams* **14(2)**, pp. 181–210.

Baumung, K., G.I. Kanel, S.V. Razorenov, D. Rusch, J. Singer, and A.V. Utkin (1997). *J. Phys. IV* **7**, pp. C3–927.

Baumung, K., G. Muller, J. Singer, G.I. Kanel, and S.V. Razorenov (2001). *J. Appl. Phys.* **89(11)**, pp. 6523–6529.

Baumung, K., J. Singer, S.V. Razorenov, and A.V. Utkin (1996b). in: *Shock compression of Condensed Matter — 1995* (eds. S.C. Schmidt and W.C. Tao) American Institute of Physics, New York, pp. 1015–1018.

Gupta, V., A.S. Argon, D.M. Parks, and J.A. Cornie (1992). *J. Mech. Phys. Solids* **40(1)**, pp. 141–180.

Gupta, V., and J. Yuan (1993). *J. Appl. Phys.* **74(4)**, pp. 2397–2404.

Kanel, G.I., S.V. Razorenov, and A.V. Utkin (1996). in: *High Pressure Shock Compression of Solids II. Dynamic Fracture and Fragmentation* (eds. L. Davison, D.E. Grady, and M. Shahinpoor), Springer-Verlag, New York, pp. 1–24.

Kanel, G.I., S.V. Razorenov, A.V. Utkin, K. Baumung, H.U. Karow, and V. Licht (1994). in: *High-Pressure Science and Technology—1993* (eds. S.C. Schmidt, J.W. Shaner, G.A. Samara, and M. Ross) American Institute of Physics, New York, pp. 1123–1126.

Trott, W.M., L.C. Chhabildas, M.R. Baer, and J.N. Castaneda (2002). in: *Shock Compression of Condensed Matter — 2001* (eds. M.D. Furnish, N.N. Thadhani, and Y. Horie) American Institute of Physics, New York, pp. 845–848.

CHAPTER 6

Polymorphic Transformations and Phase Transitions in Shock-Compressed Solids

6.1. Introduction

It is known that many solids can exist in different crystal structures depending on the pressure and temperature. The change in the crystal structure is often accompanied by a change in compressibility which, in turn, affects the evolution of compression and rarefaction waves. This circumstance opens a way to study thermodynamic parameters and kinetics of polymorphic transformations. Shock-wave loading causes irreversible heating of the material which may result in its melting upon compression or during unloading from shock-compressed states. At higher peak pressures the irreversible part of the energy of shock-wave compression becomes sufficient to vaporize a significant fraction of the material during unloading. Melting and vaporization of shocked solids occur, for example, as a result of hypervelocity collisions of meteorites with space apparatus. This makes it important to study these phenomena.

The phase diagrams of solids and properties of high-pressure phases are one of the main subjects of high-pressure science. The shock-wave investigations are not usually supported by detailed structural analysis which requires a long exposure time and, in this sense, they are not able to compete with modern quasi-static research. On the other hand, any observation of structural rearrangement of sub-microsecond duration is very interesting and intriguing by itself. It has been shown (for example, see the review paper by Kormer, 1968) that the transformation time of KBr and KCl crystals decreases sharply with increasing pressure and becomes less than 2×10^{-12} s when the peak shock pressure exceeds 15 GPa. Fast polymorphic transformations exert an essential influence on the response of materials to high-velocity impact or explosion, and this stimulates their investigation. The influence of polymorphic transformations in steel, as an effect of high-velocity impact or explosion, was studied computationally by Bertholf et al., 1975, and by Sugak et al., 1983. It was found that the polymorphism of the target material leads to a qualitative change of the flow field, so that there is a significant effect on the residual distribution of damage. Other practical reasons for studying polymorphic transformations are related to the explosive or impact production of super-hard materials such as diamond and diamond-like boron nitride.

Studies of polymorphic transformations in shock waves have been exhaustively reviewed by Duvall and Graham, 1977, and Al'tshuler, 1978. In this chapter, we shall confine our attention to the most characteristic features of the phenomenon in several of the most practically important materials: iron and steels, titanium, carbon, and boron nitride. The chapter is completed by some non-traditional observations of melting and vaporization during decompression.

The effect of polymorphic transformations with a change in volume on the profiles of compression and rarefaction waves is illustrated in Fig. 6.1 on the assumption that the deviatoric stresses are small. For pressures $p > p_1$, the initial crystal structure looses its stability and transforms into a high-pressure phase of higher density. Mixed phases appear in region 1–2, with an increase of the fraction of the high-pressure phase as the state changes from point 1 to point 2; above point 2 we deal with the Hugoniot of the high-pressure phase. The Hugoniot as whole shows strong kinks at points 1 and 2 and a weak kink at point 3 where a transition occurs from the two-shock regime to a single-shock wave. Since shock compression is not an isothermal process, the pressure in the mixed-phase region, in general, should not be constant and usually increases from point 1 to point 2 with increasing peak energy of shock compression. The excluded region 1–2–3 of the Hugoniot is not accessible to single shock-wave transitions from the initial state at point 0, and shock waves of peak pressure in the range $p_1 < p < p_3$ split into two compression waves, with the velocity of the first shock wave,

$$U_{s1} = V_0 \left(\frac{p_1}{V_0 - V_1} \right)^{1/2},$$

exceeding the Lagrangian velocity of the second wave,

$$U_{s2} = V_0 \left(\frac{p - p_1}{V_1 - V} \right)^{1/2}.$$

This situation is qualitatively similar to the case of loss of stability experienced by a shock wave at the elastic–plastic transition. No such similarity takes place for rarefaction waves.

When the high-pressure compressed phase is released, the change of state down to the onset of the inverse transformation at point 2 follows the compressibility curve for this phase. The pressure region $p_k < p < p_2$ is anomalous for release waves in the sense that the velocity of sound varies non-monotonically with decreasing pressure, and is greater along the segment $1-k$ than along the segment $2-1$. A rarefaction shock wave is therefore formed during the release wave which propagates at the velocity

$$U_{sr} = V_0 \left(\frac{p_2 - p_k}{V_k - V_2} \right)^{1/2}. \tag{6.1}$$

The waveform shown in Fig. 6.1b corresponds to a case when the compression wave is followed by unloading with a rarefaction shock that propagates in the same direction. A reflection of the compression wave from the free surface of the plate creates another unloading wave in which a rarefaction shock is also formed. Interaction of these two oppositely directed rarefaction shocks produces tension at an extremely high rate and a very smooth, mirror-like spall forms. Observation of such spall surfaces by Ivanov and Novikov, 1961, in steel was obviously the first experimental confirmation of the existence of rarefaction shocks.

Formation of the two-wave compression configuration and the rarefaction shock wave is the most obvious and convincing demonstration of a reversible polymorphic transformation in a compression pulse. The state behind the first shock wave of the two-shock configuration is the state at the beginning the polymorphic transformation into the high-pressure phase. The transformation itself occurs behind first shock, mainly in the second compression wave. The decay of the first shock and the rise time of second compression wave are governed by the transformation kinetics. Correspondingly, analysis of the unloading part of a shock compression pulse allows evaluation of the thermodynamic and kinetic parameters of the reverse transition to the low-pressure phase, if this transition occurs.

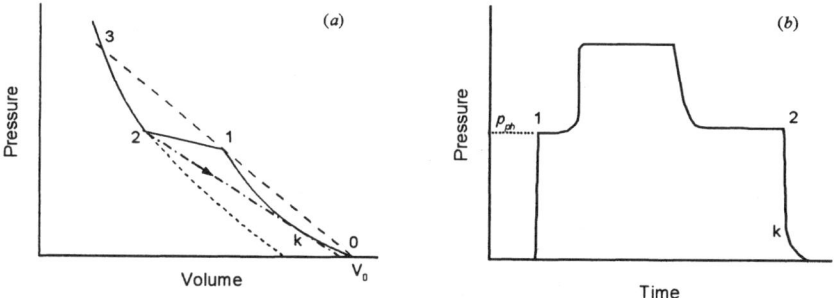

Figure 6.1. A compression pulse in a medium undergoing reversible polymorphic transformation under pressure. (a) Schematic stress-volume diagram of the process where 0–1–2–3 is the Hugoniot, 1 and 2 are the points of beginning and end of transformation, respectively. (b) Idealized stress history of a compression pulse in a material undergoing reversible transformation.

6.2. The $\alpha \leftrightarrow \varepsilon$ Polymorphic Transformation in Shock-Compressed Iron and Steels

Polymorphic transformation by shock-wave compression in a solid was first recorded by Bancroft et al., 1956, as a kink on the Hugoniot of iron at a pressure of 13 GPa. This finding was later confirmed under static compression by

Balchan and Drickamer, 1961, who observed an abrupt change in the electrical resistance of iron at 13.3 GPa and room temperature. The high-pressure phase was initially identified with the face-centered-cubic γ phase into which iron transforms at a temperature of 910 °C at normal pressure (the initial α phase of iron has a body-centered-cubic structure). However, the microstructure of iron samples shock-compressed at different initial temperatures (Johnson et al., 1962) indicated that the high-pressure phase at room temperature was different from the high-temperature γ phase. X-ray studies under static compression by Jamieson and Lawson, 1962, later showed that the low-temperature, high-pressure phase had an hexagonal-close-packed crystal structure that was identified as the ε phase of iron.

The splitting of a shock wave in iron was first observed by Bancroft et al., 1956, in unique experiments using several dozen electrical pins mounted at different distances from the sample surface. Development of methods of recording the stress and particle velocity histories in the interior of a sample has greatly simplified the detection of phase transitions in both shock and release waves. Figure 6.2 shows stress profiles in Armco iron and high-strength steel recorded with manganin pressure gauges by Anan'in et al., 1973, 1981, and by Gluzman et al., 1985. The samples were loaded by impact of aluminum flyer plates with a velocity of 1.05 or 2.06 km/s. At the lower impact velocity the wave profiles clearly demonstrate elastic–plastic compression followed by elastic–viscous unloading, as in other metals. When the compression pulse has a sufficiently high amplitude, the plastic shock wave is found to split into two shocks in the region of the α → ε transformation. The formation of a rarefaction

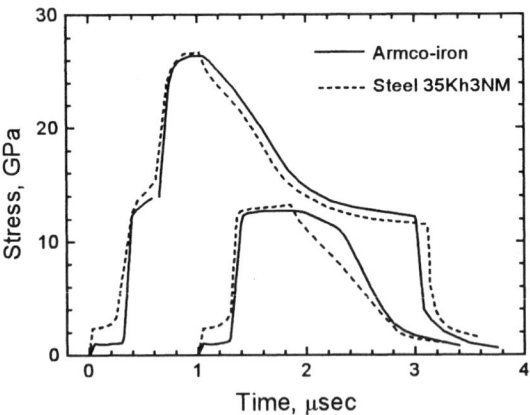

Figure 6.2. Stress profiles in Cr–Ni–Mn steel 35Kh3NM and Armco iron samples loaded by impacts of aluminum flyer plates of 7.0 and 5.0 mm thickness at velocities of 2.06 km/s and 1.05 km/s, respectively. Measurements with manganin pressure gauges were made at a distance of 10 mm from the impact surface.

shock wave due to the reverse $\varepsilon \to \alpha$ transition is also observed during the unloading process.

The wave profiles in iron and steel are generally similar. However, it is clear that the deformation of the harder steel is associated with greater (as compared with iron) deviatoric stress in the entire cycle of shock compression and release. The greater resistance to deformation of steel is probably responsible for the greater hysteresis of the parameters of the direct and inverse polymorphic transformations. The stability boundaries of the phases are obviously determined by their density or pressure, and not by the maximum stress.

Directly recorded wave profiles give the stresses behind the first plastic compression wave and before the rarefaction shock wave in iron that correspond to the beginning of the forward and reverse $\alpha \leftrightarrow \varepsilon$ transitions in Armco iron at 12.6–14 GPa and 12.3 ± 0.4 GPa, respectively. We note the small (compared to the static experiments by Giles et al., 1971) hysteresis in the pressure at the beginning of polymorphic transformations. The transition kinetics are difficult to determine because of the superposition of two relaxation processes, namely, polymorphic transformation and time-dependent plastic deformation. Experimental data at lower peak stresses suggest that the slow decrease of stress ahead of the rarefaction shock is related mostly to the viscous–elastic behavior of the material. It is actually difficult to say whether the mobility of atoms during plastic deformation provides a high transformation rate or, to the contrary, the high strain rate is provided by the high transformation rate. It seems that, under the conditions of uniaxial deformation associated with shock-wave loading, the polymorphic rearrangement of the crystal structure is one of the mechanisms of realization of plastic deformation.

Barker and Hollenbach, 1974, studied the $\alpha \leftrightarrow \varepsilon$ transformations in Armco iron at different peak stresses using high-resolution VISAR measurements of the free surface velocity histories of shocked samples. In accordance with their data, the transformation rate increases from $\sim 3 \times 10^6$ s^{-1} at a 17 GPa peak stress to $\sim 22 \times 10^6$ s^{-1} at 30 GPa. The $\alpha \to \varepsilon$ transformation is not completed at 15.7 GPa but is completed at 17.3 GPa.

Anan'in et al., 1981, recorded stress histories in shocked Armco iron at various distances from the impact surface. This provided, using the assumption of simple wave propagation, a capability to recover the histories of changing volume and other state variables during the shock compression and unloading processes. The result of this treatment of the unloading parts of measured stress histories is presented in Fig. 6.3. The figure also shows the path of the released states near the free surface obtained by Barker and Hollenbach, 1974. They determined the longitudinal stress to be 12.9 GPa behind the first plastic shock wave. This is in agreement with the data obtained employing manganin gauges. However, the stress at the beginning of the reverse $\varepsilon \to \alpha$ transition was determined by them to be 9.8 ± 0.4 GPa, a value that differs significantly from the

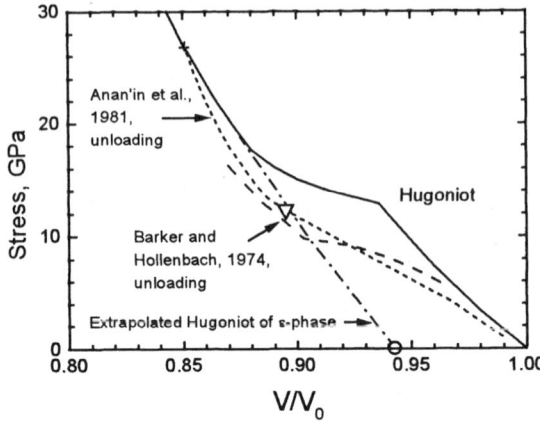

Figure 6.3. The trajectory of the changing state of Armco iron in release waves, as recovered from the stress histories in comparison with the Hugoniot of iron and the data evaluated from the free surface velocity histories recorded by Barker and Hollenbach, 1974. The triangle shows the stress measured just ahead the rarefaction shock in the stress history. The zero-pressure specific volume of the ε phase has been reported by Mao et al., 1967.

data obtained with manganin pressure gauges in the interior of the samples. The relative position in Fig. 6.3 of the trajectories of change-of-state deduced from the stress histories and the path of released states evaluated from the free surface velocity histories suggests that there is a difference between deviatoric stresses in these two cases that is due to the viscosity of the medium. It is also clear from Fig. 6.3 that the parameter values at the beginning of the reverse transition differ not only by the stress, but also by the specific volume. This is due to the effect of the rate of transformation on the values at the beginning of the $\varepsilon \to \alpha$ transition.

For comparable strain rates, the deviation of the change-of-state path in the release wave from the Hugoniot for the high-pressure phase is much greater than that for the initial phase in experiments at low impact velocities. We note that the residual hardening of iron after shock-wave treatment is very much greater when the shock-wave amplitude is sufficient for the polymorphic transformation to proceed. Hardening during the polymorphic transformation is caused by refinement of the grain structure of the material.

Figure 6.4 shows free surface velocity histories of samples of Armco iron tested at normal and elevated temperatures. Earlier similar measurements were performed by Nahme and Hiltl, 1995. The data demonstrate a decrease of the transition pressure with increasing temperature. On the other hand, Anan'in et al., 1981, made measurements with manganin pressure gauges at the tempera-

6. Polymorphic Transformations and Phase Transitions   195

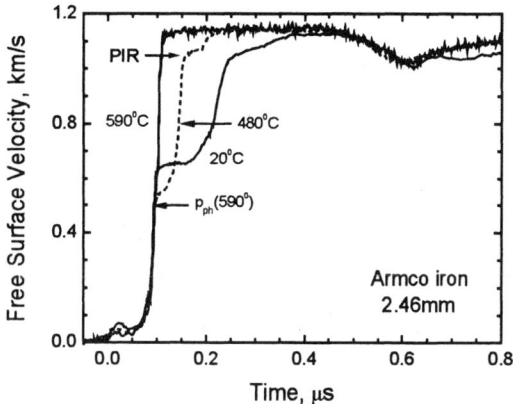

Figure 6.4. Results of measurements of free surface velocity histories of Armco iron samples at normal and elevated temperatures. The samples were 2.46 mm in thickness and were impacted by 2-mm-thick aluminum flyer plates at a velocity of 1.9 km/s.

ture of liquid nitrogen and did not find significant changes in the transition pressures. The time intervals between the first and second plastic waves in Fig. 6.4 decrease with increasing temperature as a result of an increase of the stress increments and, correspondingly, the velocity of second plastic waves. At elevated temperatures the so-called PIR wave (Barker and Hollenbach, 1974) is distinctly recorded. This fourth step in the compressive part of free surface velocity history is a result of the reflection of unloading wave at phase interface between transformed and untransformed iron.

Figure 6.5 summarizes the results of measurements of transition pressure as a function of the temperature. The points obtained from measurements of the free surface velocity histories differ slightly from data by Johnson et al., 1962, that were obtained 40 years earlier by means of observations of microstructural changes.

Figure 6.6 presents the results of experiments by Razorenov et al., 1997, with 40Kh steel in the "as-received" and quenched states. The as-received material had a ferrite–pearlite structure. The quenched steel had a martensite structure with isolated grains of the residual austenite (about 2%) and chromium carbide (about 1%) and was three times harder than the as-received one. Under shock compression, the difference between dynamic yield stresses of these two materials decreased to ~10%. It may be seen from comparison of the wave profiles in Fig. 6.6 that strengthening of the material by the heat treatment may have resulted in some deceleration of the $\alpha \rightarrow \varepsilon$ polymorphic transformation but not in any remarkable change of the transformation pressure.

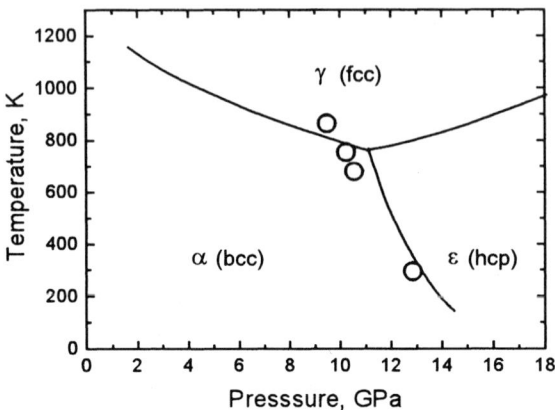

Figure 6.5. The phase diagram of iron. Lines show the data by Johnson et al., 1962; points are the results of interpretation of the free surface velocity histories shown in Fig. 6.4.

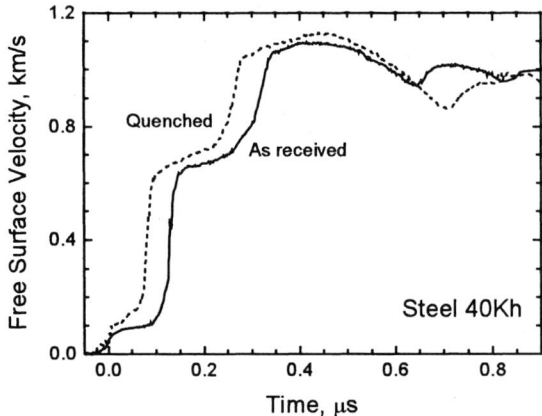

Figure 6.6. The free surface velocity histories of samples of as-received and quenched 40Kh steel of 3.95 and 3.17 mm thickness, respectively. Impacts were by 2-mm-thick aluminum flyer plates at the velocity of 1.9 km/s. Data by Razorenov et al., 1997.

6.3. The α → ω Transformation in Shocked Titanium

The α → ω phase transformation in titanium has been observed in several laboratories under both static (see Jamieson, 1963, Olinger and Jamieson, 1973, Zilbershtein et al., 1975) and dynamic (Kutsar et al., 1973) pressures. Nevertheless, there are essential contradictions in the transition pressure. The static pressure values range from 2.0 to 7.5 GPa. Factors that are responsible for this scatter were identified as pressure exposure time, sample preparation history, impurity content, and non-hydrostaticity of the applied pressure (Sikka et al., 1982). According to Zilbershtein et al., 1975, and Vohra, 1978, the $\alpha \rightarrow \omega$ transformation, being martensitic, is aided by the presence of shear forces. It has been shown that the shear strains reduce the hysteresis under static conditions. Under combined compression and shear, both forward and reverse transformations can be observed in titanium at practically the same pressure of ~2 GPa. Wave profile measurements showed shock transition pressures of 11.9 GPa (Kutsar et al., 1982), 6.0 GPa (Kiselev and Falkov, 1982) (both for the commercial grade titanium VT1-0), and 10.4 GPa (Gray, 1990) for electrolytic titanium. Examination by Kutsar et al., 1973, of Ti samples recovered after shock compression at low temperatures has shown some amount of the ω phase at the 9 GPa shock pressure. The volume decrement at the $\alpha \rightarrow \omega$ phase transformation in titanium is only ~1.2 %.

Figure 6.7 compares the free surface velocity histories of shocked samples of a high-purity titanium, commercial-grade titanium which contains (wt. %) O_2–0.15, Fe–0.10, Cr–0.018, Ni–0.015, C–0.016, Al <0.02, Cu, Zr, V, Mn <0.01, and titanium alloy Ti-6Al-2Sn-2Zr-2Cr-2Mo-Si at 4.5–6.5 GPa peak shock stress. Whereas the waveforms of commercial titanium and titanium alloys do not contain any peculiarities related to the expected polymorphic transformation, the form of plastic compression waves in pure titanium is not typical for metals: The slope of the upper part of the waveform is obviously less than the slope of the lower part that is obviously an effect of the transformation. We

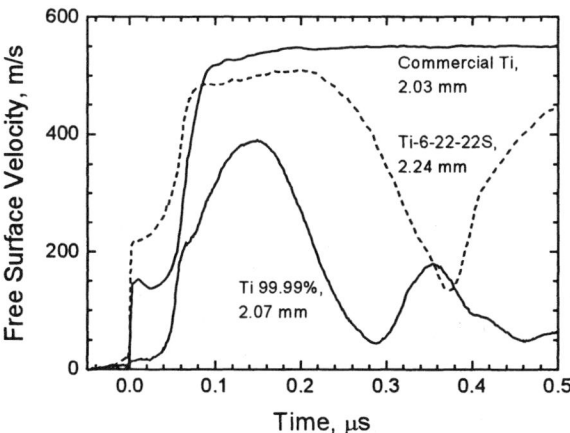

Figure 6.7. Free surface velocity histories of shocked samples of a high-purity titanium, commercial-grade titanium, and titanium alloy Ti-6Al-2Sn-2Zr-2Cr-2Mo-Si. Data by Kanel et al., 2003, and Krüger et al., 2002.

may conclude that, unlike iron, the transformation parameters of titanium are very sensitive to composition of the sample tested. The pressure at which the $\alpha \to \omega$ polymorphic transformation begins in pure titanium is estimated to be 2.37 GPa in this shot.

Figures 6.8 and 6.9 show the free-surface velocity profiles of pure titanium samples recorded by Razorenov et al., 1995, in nano- and microsecond duration ranges. A wave configuration with two shock discontinuities was never observed in these experiments. The splitting of a shock wave is observed distinctly for small propagation distances (Fig. 6.8) at 10.4 GPa peak pressure. More expressed splitting could be expected at greater propagation distances. However, the free-surface velocity profile in Fig. 6.9 at approximately the same peak pressure (11.6 GPa) on a thicker sample shows a decrease of the velocity gradient in the compression wave instead of a double-shock configuration. In both cases, the inflection occurs at the same velocity of ~440–450 m/s. Therefore, the upper part of the compression wave, which can be identified as a second plastic wave with phase transition, behaves not as second shock, but rather as a continuously spreading wave. It is natural to relate the point of inflection to the beginning of the phase transition at this peak stress. Then the pressure level of the phase transition onset in the high-purity titanium calculated from the first wave amplitude is 5.1 GPa for a peak pressure of 10.5–11.6 GPa. If this is the true value of the phase transition pressure, then profiles of shock waves of lower peak pressure cannot contain any kinks. But Figs. 6.7 and 6.9 show inflection points in the vicinity of 200 m/s in shock-wave profiles at peak pressures of ~4.5 GPa. Thus, the pressure at the inflection point depends on the peak pressure in the shock wave.

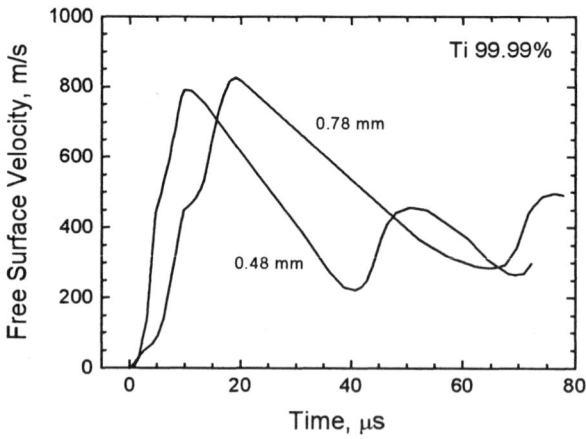

Figure 6.8. Free surface velocity histories of samples of pure titanium exposed to a pulsed proton beam. Data by Razorenov et al., 1995.

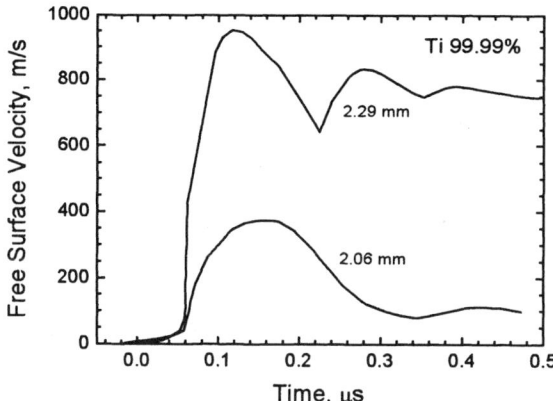

Figure 6.9. Free surface velocity histories of samples of pure titanium impacted by 0.4-mm-thick aluminum flyer plates at velocities of 0.66 and 1.25 km/s. Data by Razorenov et al., 1995.

The $\alpha \rightarrow \omega$ transformation in titanium is accompanied by a very small volume decrease. We may assume that such a small volume decrement does not form an excluded region on the Hugoniot but just decreases its curvature. For Hugoniots having only small curvature, the deviatoric stresses do not vary much within the compression wave. As a result, variations in the material viscosity should have a significant effect on the strain rate. The beginning of a polymorphic transformation which increases the compression means appearance of a bulk viscosity, whereas previous plastic deformation in the shock wave is associated only with a shear viscosity. In other words, the steepness of the lower parts of observed compression waves in titanium is governed by the shear viscosity whereas the steepness of the upper part is determined by both the shear viscosity and the kinetics of the transformation.

This hypothesis may help to explain the observed dependence of the phase transition pressure on the peak pressure as well as the discrepancy in published data on the phase transition pressure. Nevertheless, kinetic aspects of the wave evolution are still not clear. Since we did not see the establishment of a steady shock-wave structure, we can conclude that the transformation time is more than the largest observed rise time of the second wave, i.e., the phase transformation is relatively slow.

Within the framework of the martensitic phase transformation mechanism, it is natural to suppose that the transformation occurs during the plastic deformation. As a result, the transformation rate has to increase with increasing strain rate. In this case the high strain rate near the impact surface provides a high transition rate. The phase transition is accompanied by the effect of bulk viscosity that leads to increasing rise time of the wave and thus to reduction of the strain rate. A reduction of the strain rate provides a lower transition rate, which leads to further spreading of the wave profile and so on. Thus, the observed unsteady compression waves in titanium are formed due to the mutual influence of the strain rate and the phase transition rate.

Figure 6.10 demonstrates variation of the transformation pressure with temperature from 2.37 GPa at 20°C to 3.05 GPa at 465°C. The increase of transformation pressure with increasing temperature is known from quasistatic data also, although the rate of this increase, dp/dT, differs from that indicated by the shock-wave data shown in Fig. 6.10.

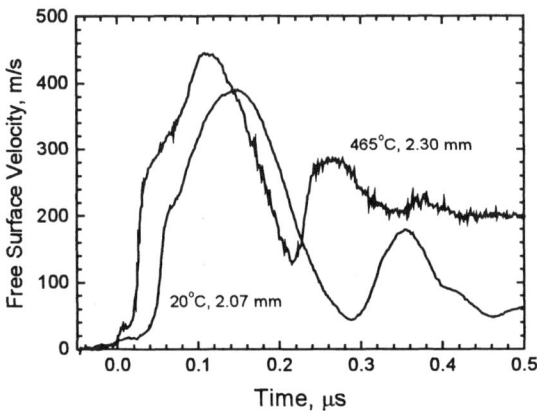

Figure 6.10. Free surface velocity histories of pure titanium samples at normal and elevated temperatures.

6.4. The Graphite to Diamond Transition under Shock Compression

Polymorphic transformations of matter under shock compression are used in industry to produce super-hard materials such as diamond and diamond-like boron nitride. Naturally, a vast volume of literature exists that is devoted to polymorphism of these materials under static and dynamic compression. In this chapter we'll discuss only briefly the appearance of the transformation in wave profiles.

The most favorable conditions for the accurate determination of transformation parameters are established in the two-wave configuration. Anan'in et al., 1978, observed a two-wave configuration in boron nitride and graphite by the step-like compression technique. The material was first pre-compressed by a shock wave to a state somewhere below the transformation pressure. After that, the pre-compressed matter was further compressed by a second shock wave.

This second wave was created as a result of reflection of first shock from a copper barrier that had been placed immediately behind the sample tested. The pressure profiles inside and on the surface of the sample were recorded by manganin pressure gauges. The step-like compression technique provides the capability to make measurements at relatively low impact velocities. Besides this, an advantage of the technique is that the two-wave configuration may be formed even in porous samples whereas, for compression by a single shock wave, the excluded region decreases and eventually disappears with increasing sample porosity.

Anan'in et al., 1978, and Razorenov et al., 1981, used the step-like compression technique for testing natural graphite and boron nitride samples at normal, low, and elevated temperatures. The samples were pressed of powders of the tested materials. The measurements have shown that the pressure of the graphite → diamond transition at an initial temperature of −196°C, +20°C, and +300°C is 23.3 ± 1.0 GPa, 19.6 ± 0.5 GPa, and 17.5 ± 1.0 GPa, respectively. The measurements therefore demonstrate quite reliably that the transformation pressure decreases with increasing temperature. Similar experiments with boron nitride have shown that, regardless of the initial sample temperature, the transformation of the initial material with hexagonal graphite-like structure into the high-pressure phase occurs at 12 ± 0.2 GPa.

Gogulya, 1989, observed the two-wave configuration under single shock compression of natural graphite and has determined the transformation pressure to be 20.0 ± 0.5 GPa. He also observed some decay of the first compression wave and estimated the initial transformation rate. Erskine and Nellis, 1992, revealed high sensitivity of the graphite → diamond transformation kinetics to the sample morphology. They compared shock response of pyrolytic graphite and highly oriented graphite single crystals. These two materials differ by disorientations of the graphite structure: the pyrolytic graphite has a mosaic structure with disorientation of the c axis of the blocks of the order of 3.5°, whereas the highly oriented graphite had only an 0.8° disorientation along the c axis. For the single crystals, a very distinct two-wave configuration has been recorded with the transformation pressure 19.6 ± 0.7 GPa and the transformation time less than 10 ns at a 30 GPa peak shock pressure. However, the density of shock-compressed carbon within the pressure range of 25–50 GPa was 5% less than the density of diamond at the same pressure and temperature. This difference was considered as evidence of transformation of graphite to a disordered or distorted diamond-like structure. In the case of pyrolytic graphite, the wave profiles demonstrated much slower transformation and the pressure at which the transformation began varied from shot to shot between 24 and 42 GPa. Undoubtedly, the difference in response of the pyrolytic and highly oriented graphite is the result of different disorientations of their structures. However, it still is not clear why the highly disoriented samples, which were pressed of graphite powder, show the same transformation pressure and rate as highly oriented single crystals demonstrate.

6.5. On the Possibility of Polymorphic Transformations Occurring in the Negative-Pressure Region

Since many solids transform to a denser crystal structure under compression, it may be expected that, under certain conditions, transformation to a less dense structure could occur in some materials under tension. This could be expected, for example, in cobalt. At zero pressure cobalt can exist in two modifications. The low-temperature modification has the hexagonal-close-packed (hcp) crystal structure. At the temperature $T_f = 423\,°C$ (or $447\,°C$ according to other data) the hcp cobalt transforms to the high-temperature face-centered-cubic (fcc) phase (Kaufman, 1963). This transformation is accompanied by a volume increment of 0.026 cm^3/mol or 0.35%. The transformation temperature increases with pressure at a rate of $dT_f/dp = 60$ K/GPa so, under tension, the transition may occur at a lower temperature.

Razorenov et al., 2002, investigated the response of cobalt to shock compression and rarefaction at normal and elevated temperatures. It was expected that the hcp → fcc polymorphic transformation should exhibit itself in the free surface velocity history. The results of measurements are shown in Figs. 6.11 and 6.12.

It was supposed that the volume increase in the course of the hcp → fcc transformation should result in formation of a rarefaction shock wave as observed earlier for iron. The corresponding tensile stresses of the beginning the transformation were estimated in the following way.

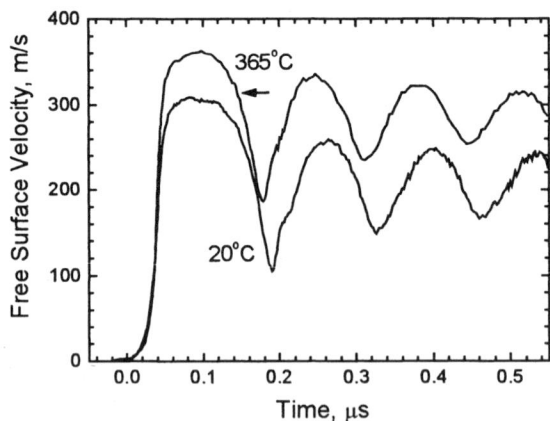

Figure 6.11. The free surface velocity histories of 1-mm-thick cobalt samples impacted by 0.4-mm-thick aluminum flyer plates at the room temperature and at 365 °C. The arrow shows the part of waveform where an anomaly was expected.

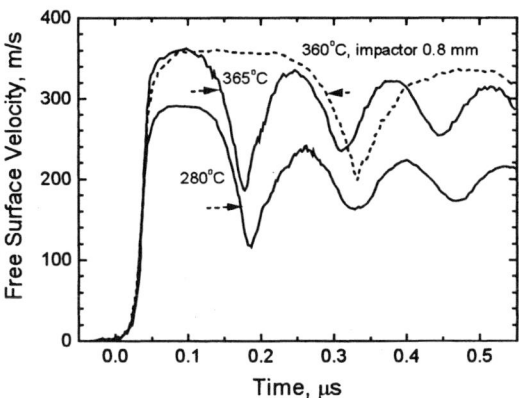

Figure 6.12. The free surface velocity histories of cobalt samples impacted by aluminum flyer plates 0.4 mm or 0.8 mm in thickness at elevated temperatures.

In the linear approach, a relationship between the pressure and the temperature along the line of phase equilibrium and along the isentrope are

$$T_f = T_{f0} + \frac{dT_f}{dp} p$$

$$T_S = T_0 + \left(\frac{\partial T}{\partial p}\right)_S p, \qquad (6.2)$$

where T_f is the temperature on the phase equilibrium line, T_{f0} is the transformation temperature at zero pressure, T_S is the temperature on the expansion isentrope, and T_0 is the test temperature. The subscript S means differentiation along an isentrope. Accounting for the thermodynamic identity

$$\left(\frac{\partial T}{\partial V}\right)_S = -\frac{\Gamma}{V} T,$$

the second relationship (6.2) becomes

$$T_S = T_0 + \frac{\Gamma T_0}{K_S} p, \qquad (6.3)$$

where Γ is the Grüneisen coefficient and K_S is the isentropic bulk modulus. The beginning of transformation during isentropic expansion corresponds to the point of intersection of the isentrope and the phase equilibrium line. Then the condition $T_f = T_S$ gives us the pressure at the intersection point:

$$p = K_S \frac{T_{f0} - T_0}{\Gamma T_0 - K_S (dT_f / dp)}.$$

In accordance with this estimate, the state of cobalt at the initial test temperature of 365°C should, under tension, pass the phase equilibrium line at a pressure of −1.0 to −1.05 GPa. This corresponds to approximately a 50 m/s decrement of the free surface velocity. However, no anomaly is observed in this part of the wave profiles in Fig. 6.11. Obviously the volume change at the transformation is too small or the transformation is too slow to produce an appreciable effect in the wave structure formed on the way from the spall plane to the sample surface. Probably a larger distance of propagation of the tensile wave is needed to accumulate the effect of anomalous compressibility in the region of the polymorphic transformation. However, doubling the impactor thickness (see Fig. 6.12) did not result in formation of rarefaction shock or any other anomaly.

One should admit that it is actually not yet clear how the transition under tension manifests itself in the waveform. Zaretsky and Kaluzhny, 1996, tried to establish a correlation between the shock-induced $\gamma \rightarrow \alpha'$ martensitic transformation in stainless steel (see also references in this paper) and the free surface velocity histories at the normal and lowered temperatures. They have found that, whereas the tensile stress required to produce spall increases the transformation temperature from 214 K to 240 K, no peculiarity that could be related to the transformation appears in the free surface velocity history.

Even if we do not yet have convincing records of polymorphic transformation under high-rate tension, our opinion is that it would be very interesting and important to continue the search for evidence of such phenomena. In this regard, it could be mentioned that, in its beginning, high pressure physics dealt with pressures of only a few GPa but revealed numerous pressure-induced transformations and other effects. With modern experimental techniques, measurements are available at negative pressures up to 30–50% of their ultimate values, that is up to tensile stresses of ~20 GPa. Polymorphic transformation of stressed material may put a natural limit on the ultimate strength of a material, as was shown, for example, by Roundy and Cohen, 2001, for diamond. Besides the new physics that may be expected in the negative pressure range, the investigations may finally give us new technologies to, for example, develop super-light materials.

6.6. Melting of Shock-Compressed Metals During Decompression

Shock compression is accompanied by heating of the medium. At a certain peak pressure, the entropy increase in the shock wave results in melting of solids during isentropic unloading. The shock pressure leading to material melting in the release wave is of general interest for the verification of equations of state, for practical applications, and for the shock-wave experiments themselves. The onset of melting produces a qualitative change in the effects of hypervelocity impact and other intense attacks upon materials and structures. In this sense it is often unimportant whether the melting occurs at shock compression or during unloading; the main question is just whether or not it occurs.

It is known that the reflection of strong shock waves from the surface of a solid body can lead to ejection of material (Asay et al., 1976). Anrdiot et al., 1984, have found out that the effect is sharply intensified when post-shock melting is approached during release. The development of surface instabilities at melting was also observed by Belyakov et al., 1967, and Werdiger et al., 1996, who took x-ray shadowgraphs of samples after shock compression and release. When a solid–liquid transition begins on the release isentrope, the reflectivity of optically polished surfaces drops drastically. Chapron et al., 1988, suggested using this phenomenon to detect melting during release of shock-loaded materials.

In this section, results of measurements by Baumung et al., 1997, and Kanel et al., 1998, of the threshold pressure leading to melting of aluminum, copper, molybdenum, and titanium in the release wave are presented. The method utilizes the particular properties of experimental capabilities at the Karlsruhe Light Ion Facility, KALIF, and an improved line-imaging laser-Doppler velocimeter that can operate in either the VISAR or ORVIS mode. This instrument allows measurement of the radial velocity profile over an ~8-mm-long line across the ion beam focus. The facilities are described in the paper by Baumung et al., 1996. According to the power density profile of the beam cross-section, the radial velocity distribution of the ablatively accelerated flier foil is bell-shaped showing an ~25% decrease over a 4-mm distance from the beam axis.

Figure 6.13 shows the arrangement of the targets used for the shock melting experiments. Shock waves in the targets were created by impact of flyers. In the experiments performed, aluminum foil fliers of 40 to 75 µm initial thickness were accelerated by the ablation pressure. Given the material ablation of 25– 30 µm up to the time of impact, the final solid impactor thickness was ~10–50 µm. The upper edge of the target was positioned ~1 mm below the beam axis in order to allow a proper extrapolation of the flier velocity distribution measured in the upper part of the field of view into the region hidden by the target. The impact velocity was adjusted by varying the initial gap between the flier and the target and the flier thickness. All samples were 20–70-µm-thick foils.

Using a line-imaging laser Doppler velocimeter, the velocity history was recorded simultaneously along a 4-mm distance of both the flier and the target. When an appropriate velocity interval is chosen, the pressure threshold for shock melting is contained in the pressure interval observed. The loss of intensity of the reflected laser light was considered indicative of material spraying that occurs upon decompression of the shocked state to that of a solid + liquid mixture.

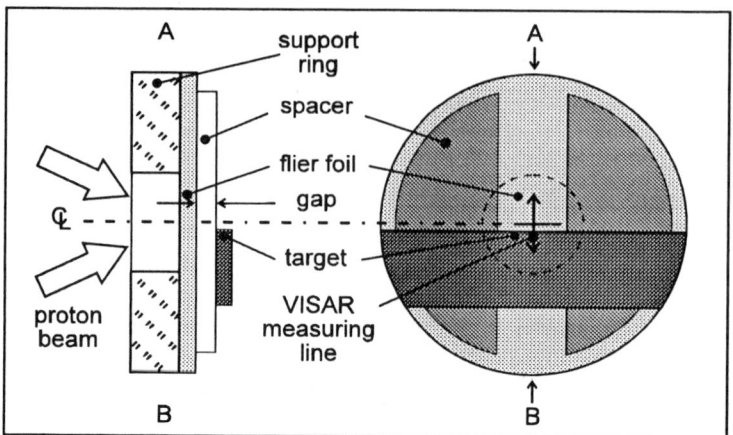

Figure 6.13. Schematic of the target arrangement. Radial and axial dimensions are not at the same scale. The total diameter is 40 mm, the flier foils were 50–75-μm thick, the gaps were varied between 50 and 130 μm, and the target was 50 μm thick.

Figure 6.14 shows a line-imaging record in the VISAR mode of an experiment where the melting boundary is seen. The upper part of the interferogram shows the flier acceleration and the lower part is related to the free surface of the target. In the VISAR mode, the interference fringes represent counts of constant velocity. The interferogram shows a clear separating line between the upper dark zone where the reflectivity is lost and the section below where the emergence of two fringes indicates a vertical velocity gradient. Taking into account the initial phase of the fringe, the threshold velocity is 5.1 ± 0.2 km/s corresponding to a pressure of 62 ± 3 GPa. In another shot we observed the loss of reflectivity at an impact velocity of 5.3 km/s so, using the Hugoniot of aluminum and presuming

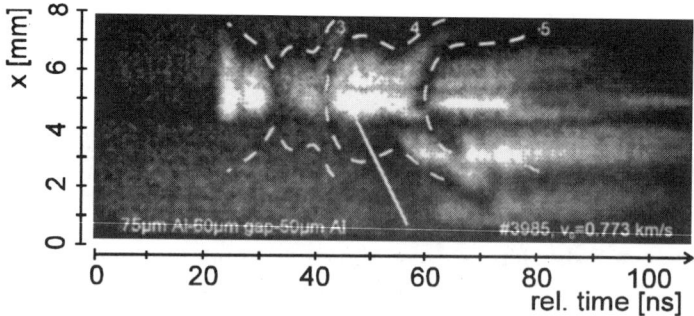

Figure 6.14. A line-imaging VISAR record of an impact experiment with melting of the aluminum target. The lower half shows the target. In its upper part (2.5–4.0 mm) the impact causes a loss of reflectivity whereas, in the adjacent section (1.0 to 2.5 mm), two fringes appear indicating a velocity gradient. The velocity-per-fringe constant was 0.773 km/s and the initial jump of flier velocity corresponds to ~3.5 lost fringes.

the loss of reflectivity to be characteristic of material melting, we find a threshold pressure for melting in the range of 62–65 GPa.

Figure 6.15 shows an example of the line-imaging ORVIS interferogram for a 50-µm-thick titanium sample. Whereas the VISAR mode gives directly a clear view of the time dependence of the velocity field, the ORVIS mode provides more precise velocity measurements. For both modes, the velocity is calculated through the number of fringes recorded for the given spatial point for the time interval up to the moment considered. In the line-imaging ORVIS mode, the interference fringes have a slope that is determined by both the acceleration and the spatial velocity gradient. If the direction of the fringe deflection coincides with the velocity gradient, the fringe slope increases; if not, it decreases. The impact velocity in this shot varied between ~5.2 and ~6.3 km/s. At impact velocities of 5.6 km/s or less, both reflectivity and interference contrast are maintained during the entire recording time. This permitted measurement of the target free surface velocity at the shock front and, thereby, recovery of the obscured parts of the flyer velocity distribution. Between ~5.6 and 6.0 km/s (shock pressures of 82.5–90.8 GPa) the partially maintained reflectivity and interference contrast decreases continuously to zero.

Figure 6.16 displays interferograms of two experiments using 20-µm-thick molybdenum samples and aluminum impactors of 40 µm initial thickness. In the range of 7.6–8.0 km/s impact velocity (shot #4143) a decrease of the reflectivity and interference contrast appear, but fringes remain observable for a relatively long period. In shot #4163, the impact velocity varies from 8.8 km/s near the bottom of the interferogram to 9.8 km/s at the upper border of the target foil. In this range, a continuous reduction of the reflectivity and interference contrast is recorded. Nevertheless, some weak fringes can be observed over the entire range for a short time after shock break-out. The intense luminosity of the flyer in the last third of the interferogram is caused by burn-through of the ablation plasma. In another similar shot at an impact velocity of ~9.4 km/s, a weak fringe was recorded for a few nanoseconds. Therefore, 9.4 ± 0.4 km/s is a best estimate of

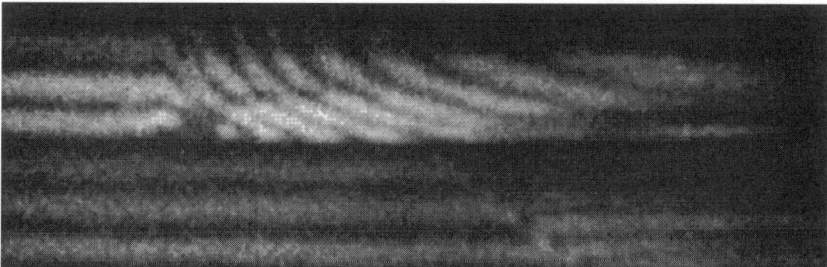

Figure 6.15. The ORVIS interferogram of an experiment with a 50-µm-thick titanium sample impacted by an aluminum flyer of 50 µm initial thickness. The velocity increase corresponds to the negative fringe slope.

the critical impact velocity for molybdenum, corresponding to a pressure of 252 ± 16 GPa.

Figure 6.17 shows one of interferograms of the shots with 45-μm-thick copper foils and aluminum impactors of 50 μm initial thickness. Weak fringes remain clearly visible for more than 10 ns in the lower part of the interferogram up to impact velocities of 6.6 km/s. Their initial negative slope indicates that there is still some notable tensile strength. The impact velocity of 6.6 km/s corresponds to a shock pressure of 136 GPa.

The results of the experiments performed show a good agreement between the threshold pressure for loss of reflectivity and the shock pressure of melting during release calculated by McQueen et al., 1970, for aluminum (64 ± 2 GPa and 65 GPa, respectively), copper (136 ± 5 GPa and 137 GPa), and molybdenum (252 ± 16 GPa and 230 GPa) whereas there is a large discrepancy in the case of titanium (86.5 ± 4 GPa measured and 120 GPa following from the EOS). The good agreement means that the release process occurring in shock-compressed metals is isentropic or nearly isentropic in our conditions. Obviously, the $\alpha \rightarrow \omega$ phase transition in shock-compressed titanium is a reason for the discrepancy with the equation of state data. The data obtained herein are related to onset of melting. Since, under shock-wave conditions, the material is heated as a result of mechanical work, any imperfection in the material creates conditions for localization of the energy deposition. The volume increment at melting in such hot spots creates conditions for nucleation of surface instabilities which appear in the observed rapid decrease of the surface reflectivity.

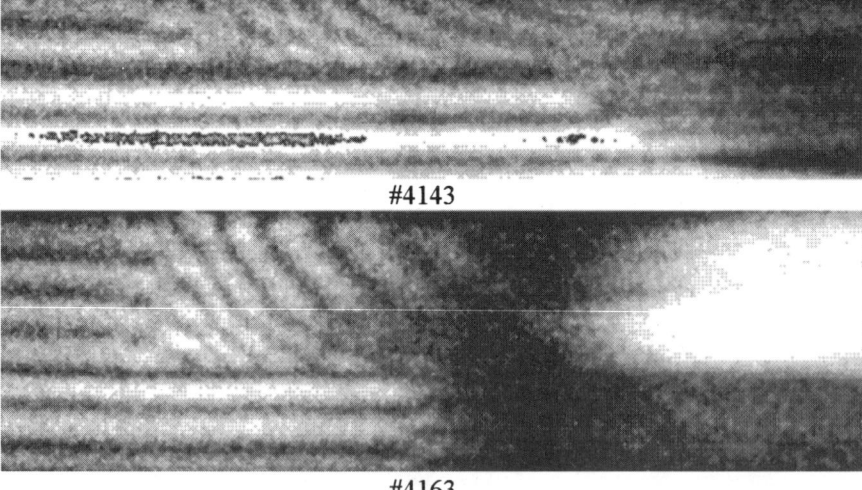

Figure 6.16. Interferograms of experiments with molybdenum at a velocity-per-fringe constant of 1261 m/s.

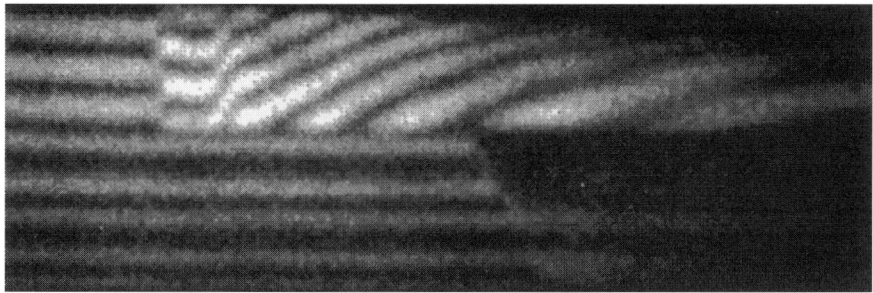

Figure 6.17. ORVIS interferogram of an impact experiment with a copper foil target.

6.7. Measuring Unloading Isentropes and Vaporization of Shock-Compressed Polymers

Experimental isentrope data are most important in the region of vaporization. Usually, the unloading isentrope is recovered from a series of experiments in which the shock-wave parameters are measured in plates of standard low-impedance materials placed behind the sample. The specific internal energy and specific volume are calculated from the measured $p(u)$ release curve, which corresponds to the Riemann integral.

The thick-foil method of Remiot et al., 1996, provides several experimental points on the isentrope in one shot. When a foil of higher shock impedance than the material studied is placed on its surface, the release process occurs by steps having a duration corresponding to the time required for two shock-wave transits of the foil. The velocity during the different steps, combined with a knowledge of the Hugoniot of the foil, allows us to determine a few points on the isentropic unloading curve. However, the method becomes insensitive when the low-pressure range of vaporization is reached in the course of the unloading. The isentrope in this region can be measured by recording the smooth acceleration of a thin foil witness plate. With the mass of the foil known, measurements of the foil acceleration will give us the vapor pressure.

Earlier, Bushman et al., 1993, found evidence of vaporization of PMMA during unloading after shock compression to a 40 GPa peak pressure. Figure 6.18 shows results of their measurements of the unloading isentropes. The data are presented in semi-logarithmic coordinates in order to demonstrate the behavior in low-pressure vaporization region. The measurements show that the particle velocity increment in the unloading wave much exceeds that in the shock wave.

In explosively driven experiments conducted by Kanel et al., 2000, shock waves with peak pressures in a range up to 34.5 GPa were created in PMMA sample plates. The rear sample surfaces were covered by 25–50-μm-thick aluminum or titanium foils and the velocity histories of these surfaces were recorded.

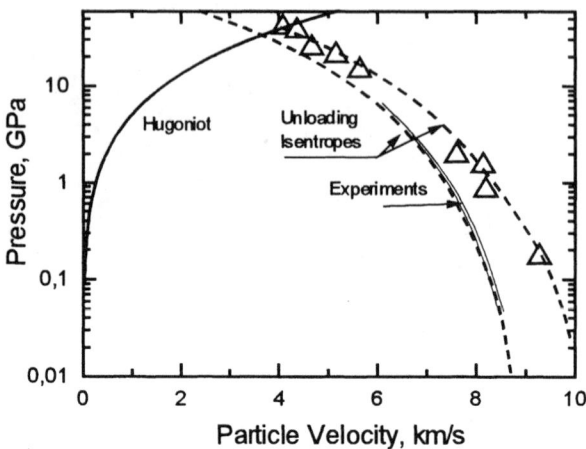

Figure 6.18. Unloading data for shock-compressed PMMA. The triangles present the data by Bushman et al., 1993, the dashed lines were calculated using a modified EOS, and the thin lines at low pressures are the results of our experiments. Calculations have been done starting at various initial shock pressures.

Figure 6.19 shows the foil velocity histories measured in experiments with PMMA at various peak pressures. According to computer simulations, a final constant velocity of the rear foil should be established after ~200 ns of the wave reverberation process. This behavior was observed in the experiments at lower shock pressures. However, the results of measurements at the highest peak pressure demonstrate a much longer time for the velocity increase and a somewhat higher final velocity. This finding is evidence of vaporization of the shock-heated PMMA at a peak pressure of 34.5 GPa.

Figure 6.20 illustrates the interpretation of the measured velocity histories. The upper parts of the velocity profiles have been approximated by a smooth function $u_f(t)$ with monotonically decreasing first and second derivatives. Knowing the foil mass, the vapor pressure was evaluated using Newton's law, $p(t) = \rho \delta \, du_f(t)/dt$, where ρ and δ are the foil density and thickness. The values of the pressure, p, and the particle velocity, $u = u_f$, taken at the same time, t, correspond to a point on the unloading isentrope of the shock-compressed PMMA.

The results of this data analysis are shown in Fig. 6.18 by thin lines in the low pressure part of the $p-u$ diagram. It can be seen that the experiments with aluminum and titanium witness foils of different mass are in a good agreement. The thin-foil data show the same trend as the data of Bushman et al., 1993, and in this sense may be considered as quite reasonable.

6. Polymorphic Transformations and Phase Transitions   211

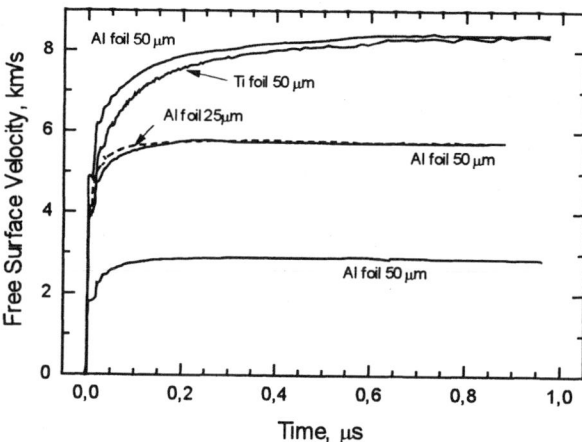

Figure 6.19. The velocity histories of 25 or 50-μm-thick aluminum and titanium foils covering PMMA sample plates. The peak shock pressures in the PMMA samples were 9 GPa, 23.8 GPa, and 34.5 GPa.

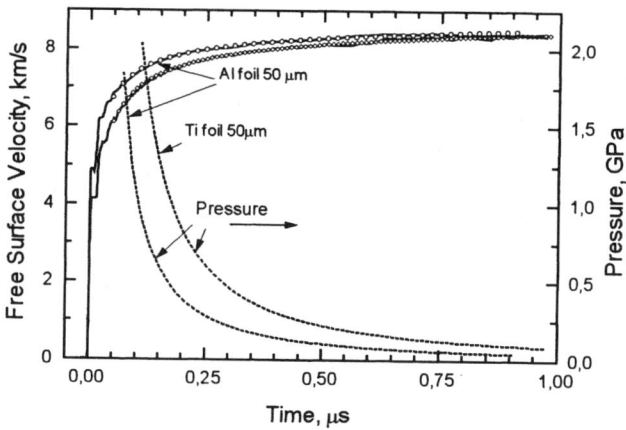

Figure 6.20. Interpretation of the measured velocity history. The points are the approximation and the dashed lines show the pressure histories calculated from the foil acceleration.

The caloric Mie–Grüneisen equation of state (EOS), generalized to account for gas-phase states, was used to describe the experimental data on acceleration of thin foils by shock-compressed PMMA. Figure 6.18 shows calculated unloading isentropes and Fig. 6.21 presents the results of computer simulations of the shock-wave experiments. The simulations satisfactorily reproduce the experimental data at high pressure but do not agree with the experimental data at lower shock amplitudes. The good agreement of the simulated and measured foil velocity histories at 34.5 GPa shock pressure may be considered to be evidence of the reliability of the thin foil acceleration method for investigation of EOS properties in the vaporization region. There is probably more than just vaporization occurring in PMMA after shock compression up to high pressures. Hauver, 1965, observed a rapid increase in the shock polarization and dielectric constant at shock pressures of 26.7 GPa and higher and the relaxation time of polarization suddenly decreased. These results may be considered as evidence of some physical or chemical transformation in the shock-compressed PMMA. In this case we should use a different EOS for the material in its initial and transformed states.

Figure 6.22 presents results of measurements of acceleration of aluminum witness foils by shock-compressed polyethylene in comparison with similar data for PMMA. Although the foil acceleration was recorded for the shot with polyethylene for at least 0.6 µs, the magnitude of the acceleration was much less than in the case of PMMA at the same impact conditions.

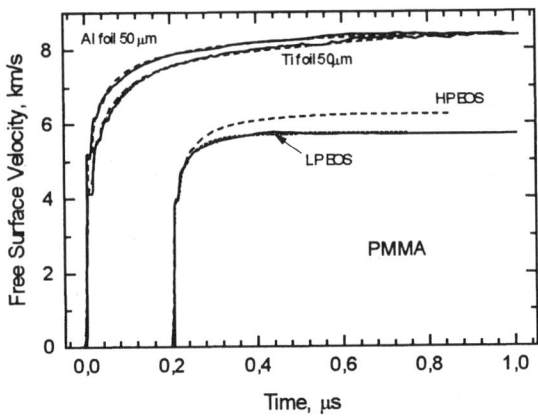

Figure 6.21. Comparison of measured (solid lines) and simulated (dashed and dot-dashed lines) velocity histories of aluminum and titanium witness foils at incident shock pressures of 34.5 GPa (top curves) and 23.8 GPa (lower curves). The high-pressure (HP EOS) and low-pressure (LP EOS) versions of the equation of state were used in simulations of the 23.8 GPa shot.

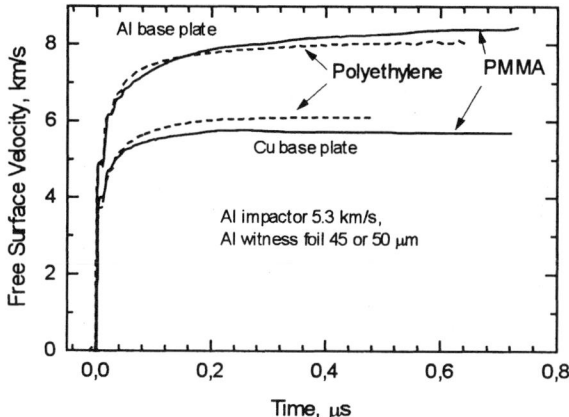

Figure 6.22. Acceleration histories of aluminum witness foils accelerated by shock-compressed polyethylene and PMMA. The foil thickness was 50 μm in shots with PMMA samples and 45 μm in the case of polyethylene. The peak shock pressure was varied by using base plates of different dynamic impedances at the same impact velocity of 5.3 km/s.

References

Al'tshuler, L.V. (1978). "Phase transitions in shock waves (Review)," *J. Appl. Mech. Tech. Phys.* **19(4)**, pp. 496–505. [trans. from *Zh. Prikl. Mekh. Tekh. Fiz.* **(4)**, pp. 93–103 (1978).]

Anan'in, A.V., A.N. Dremin, and G.I. Kanel (1973). "Structure of shock waves and rarefaction waves in iron," *Comb. Expl. Shock Waves* **9(3)**, pp. 381–385. [trans. from *Fiz. Goreniya Vzryva* **9(3)**, pp. 437–443 (1973).]

Anan'in, A.V., A.N. Dremin, and G.I. Kanel (1981). "Polymorphous transformations of iron in shock wave. *Comb. Expl. Shock Waves* **17(3)**, pp. 320–326. [trans. from *Fiz. Goreniya Vzryva* **17(3)**, pp. 93–102 (1981).]

Anan'in, A.V., A.N. Dremin, G.I. Kanel, and S.V. Pershin (1978). "Investigation of the structure of shock waves in boron nitride and graphite in the region of polymorphous transformation," *J. Appl. Mech. Tech. Phys.* **19(3)**, pp. 372–376 [trans. from *Zh. Prikl. Mekh. Tekh. Fiz.* **19(3)**, pp. 112–116 (1978)].

Anrdiot, P., P. Chapron, V. Lambert, and F. Olive (1984). "Influence of melting on shocked free surface behavior using Doppler laser interferometry and X-ray densitometry," in: *Shock Waves in Condensed Matter—1983* (eds. J.R. Asay, R.A. Graham, and G.K. Straub) North-Holland, Amsterdam, pp. 277–280.

Asay, J.R., L.P. Mix, and F.C. Perry (1976). "Ejection of material from shocked surface," *Appl. Phys. Lett.* **29**, pp. 284–287.

Balchan, A., and H.G. Drickamer (1961). "High-pressure electrical resistance cell and calibration points above 100 kbar," *Rev. Sci. Instr.* **32(3)**, pp. 308–314.

Bancroft, D., E.L. Peterson, and S. Minshall (1956). "Polymorphism of iron at high pressure," *J. Appl, Phys.* **27(3)**, pp. 291–298.

Barker, L.M. and R.E. Hollenbach (1974). "Shock wave study of phase transition in iron," *J. Appl. Phys.* **45(11)**, pp. 4872–4887.

Baumung, K., H.J. Bluhm, B. Goel, P. Hoppe, H.U. Karow, D. Rush, V.E. Fortov, G.I. Kanel, S.V. Razorenov, A.V. Utkin, and O.Yu. Vorobjev (1996). "Shock-Wave Physics Experiments with High-Power Proton Beams," *Laser and Particle Beams* **14(2)**, pp. 181–210.

Baumung, K., G.I. Kanel, S.V. Razorenov, D. Rush, J. Singer, and A.V. Utkin (1997). "Shock-melting pressures from non-planar impacts," *Int. J. Impact Eng.* **20(1–5)**, pp. 101–110.

Belyakov, L.V., V.P. Valitski, N.A. Zlatin, and S.M. Mochalov (1967). "On melting of lead in shock wave," *Sov. Phys.–Dokl.* **11(9)**, pp. 808–810. [trans. from *Dokl. Akad. Nauk SSSR* **170**, pp. 540–543 (1966).]

Betholf, L.D., L.D.Buxton, B.J. Thorne, et al. (1975). "Damage in steel plates from hypervelocity impact. II. Numerical results and spall measurement," *J. Appl. Phys.* **46(9)**, p. 3776.

Bushman, A.V., M.V. Zhernokletov, I.V. Lomonosov, Yu.N. Sutulov, V.E. Fortov, and K.V. Khrishchenko (1993). "Investigations of Plexiglass and Teflon in waves of secondary shock compression and isentropic unloading. The equation of state of polymers at high energy densities," *Phys.–Dokl.* **38(4)**, pp. 165–167 [trans. from *Dokl. Akad. Nauk SSSR* **329(5)**, pp. 581–584 (1993)].

Chapron, P., P. Elias, and B. Laurent (1988). "Experimental determination of the pressure inducing melting in release for shock-loaded metallic samples," in: *Shock Waves in Condensed Matter—1987* (eds. S.C. Schmidt and N.C. Holmes) North-Holland, Amsterdam, pp. 171–173.

Duvall, G.E., and R.A. Graham (1977). "Phase transitions under shock wave loading," *Rev. Mod. Phys.* **49(3)**, pp. 523–581.

Erskine, D.J. and W.J. Nellis (1992). "Shock-induced martensitic transformation of highly oriented graphite to diamond," *J. Appl. Phys.* **71(10)**, p. 4882.

Giles, P.M., M.H. Longen, and A.R. Marder (1971). "High-pressure martensitic transformation in iron," *J. Appl. Phys.* **42(11)**, pp. 4290–4296.

Gluzman, V.D., G.I. Kanel, V.F. Loskutov, V.E. Fortov, and I.E. Khorev (1985). "Resistance to deformation and fracture of 35Kh3NM steel under conditions of shock loading," *Strength of Materials* **17(3)**, pp. 1093–1098. [trans from *Problemy Prochnosti*, **17(8)**, pp. 52–57 (1985).]

Gogulya, M.F. (1989). "Shock structure and parameters under dynamic loading of natural graphite in polymorphic transformation domain," *Comb. Expl. Shock Waves* **25(1)**, pp. 87–95 [trans from *Fiz. Goreniya Vzryva* **25(1)**, pp. 95–104 (1989)].

Gray III, G.T. (1990). "Shock recovery experiments: an assessment," in: *Shock Compression of Condensed Matter—1989* (eds. S.C.Schmidt, J.N. Johnson, and L.W. Davison) North-Holland, Amsterdam, pp. 407–414.

Hauver, E. (1965). "Shock-induced polarization in plastics. II. Experimental study of plexiglas and polystyrene," *J. Appl. Phys.* **36(7)**, pp. 2113–2118.

Ivanov, A.G., and Novikov, S.A. (1961). "About the rarefaction shock waves in steel," *Sov. Phys.–JETP* **13**, pp. 1321–1323 [trans. from *Zh. Eksp. Teor. Fiz.* **40(6)**, pp. 1880–1881 (1961)].

Jamieson, J.C. (1963), *Science* **140**, p. 72.

Jamieson, J.C., and A.W. Lawson (1962). "X-ray diffraction studies in the 100-kbar pressure range," *J. Appl. Phys.* **33(3)**, pp. 776–780.

Johnson, P.C., B.A. Stein, and R.S. Davis (1962). "Temperature dependence of shock-induced phase transformation in iron," *J. Appl. Phys.* **33(2)**, pp. 557–564.

Kanel, G.I., K. Baumung, D. Rush, J. Singer, S.V. Razorenov, and A.V. Utkin (1998). "Melting of shock-compressed metals in release," in: *Shock Compression of Condensed Matter—1997* (eds S.C. Schmidt, D.D. Dandekar, and J.W. Forbes) American Institute of Physics, New York, pp. 155–158.

Kanel, G.I., V.E. Fortov, K.V. Khishchenko, A.V. Utkin, S.V. Razorenov, I.V. Lomonosov, T. Mehlhorn, J.R. Asay, and L.C. Chhabildas (2000). "Thin foil acceleration method for measuring the unloading isentropes of shock-compressed matter," in: *Shock Compression of Condensed Matter—1999* (eds. M.D. Furnish, L.C. Chhabildas, and R.S. Hixson) American Institute of Physics, New York, pp. 1179–1182.

Kanel, G.I., S.V. Razorenov, E.B. Zaretsky, B. Herrman, and L. Meyer (2003). "Thermal "softening" and "hardening" of titanium and titanium alloy at high strain rates of a shock-wave deforming," *Phys. Solid State*, **45(4)** pp. 656-661.

Kaufman, L. (1963). in: *Solids under Pressure* (eds. W. Paul and D.M. Warschauer) McGraw-Hill, New York.

Kiselev, A.N., and A.A. Falkov (1982). "Phase Transformation in Titanium in Shock Waves," *Fiz. Goreniya Vzryva* **18**, p. 105 (1982). (in Russian)

Kormer, S.B. (1968). "Optical investigations of properties of shock-compressed condensed dielectrics," *Sov. Phys.–Usp.* **11(4)** pp. 229–254 (1968) [trans. from *Usp. Fiz. Nauk* **94(4)**, pp. 641–687 (1968)].

Krüger, L., G. I. Kanel, S.V. Razorenov, L. Meyer, and G.S. Bezrouchko (2002). "Yield and strength properties of the Ti-6-22-22S alloy over a wide strain rate and temperature range," in: *Shock Compression of Condensed Matter—2001* (eds. M.D. Furnish, N.N. Thadhani, and Y. Horie) American Institute of Physics, New York, pp. 1327–1330.

Kutsar, A.R., V.N. German, and G.I. Nasova (1973). "($\alpha \rightarrow \omega$)-transformation in titanium and zirconium in shock waves," *Sov. Phys.–Dokl.* **131(3)**, pp. 317–320 [trans. from *Dokl. Akad. Nauk SSSR* **213(1)**, pp. 81–84 (1973)].

Kutsar, A.R., M.N. Pavlovsky, and V.V. Komissarov (1982). "The Observation of Two-Wave Configuration of Shock Wave in Titanium." *JETP Lett.* **35(3)**, pp. 108–112 [trans. from *Pis'ma Zh. Eksp. Teor. Fiz.* **35(3)**, pp. 91–94 (1982)].

Mao, H.K., W.A. Bassett, and T. Takahashi (1967). "Effect of pressure on crystal structure and lattice parameters of iron up to 300 kbar," *J. Appl. Phys.* **38(1)**, pp. 272–278.

McQueen, R.G., S.P. Marsh, J.W. Taylor, J.N. Fritz, and W.J. Carter (1970). "The equation of state of solids from shock wave studies," in: *High-Velocity Impact Phenomena* (ed. R. Kinslow) Academic Press, New York, pp. 293–417 (see also Appendices, pp. 518–568).

Nahme, H., and M. Hiltl (1995). "Dynamic properties and microstructural behavior of shock-loaded Armco iron at high temperatures," in: *Metallurgical and Materials Applications of Shock-Wave and High-Strain-Rate Phenomena* (eds. L.E. Murr, K.P. Staudhammer and M.A. Meyers) Elsevier, New York, pp. 731–738.

Olinger, B., and J.C. Jamieson (1973). *High Temp.–High Press.* **5**, p. 123.

Razorenov, S.V., A.A. Bogach, and G.I. Kanel (1997). "Influence of heat treatment and polymorphous transformation on the dynamic rupture resistance of 40X steel," *Phys. Met. Metall.* **83(1)**, pp. 100–103 [trans. from *Fiz. Metall. Metalloved.* **93(1)**, pp. 147–152 (1997)].

Razorenov, S.V., G.I. Kanel, E. Kramshonkov, and K. Baumung (2002). "Shock compression and spalling of cobalt at normal and elevated temperatures," *Comb. Expl. Shock Waves* **38(5)**, pp. 598–601 [trans from *Fiz. Goreniya Vzryva* **38(5)**, pp. 119–123 (2002)].

Razorenov, S.V., G.I. Kanel, and A.A. Ovtchinnikov (1981). "Recording of shock waves by manganinin gauges and the pressure of graphite–diamond transformation at elevated temperature," in: *Detonation. Proceedings of the Second All-Union Workshop on Detonation*, Inst. of Chem. Physics, Chernogolovka, pp. 70–74. (in Russian).

Razorenov, S.V., A.V. Utkin, G.I. Kanel, V.E. Fortov, A.S. Yarunichev, K. Baumung, and H.U. Karow (1995). "Response of high-purity titanium to high-pressure impulsive loading," *High Press. Res.* **13(6)**, pp. 367–376.

Remiot, C., J.M. Mexmain, and L. Bonnet (1996). "Precise method to determine points on isentropic release curve on copper," in: *Shock Compression of Condensed Matter—1995* (eds: S.C. Schmidt and W.C. Tao) American Institute of Physics, New York, pp. 955–958.

Roundy, D., and M.L. Cohen (2001). "Ideal strength of diamond, Si, and Ge," *Phys. Rev. B* **64**, pp. 212103–212103.

Sikka, S.K., Y.K. Vohra, and R. Chidambaram (1982). "Omega phase in materials," *Prog. Mater. Sci.* **27**, p. 245–310.

Sugak, S.G., G.I. Kanel, V.E. Fortov, A.L. Ni, and V.G. Stelmakh (1983). "Numerical modeling of the action of an explosion on an iron slab," *Comb. Expl. Shock Waves* **19(2)**, pp. 239–246 [trans. from *Fiz. Goreniya Vzryva* **19(2)**, pp. 121–128 (1983)].

Vohra, Y.K. (1978). *J. Nucl. Mat.* **75**, p. 288.

Werdiger, M., B. Arad, Z. Henis, Y. Horowitz, E. Moshe, S. Maman, A. Ludmirsky, and S. Elizer (1996). "Asymptotic Measurements of Free Surface Instabilities in Laser-Induced Shock Waves," *Laser Particle Beams* **14(2)**, pp. 133–147.

Zaretsky, E., and M. Kaluzhny (1996). "Fracture threshold and shock induced strengthening of stainless steel," in: *Shock Compression of Condensed Matter—1995* (eds. S.C. Schmidt and W.C. Tao) American Institute of Physics, New York, pp. 627–630.

Zilbershtein, V.A., N.P. Christotina, A.A. Zharov, N.S. Grishina, and E.I. Estrin (1975). *Phys. Met. Metall.* **39(2)**, pp. 208–217. [trans. from *Fiz. Metall. Metalloved.* **39**, pp. 445–457 (1975).]

CHAPTER 7

Equations of State and Macrokinetics of Decomposition of Solid Explosives in Shock and Detonation Waves

7.1. Introduction

A natural, general goal of studies of detonation is to provide means for predicting explosion phenomena. In modern understanding, complete predictability is achieved when the phenomena considered can be simulated using a computer. In the calculations, properties of explosive materials are characterized by the equation of state of the unreacted high explosive (HE) and of the detonation products, and by macrokinetic equations that describe the rate of chemical reaction of the HE to form the detonation products. Other practical goals of the investigations are to reveal microscopic mechanisms of initiation and evolution of the HE energy release, identifying methods of controlling the sensitivity of HE to intense impulsive loading, and to provide a theoretical basis for solving the safety problems of explosive handing.

In this chapter we discuss ways of obtaining data on equations of state and macrokinetics of HE decomposition. Since the kinetic information is always limited, a comprehensive understanding of decomposition mechanisms is needed in order to generalize the experimental data by developing a robust constitutive relationship. With this goal in mind, we present a brief overview of observations and macroscopic-level mechanisms of ignition of exothermic processes in various kinds of explosive materials. For those readers who are unfamiliar with detonation phenomena, we present a short exposition of the modern theory of detonation. There are a number of specialized books that discuss in detail various aspects of detonation phenomena and the effects of explosions. Among relatively recent publications we mention the monographs by Johansson and Persson, 1970, Baum et al., 1975, Mader, 1979, and Chéret, 1993.

7.2. General Structure of Plane Steady Detonation Waves

The first theoretical treatment of detonation was given by Chapman, 1899, and Jouguet, 1905. The Chapman–Jouguet (CJ) theory represents the detonation wave as a steady shock wave at which the explosive undergoes an instantaneous chemical transformation into detonation products. An important result of the Chapman–Jouguet (CJ) theory is that a minimum velocity of steady detonation exists for any given explosive material. This result is known as the Chapman–Jouguet condition. For the minimum detonation velocity, the Rayleigh line is tangent to the detonation adiabat in pressure–volume coordinates. The detonation adiabat is actually the Hugoniot for a shock transition from HE in its initial unreacted state to states of the fully reacted detonation products. The point of tangency is known as the Chapman–Jouguet point and the corresponding state is the CJ state.

The modern theory of detonation is known in Russian literature as the Zel'dovich theory and in international literature as the Zel'dovich–von Neumann–Döring (ZND) theory. The refined theory accounts for the chemical reaction process and proves that a self-supporting steady detonation can propagate only with the CJ minimum velocity of detonation. In accordance with Zel'dovich and Kompaneets, 1955, the detonation transformation of explosives occurs as an effect of a shock wave that initiates the exothermic chemical reaction. The energy released as a result of the reaction sustains the process as a whole. The self-sustaining detonation has a steady velocity that is independent of the initiating impulse and is determined solely by the energy released in the chemical reaction (the heat of explosion) and the equation of state of the detonation products. As shown in Fig. 7.1, the complete detonation waveform includes a shock discontinuity, a chemical reaction zone (often called the "chemical peak" or "ZND spike") of constant width, and a non-stationary region of expanding detonation products. Since the leading portion of the detonation wave, including the shock discontinuity and the reaction zone, is steady, it is described by the Rankine–Hugoniot conservation equations and the equation of state:

$$\mathscr{E}_{CJ} = \frac{p_{CJ}}{2}(V_0 - V) + \mathscr{E}_0 + Q_V, \tag{7.1}$$

$$V_{CJ} = V_0 \frac{D - u_{CJ}}{D}, \tag{7.2}$$

$$p_{CJ} = \rho_0 D u_{CJ}, \tag{7.3}$$

$$p_{CJ} = p_{CJ}(V_{CJ}, \mathscr{E}_{CJ}), \tag{7.4}$$

where the subscript CJ indicates the energy, $\mathscr{E}$, the pressure, p, the specific volume, V, and the particle velocity, u, at the end of reaction zone. The subscript 0 indicates the same parameters in the initial state at zero pressure, D is the detonation wave speed, and Q_V is the heat of detonation.

At the end of the reaction zone, the condition that the particle velocity relative to the detonation wave front must be equal to the local sound velocity (the Chapman–Jouguet condition) is satisfied. On the pressure–volume diagram shown in Fig. 7.1a, the CJ condition is expressed in that the Rayleigh line defined by the equation

$$p = \rho_0^2 D^2 (V_0 - V), \tag{7.5}$$

and designated OGA in the figure, is tangent to both the detonation adiabat and the detonation-product isentrope (the curve S in the figure). The motion of the material behind the Chapman–Jouguet plane is supersonic relative to the detonation front, so no disturbance from this region can penetrate the reaction zone to influence the detonation-front parameters. The flow within the reaction zone is subsonic, thereby ensuring arrival of the released energy at the detonation wave front. The structure of the reaction zone is defined by the kinetics of transformation of the explosive into the detonation products. Thus, measurements of the profiles of steady detonation waves yield information on the reaction rate and provide reference points for determining the shock compressibility of the initial unreacted explosive and the equation of state of the detonation products.

The existence of a chemical reaction zone in detonation waves in high explosives has been confirmed by numerous experiments using different experimental techniques including various versions of the magnetoelectric method of measuring particle velocity histories, manganin pressure gauges, and laser-Doppler interferometer velocimeters, as well as methods based on recording the attenuation of the shock wave excited in a reference barrier by the detonation of the explosive under investigation. Figures 7.2–7.5 show typical examples of such measurements.

Figure 7.2 presents the pressure history recorded with a manganin gauge at the interface between a detonating TNT sample of 1.06 g/cm³ density and a Teflon barrier. Manganin pressure gauges and magnetoelectric particle velocity

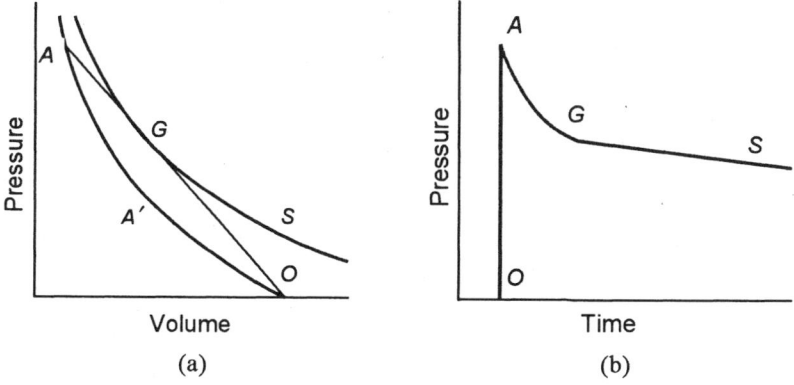

Figure 7.1. Steady detonation-wave structure. (a) The curve OA'A is the Hugoniot of the unreacted explosive and the curve GS is the detonation-product isentrope. The point G is the Chapman–Jouguet point. (b) The portion of the waveform OAG is the "chemical spike" or "reaction zone" of the detonation wave and GS is the region of expanding detonation products.

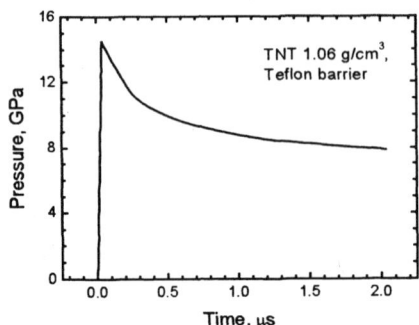

Figure 7.2. Pressure profile of a steady detonation in milled TNT of 1.06 g/cm^3 density measured with a manganin gauge at the interface with a Teflon barrier.

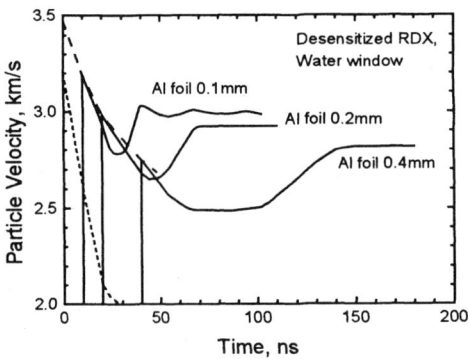

Figure 7.3. Results of measurements of the detonation wave structure in an RDX + Wax HE composition of 1.6 g/cm^3 density. Measurements with an ORVIS laser-Doppler velocimeter using a water window and aluminum foil reflectors of various thickness. The dashed line shows the extrapolation to zero foil thickness and the dotted line shows an estimate of the real detonation-wave structure inside the HE.

gauges provide very demonstrative data and are effectively used when the reaction time exceeds ~50 ns. Shorter time intervals can be resolved by use of laser interferometer techniques. Figure 7.3 shows velocity histories of the surface of foil plates placed at the boundary between a sample of desensitized RDX and a water window (Utkin and Kanel, 1986). In these experiments, the foil thickness was comparable to, or greater than, the reaction zone thickness in the detonation wave. The use of relatively thick foils somewhat complicates the recorded waveform: The second velocity rise in the $u_p(t)$ profiles results from wave reverberation in the foil between the water window and the explosive. On the other hand, measurements with thick foils are more certain and less sensitive to inhomogeneities in the HE structure. The extrapolation of peak values of velocity to zero thickness of the foil makes it possible to calculate the particle velocity in the explosive at the beginning of chemical reaction zone, thereby refining the Hugoniot of the unreacted HE in the region of the detonation pressure.

Results of measurements of the shock-wave velocity as a function of distance in a PMMA plate placed against the downstream face of a charge of desensitized HMX are shown in Fig. 7.4 (Al'tshuler et al., 1980). The data reflect the detonation wave structure and make it possible to recover the pressure history at the interface between the HE and a barrier using, for example, the method of characteristics. The measurements of the shock front velocity on the bases of 0.2 mm near the contact surface yield information on the state variables of the partially reacted explosive 5–10 ns after the start of the reaction.

Another similar indicator measurement technique is based on recording the thermal radiation from a shock-wave front in transparent barrier material (Voskoboinikov and Gogulya, 1984). Since the brightness of thermal radiation is proportional to approximately the fourth power of the shock pressure, the technique provides for high precision of measuring shock-wave attenuation in the barrier as a function of time. The total resolution of the technique when measuring the detonation wave structure is estimated to be 5–10 ns. Some examples of application of the brightness indicator technique are shown in Fig. 7.5.

For orientation, Table 7.1 presents typical values of the peak detonation pressure, p_{peak}, the Chapman–Jouguet pressure, p_{CJ}, and the reaction zone duration, Δt, for several important high explosives. It is clear from the table that, with practically equal parameters in the Chapman–Jouguet plane, the reaction time in the detonation wave depends on the explosive structure, as illustrated by the example of cast and pressed TNT. The reaction time increases as the detonation pressure decreases along with the density of the explosive charge.

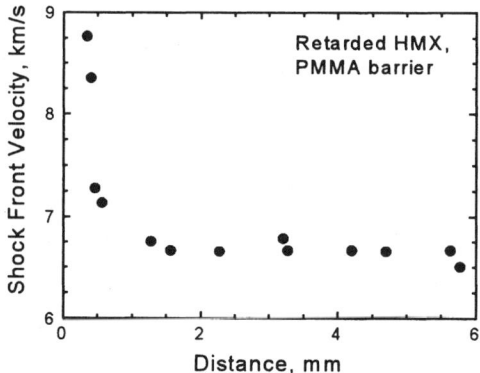

Figure 7.4. Attenuation of a shock wave in a PMMA barrier placed behind a detonating HMX + wax charge. Data from a paper by Al'tshuler et al., 1980.

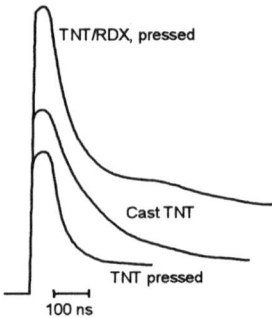

Figure 7.5. Brightness histories of radiation from a shock-wave front in a CCl_4 indicator placed behind a detonating explosive charge. Measurements made by Voskoboinikov and Gogulya, 1984.

Table 7.1. Detonation Parameters of Some Selected High Explosives.

High Explosive	ρ_0 g/cm^3	p_{peak} GPa	p_{CJ} GPa	Δt ns	Reference
Desensitized RDX	1.41	28.8	20.6	60±5	Utkin and Kanel, 1986
	1.60	38.4	26.9	25±5	
	1.67	43.0	29.2	15±3	
Pressed TNT	1.60	23.7	20.3–20.9	41–90	Voskoboinikov and Gogulya, 1984
	1.64	27.5	22.0	80	Sheffield et al., 1984
Cast TNT	1.60	24.2	20.3–20.9	140–218	Voskoboinikov and Gogulya, 1984
Desensitized HMX	1.77	–	31.2	40	Al'tshuler et al., 1980
Desensitized PETN	1.65	–	24.5	35	Al'tshuler et al., 1980
PBX-9502	1.88	37.5	28.5	280	Sheffield et al., 1984
Pressed CP	1.75	29.0	20.0	80	Sheffield et al., 1984

Recently Ashaev et al., 1988, and later Utkin et al., 2000b, observed detonation wave profiles without a spike in several high-density explosives. Utkin et al. also traced evolution of the detonation wave structure with increasing HE density: As the density is increased, the magnitude of the chemical reaction zone spike decreases to almost zero. Then, with a further increase of density, the pressure decay in the chemical reaction zone is replaced by its growth. Similar evolution of wave profiles also occurs with increasing pressure as found by Erskine et al., 1990, in experiments with overdriven piston-supported detonations. They interpreted this finding as evidence of crossing of the product and reactant Hugoniots at elevated pressures. This assumption does not seem natural but is not improbable because the unreacted explosive and the detonation products are materials with different compressibilities. However, it is not yet clear what controls the detonation parameters in this case.

In order to understand the way in which a self-supporting steady detonation is realized, let us consider the influence of gradients of the kinematic parameters behind a shock discontinuity on amplification or decay of the discontinuity. Following Nunziato, 1973, and Kanel, 1977, variations of the pressure, p, and particle velocity, u_p, at a shock discontinuity are related by the equations

$$\left.\frac{dp}{dt}\right|_D = \frac{\partial p}{\partial t} + U_s \frac{\partial p}{\partial h} = \frac{\partial p}{\partial t} - \rho_0 U_s \frac{\partial u_p}{\partial t}$$

$$\left.\frac{du_p}{dt}\right|_D = \frac{\partial u_p}{\partial t} + U_s \frac{\partial p}{\partial h} = \frac{\partial u_p}{\partial t} + \rho_0 U_s \frac{\partial V}{\partial t},$$
(7.6)

where $d(\cdots)/dt|_D$ is the time derivative along the shock front trajectory, $X_D(t)$, $\partial(\cdots)/\partial t$ is the partial time derivative at fixed Lagrangian (material) position coordinate

$$h = \int_0^x \frac{\rho}{\rho_0} dx,$$

and U_s is the shock-front velocity. Equations (7.6) have been written to account for conservation of mass and momentum in the flow behind the shock front. By eliminating $\partial u/\partial t$ and accounting for conservation of momentum through the shock wave, $p = \rho_0 U_s u_p$, Eqs. (7.6) are transformed to the single equation

$$2 \left.\frac{dp}{dt}\right|_D = \frac{\partial p}{\partial t} - \rho_0^2 U_s^2 \frac{\partial V}{\partial t}$$
(7.7)

that describes the relationship between variations of pressure and specific volume behind the shock front. It follows from Eq. (7.7) that the shock wave is steady when the state of the material behind it is changing along the Rayleigh line, $p = \rho_0^2 U_s^2 (V_0 - V)$. The shock wave accelerates if the pressure is above the Rayleigh line, or decays if the pressure is below the Rayleigh line in the course of a process occurring immediately behind the shock-wave front.

The specific volume of the reacting mixture in Eq. (7.7) is a sum of volumes of the components:

$$V(p, \alpha) = \alpha V_{DP}(p) + (1 - \alpha) V_{HE}(p)$$

$$\frac{\partial V}{\partial \alpha} = V_{DP} - V_{HE},$$
(7.8)

where V_{HE} is the specific volume of the unreacted HE, V_{DP} is the specific volume of the detonation products, and, α is the mass fraction of the detonation products in the mixture. It is assumed for simplicity of estimation that changes of volume of the components of the mixture occur isentropically. This automatically means that the lifetime of possible intermediate products is negligible, heat transfer between the initial and final products is insignificant, and the products themselves are immiscible. Equations (7.7) and (7.8) are transformed to a relationship between the rate of local pressure change, the chemical reaction rate, and amplification of shock wave:

$$2\frac{dp}{dt}\bigg|_D = \frac{\partial p}{\partial t} + \rho_0^2 U_s^2 \left(\frac{\partial V}{\partial p}\frac{\partial p}{\partial t} + \frac{\partial V}{\partial \alpha}\frac{\partial \alpha}{\partial t}\right)$$

or

$$2\frac{dp}{dt}\bigg|_D = \frac{\partial p}{\partial t}\left(1 - \frac{U_s^2}{a^2}\right) + \rho_0^2 U_s^2 (V_{DP} - V_{HE})\frac{\partial \alpha}{\partial t}, \qquad (7.9)$$

where $a^2 = -V_0^2 (\partial p/\partial V)_\alpha$ is the squared sound speed in the Lagrangian coordinates. Then the condition that the wave be steady becomes

$$\frac{\partial p}{\partial t} = -\rho_0^2 D^2 (V_{DP} - V_{HE})\frac{a^2}{a^2 - U_s^2}\frac{\partial \alpha}{\partial t}$$

$$\frac{\partial \alpha}{\partial t} \geq 0. \qquad (7.10)$$

Thus, the shock pressure increases when the reaction rate exceeds some value that is determined, in particular, by the rate of change of the local pressure:

$$\frac{\partial \alpha}{\partial t} > -\frac{\partial p}{\partial t}\frac{a^2 - U_s^2}{\rho_0^2 a^2 U_s^2 (V_{DP} - V_{HE})}. \qquad (7.11)$$

In other words, the magnitude of the shock pressure may grow as the wave propagates even when the local pressure behind the shock-wave front is decreasing. The higher the reaction rate, the higher the admissible rate of the pressure decrease.

For steady waves it follows from Eq. (7.10) that, for the subsonic regimes of flow behind the shock front ($U_s < a$), the pressure is decreasing when the reaction is accompanied by an increase of the volume of the mixture, or it is increasing when the volume decreases, or the local pressure is maintained constant when the volume remains unchanged during the reaction process. As the sonic regime is approached ($a \to U_s$) a finite pressure gradient may persist if the reaction rate tends to zero or if the specific volumes of the components are equal to each other. On the other hand, if the reaction rate drops to zero while the flow is subsonic, stability of this part of the wave may persist only if the pressure is maintained constant. In the case when a non-zero reaction rate is maintained as the subsonic regime is approached, for $V_{DP} > V_{HE}$ stability is possible if $\partial p / \partial t \rightarrow -\infty$, whereas for $V_{DP} < V_{HE}$, stability requires $\partial p / \partial t \rightarrow \infty$.

Let us try to image the mechanism by which the Chapman–Jouguet condition of steady detonation is achieved when $V_{DP} > V_{HE}$. Let an intermediate state of the reacting mixture appear above the Rayleigh line at some moment in time. In accordance with Eq. (7.7), the excess pressure finally reaches the shock discontinuity and causes its propagation velocity to increase. As a result, the flow becomes subsonic during the whole time of reaction and is maintained subsonic at the moment the reaction is completed. In the subsonic flow, an unloading wave can propagate from the region of expanding products into the reaction zone and forward to the shock discontinuity. In this way the wave propagation velocity recovers to the detonation velocity.

If an intermediate state of the reacting mixture appeared below the Rayleigh line, this should cause a decrease of the wave propagation velocity. As a result, the final state of the detonation products lies above the Rayleigh line. In the course of the reaction, the state of the material passes through the point at which the local Lagrangian sound speed in the reacting mixture is equal to the propagation speed of the shock wave. As discussed earlier, it is necessary in this case that $\partial p / \partial t \rightarrow -\infty$ for steady flow. If, however, the pressure is decreasing at a more probable finite rate, then the mechanism of wave amplification by excessive pressure becomes effective in accordance with Eq. (7.7). In this way the detonation process comes back to the Chapman–Jouguet condition.

Let us now suppose that the Hugoniot of the unreacted HE intersects the adiabat of the detonation products. This situation is illustrated schematically in Fig. 7.6, where all states of the detonation products are presented for simplicity by a single $p(V)$ curve, S_p, and states of the initial unreacted HE are represented by the curve H_i. At small HE density which corresponds to the specific volume V_{00}, the normal detonation wave with a chemical peak is realized since the intersection of Hugoniots occurs at a higher pressure than that in the CJ plane for this HE density. As the HE density increases, the relative spike amplitude (the difference between the state at the point P_{00} at the shock discontinuity and the CJ state at the point CJ_{00}) should decrease, as is observed in experiments by Utkin et al., 2000b. At some threshold value V_{0c} of the initial specific volume of the HE, the point of Chapman–Jouguet tangency reaches the point of intersection of the Hugoniots. At this point, the chemical spike of the detonation wave is transformed to a plateau: As seen from Eq. (7.10), the process of detonation transformation at $V_{DP} = V_{HE}$ proceeds to completion at a constant pressure and points P_{0c} and CJ_{0c} coincide.

226 Shock-Wave Phenomena and the Properties of Condensed Matter

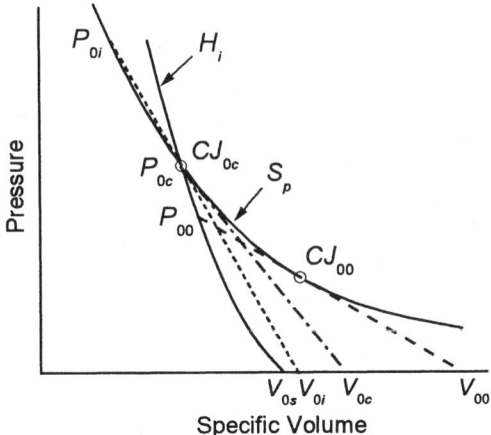

Figure 7.6. Diagrams of states in detonation waves in the case of intersecting Hugoniots of the unreacted HE and its detonation products. For simplicity all states of the detonation products are presented by a single $p(V)$ curve, S_p, and states of the unreacted HE are presented by a curve H_i.

Let us consider in more detail what occurs with further increase of the HE density. As follows from Eq. (7.9), the initiating shock wave should increase in amplitude until the pressure behind its front reaches the value P_{0c} that corresponds to the point of intersection of the Hugoniots. As the state approaches this point, the difference of specific volumes of unreacted HE and detonation products, $V_{DP} - V_{HE}$, decreases to zero. As a result, the energy release process ceases to influence the evolution of the shock wave. After transition via the intersection point, the specific volume difference $V_{DP} - V_{HE}$ changes sign and the reaction process begins to affect the evolution of the shock wave in the opposite way.

Since all isentropes of mixtures of unreacted HE and detonation products intersect the Rayleigh line, in this regime the flow is subsonic everywhere in the reaction zone. As a result, an unloading wave can propagate from the region of expanding products into the reaction zone and decrease the pressure there. This circumstance is valid for all shock pressures above P_{0c}. Thus, there is a mechanism which causes the pressure in the entire reaction zone to decrease or increase to the P_{0c} value. In other words, the intersection of the Hugoniot of the unreacted HE and the Hugoniot of the detonation products determines the detonation parameters and thus replaces the Chapman–Jouguet condition for HE of high density $\rho_0 > \rho_{0c}$ ($V_0 < V_{0c}$). It may be expected that the pressure in the chemical reaction zone of such dense explosives is constant or is slightly varying as a result of secondary reactions or heat exchange between the reacted and unreacted particles. It may also be expected that the intersection should weaken the dependence of the detonation speed upon the HE density.

7.3. Detonation Failure Diameter

A critical diameter exists such that explosive charges of smaller diameter prove incapable of self-sustaining detonation. This phenomenon was explained by Khariton, 1947, based on the fact that some time is needed to complete the energy release in a detonation wave. According to the Khariton criterion, the critical detonation conditions are defined by equality of the reaction time of the compressed HE and the time required for lateral expansion. Therefore, the value of the failure diameter is directly related to the kinetics of the reaction producing the detonation products. Measurements of the detonation failure diameter may be used to make a semi-quantitative estimate of the reaction kinetics or as tests for verification of computational models of the decomposition process.

Considerable non-uniaxiality of the detonation process in small-diameter charges hinders the development of a consistent theory of this effect. The relationship between the decomposition kinetics and the detonation-failure diameter is complicated by the fact that the detonation velocity and pressure decrease as the HE charge diameter approaches the value at which failure occurs. Investigations of detonation failure in homogeneous (liquid and gaseous) and inhomogeneous (solid) explosives have demonstrated a qualitative difference in the mechanisms of the phenomenon for these two types of explosive (Dremin et al., 1970a, Engelke and Sheffield, 1998). Different mechanisms are exhibited in the differing behavior of detonation velocity with decreasing charge diameter. Owing to the strong dependence of the reaction time on the temperature of the shock-compressed explosive, detonation failure occurs suddenly in homogeneous explosives when the charge diameter approaches the threshold value even though its velocity changes very little. Inhomogeneous explosives show a smooth decrease of the detonation speed as the charge diameter decreases to its failure-threshold value.

Much as for the duration of the chemical reaction zone of a plane detonation wave, the detonation-failure diameter depends on the physical structure and porosity of the explosive charge. This can be seen in Fig. 7.7 based on the data by Bobolev, 1947, and Fig. 7.8 based on the data by Apin and Stesik, 1955.

Figure 7.9 shows data by Kurbangakina, 1969, that demonstrate the effect of mechanical inclusions on the failure diameter of liquid HE. One can see that introducing even a small fraction of fine rigid particles into a homogeneous explosive (liquid or gelled nitromethane) leads to a sharp decrease of the failure diameter. Aeration of a liquid explosive yields the same results.

228   Shock-Wave Phenomena and the Properties of Condensed Matter

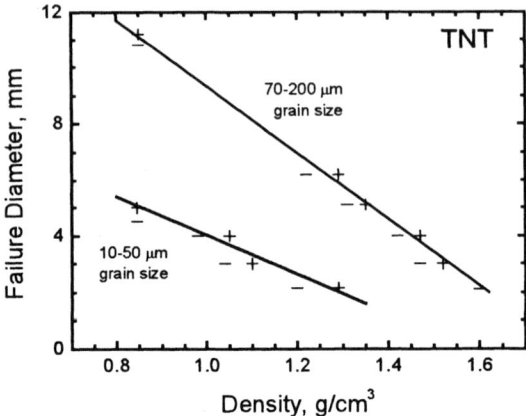

Figure 7.7. Detonation failure diameter of TNT as a function of the density and the grain size. Data by Bobolev, 1947.

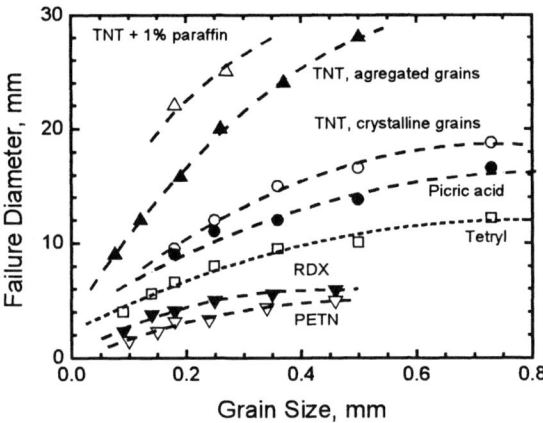

Figure 7.8. Detonation failure diameter as a function of grain size for different solid explosives of 1.0 g/cm^3 density. Data by Apin and Stesik, 1955.

Belyayev and Kurbangalina, 1960, studied the effect of temperature on the failure diameter of solid and liquid explosives. The results of measurements presented in Fig. 7.10 indicate that the detonation parameters of porous solid explosives are much less sensitive to the temperature than those of homogeneous liquids. Note further that, even at temperatures very close to the temperature of thermal ignition at normal pressure, the detonation failure diameter of liquid TNT is several times larger than that of solid porous TNT of the same density at room temperature.

7. Solid Explosives in Shock and Detonation Waves   229

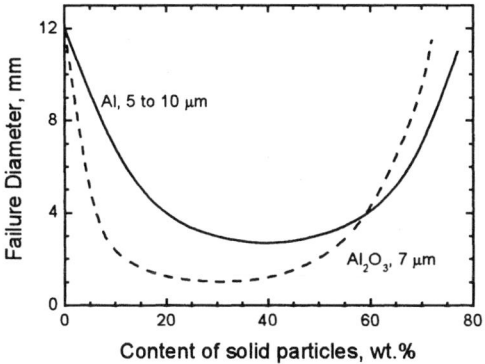

Figure 7.9. Detonation failure diameter of a mixture of nitromethane with solid particles as a function of the weight fraction of the inclusions. The particle material and grain size are noted on the curves.

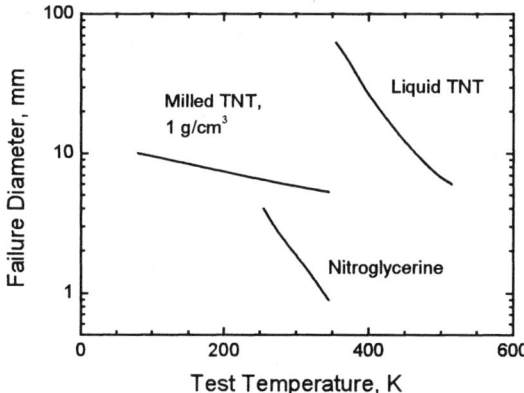

Figure 7.10. Detonation failure diameter of as a function of temperature for solid and liquid explosives.

7.4. Initiation of Detonation by Shock Waves

The shock-wave initiation process is controlled by the rate of release of the chemical energy stored in the HE. Depending on the HE composition and its physical structure, the energy release may occur in different ways. The rate of homogeneous chemical decomposition of a single-component HE at absolute temperature T is described by the equation (Emanuel and Knorre, 1974)

$$\frac{d\alpha}{dt} = -\alpha^n \nu e^{-\mathscr{E}_a/RT}, \tag{7.12}$$

where α is the reactant concentration, n is the reaction order, $\mathscr{E}_a$ is the reaction activation energy, and v is a constant. If Q is the heat of detonation, the rate of heat release, $\dot{q}$, is given by the equation

$$\dot{q} = Q\alpha^n v e^{-\mathscr{E}_a/RT}. \tag{7.13}$$

The energy release rate increases rapidly with temperature. Under adiabatic conditions, the released heat further increases the temperature of the reactant, thus accelerating the reaction. Therefore, the reaction rate increases exponentially, leading to a thermal explosion. The delay time, or adiabatic induction period, of the thermal explosion is expressed by the equation

$$\tau_{ad} = \frac{c\rho R T_0^2}{QEv} e^{\mathscr{E}_a/RT}, \tag{7.14}$$

where c and ρ stand for the specific heat capacity and density of the explosive, and T_0 is the initial temperature (Todes, 1939).

Shock-wave initiation of detonation following the mechanism of adiabatic thermal explosion is observed in homogeneous explosives. The shock wave heats the explosive and triggers a relatively slow decomposition reaction. Since the process occurs under essentially adiabatic conditions, the heat liberated by the exothermic reaction increases the HE temperature and, in this way, accelerates the reaction. As a result, a thermal explosion develops near the charge boundary where the explosive is in a shock-compressed state for the longest period of time. As a result, after the induction time elapses, the pressure near the charge boundary drastically increases, a detonation wave is formed in the compressed material, catches up with the front of the initiating shock wave, and continues propagating through the unreacted explosive (Fig. 7.11).

Measurements of the delay time for initiation of detonation of shock-compressed homogeneous explosives are used to determine the constants of thermal kinetics of decomposition of materials under such conditions. For this purpose, the dependence of the delay time on the intensity of the initiating shock wave is determined, the temperature of the shock-compressed material is calculated using its equation of state, and the constants sought are found from Eq. (7.14) using the theory of adiabatic thermal explosion. The results of such an analysis for nitromethane and PETN are given in Table 7.2. Also listed in this table are the appropriate constants measured under isothermal conditions at atmospheric pressure.

The data of Table 7.2 show that a significant decrease of the activation energy occurs under conditions of shock-wave compression as compared with isothermal conditions. Note that the accuracy of determining the constants of thermal kinetics of decomposition of explosives in shock waves is very sensitive

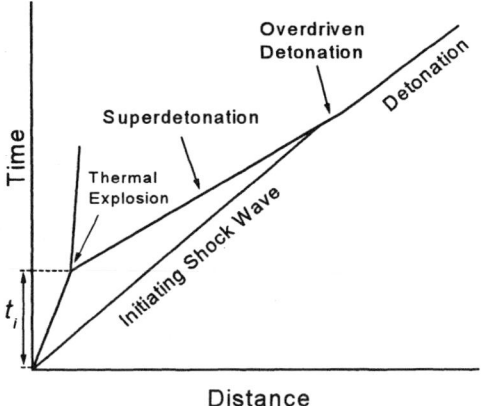

Figure 7.11. Shock-wave initiation of a detonation wave in a homogeneous explosive. The time t_i is the induction time of adiabatic thermal explosion.

to errors of the equation of state. In view of this, Shvedov and Koldunov, 1980, who performed such measurements for tetranitromethane, question the effect of shock compression on the kinetics of thermal decomposition of explosives. On the other hand, one should take into account that, in different temperature ranges, the reaction kinetics differ because the overall reaction proceeds by several successive stages. This is reflected in the difference of effective activation energies for different temperature ranges. The studies of thermal explosion under high hydrostatic pressures by Sophy, 1966, and Shipitsyn, 1980, demonstrated a decrease of the ignition temperature and activation energy for RDX, TNT, and HMX with an increase of pressure to 5–10 GPa. For instance, at a pressure of 10 GPa, the activation energy in RDX is found to be 10 ± 2 kcal/mole, whereas it is 38 kcal/mole at atmospheric pressure. On the other hand, Lee et al., 1970, found no indication of a change of the activation energy for a number of explosives at pressures of up to 5 GPa.

Most practically important explosives are inhomogeneous, having various defects such as pores, cracks, inclusions, grain boundaries, and crystallographic defects. The physical inhomogeneity leads to non-uniform distribution of energy under extremely rapid compression of the explosive by a shock wave. Shock compression of inhomogeneous explosives generates so-called "hot spots" or nuclei in which a decomposition reaction originates (Campbell et al., 1961). The less-heated mass of explosive then burns in the reaction waves propagating from the nuclei.

The hypothesis of initiation of explosion in nuclei was proposed and substantiated by Bowden and Yoffe, 1952, during their investigations of the initiation of explosion of condensed explosives by mechanical impact. They regarded

Table 7.2. Kinetic constants of some high explosives from shock-wave measurements

Explosive	$\mathscr{E}_a$ kcal/mole	ν 1/s	p GPa	T K	Reference
		Shock compression*			
PETN	6.4	5×10^6	8.5–20	—	Dick, 1986
Nitromethane	23	3×10^9	6.0–10	—	Hardesty, 1976
Nitromethane	18–38	5.5×10^{10} -5.0×10^{13}	7.0–13	650–1400	Chaiken, 1978
		Isothermal conditions			
PETN	47	6×10^{19}	—	—	Dick, 1986
Nitromethane	53.6	4×10^{14}	—	—	Campbell et al., 1961
Nitromethane	59	4×10^{15}	—	—	Dick, 1986

*p and T indicate the range of pressures and temperatures in which the measurements were made.

the adiabatic compression of gas inclusions, friction between explosive particles and impurities, and viscous heating of the explosive during high-velocity deformation, as the main mechanisms of formation of reaction-ignition nuclei at impact. The concept of hot spots was introduced into the description of the process of shock-wave initiation of inhomogeneous explosives because the initiating shock waves did not produce an increase of the average temperature that was sufficient to cause the fast decomposition observed. Of course, the heterogeneous character of the process does not exclude contributions from homogeneous heating to the bulk decomposition of the explosive. However, homogeneous heating does not appear to be decisive for initiation of most explosives.

Without giving a detailed description of the mechanism by which decomposition nuclei are formed, the general scheme of shock-wave initiation of reaction in inhomogeneous explosives appears to be as follows. Owing to the physical inhomogeneity of the explosive material, part of the shock-compression energy is localized in individual hot spots, which leads to an exothermic decomposition reaction of the explosive excited in these spots. The heat released is partly dissipated in the material surrounding the nucleus and partly remains in the nucleus and causes its temperature to rise, thereby accelerating the decomposition process. The development of the process is governed by the ratio of the rate of heat release by the exothermic reaction and that of heat conduction into the surrounding material. If the heat release is predominant, the nucleus temperature and the reaction rate increase progressively and thermal ignition occurs. Otherwise, the reaction in the nucleus ceases. Since the thermal losses are proportional to the surface area of the nucleus whereas the amount of energy released is proportional to the temperature and volume of the nucleus, it is clear that flashes are possible in nuclei whose size exceeds some limiting value for the given temperature.

According to Merzhanov, et al., 1963, the critical relation between the radius r of a spherical nucleus, its temperature, T_1, and the temperature, T_0, of the explosive mass surrounding the nucleus takes the form

$$\frac{r_c^2 Q \mathscr{E}_a \nu}{\lambda R T_c^2} e^{-\mathscr{E}_a/RT_c} \approx 12.1 \cdot \left[\ln \frac{\mathscr{E}_a}{R T_c^2}(T_c - T_0) \right]^{0.6}, \qquad (7.15)$$

where λ is the thermal conductivity. For all $T_1 > T_c$ and $r > r_c$, an explosion occurs after the induction period elapses. The induction period of nucleus self-ignition is close to the adiabatic induction period expressed by Eq. (7.14).

From general considerations it follows that the size of the hot spots formed by the shock wave must be approximately proportional to the size of inhomogeneities occurring in the unreacted explosive. As a rule, one must consider the emergence of a spectrum of nuclei. For a nucleus to be capable of igniting the surrounding explosive, its dimensions must exceed the limiting size in accordance with the thermal explosion theory. The lower the nucleus temperature, the larger must be its size to provide for stable ignition of the explosive. It is intuitively clear, and shown by estimates made for different mechanisms of hot spot formation, that the temperature of the sites increases with the shock-wave amplitude. Therefore, a larger number of nuclei capable of igniting the surrounding explosive are formed by a stronger shock wave. Further on, the chemical reaction develops in the form of layer-by-layer combustion at a rate that depends on the pressure according to the equation $u_b = b_0 + b_1 p^{b_2}$, where b_2 is of the order of 0.8–1.2 (Andreev, 1966). As the sites ignite, the combustion surface grows and the volume rate of the chemical reaction increases. After the burning sites merge, the regressive stage of combustion begins. During this stage, the volume rate of the process is approximately proportional to the mass fraction of unreacted explosive to the power 2/3.

The presence of a spectrum of nuclei has a qualitative effect on the dynamics of shock-wave initiation of detonation of inhomogeneous explosives. Unlike the liquid and gaseous explosives, the inhomogeneous explosives do not exhibit a pronounced period of induction. The initiating shock wave grows to a detonation relatively smoothly, without shocks and supercompressions. Figure 7.12 shows the evolution of pressure profiles of waves in cast TNT when detonation is initiated by shock compression pulses of different amplitudes and durations (Kanel, 1978). The evolution of the process of explosive transformation results in an increase of pressure behind the shock and continuous amplification of the shock wave up to the detonation mode. In experiments with a short initiating pulse one observes amplification of the shock wave until it starts interacting with the rarefaction wave overtaking from the rear, followed by some attenuation due to the effect of this rarefaction wave. After that, as the energy release develops, the pressure peak transforms into a weak maximum and shock-wave amplification is resumed. When the pulse duration is reduced, the initiating shock wave may attenuate to the point that the initiation process ceases.

Important results have been obtained from investigations of the effect of the loading rate on the process of initiation of solid explosives by Kipp et al., 1981, Setchell, 1981, and Bordzilovskii and Karakhanov, 1985. Figure 7.13 gives a comparison of the wave profiles measured by Setchell, 1981, when initiating the detonation of PBX-9404 by shock and quasi-isentropic (ramped) compression waves. These experiments have shown that the reaction initiation is suppressed by broadening of the compression wave. No clear-cut features of energy release are observed until a compression shock is formed. Following the formation of a shock wave, its amplification is observed although the transition to detonation is considerably delayed as compared to the case of initiation by a shock wave.

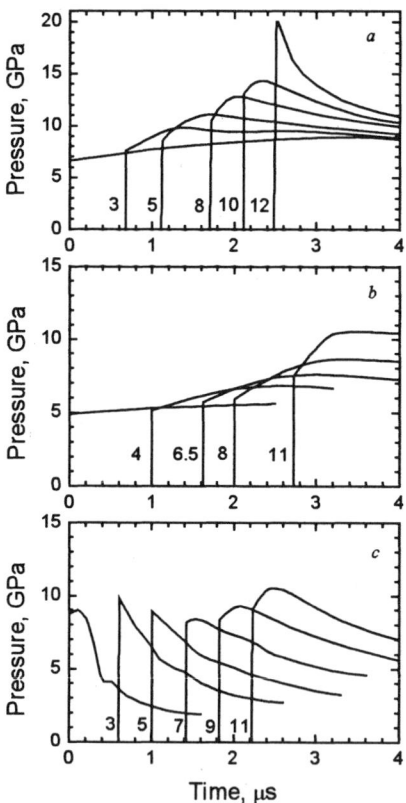

Figure 7.12. Evolution of wave profiles during the shock-to-detonation transition in cast TNT. Graphs a and b show the case of initiation by a shock pulse of initially square profile; graph c corresponds to impact by a thin flyer plate. The numbers denote distances, in mm, into the HE samples from the impacted face.

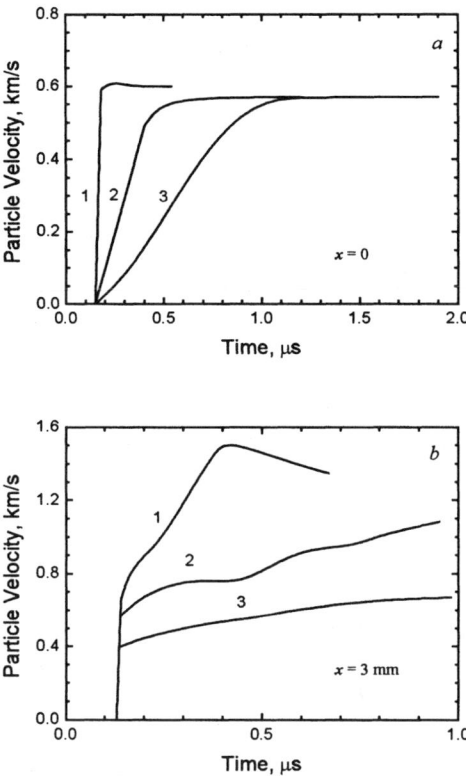

Figure 7.13. Effect of the loading rate on the process of initiation of PBX-9404 (data by Setchell, 1981). (a) Initiating pulses. (b) Wave profiles at 3 mm propagation distance into the HE sample.

Investigation of initiation of detonation by a series of shock waves of rising amplitude has shown that preliminary stimulation by a weak compression wave leads to a reduction of the rate of decomposition of the explosive in the initiating shock wave. This effect is attributed to an increase of homogeneity of the material upon its pre-compaction by a weak compression wave, deactivation of some of the nucleation sites, and reduction of the temperature and size of the hot spots formed in the initiating shock wave (Campbell et al., 1961).

These considerations indicate that the shock-wave amplitude is one of the main factors determining the mean rate of energy release in compressed explosives. Study of the regularities of shock-wave initiation of detonation enables one to characterize the sensitivity of explosives to such effects and to obtain information about the macroscopic kinetics of decomposition of the explosive at initiation sites and about the influence of various structural and thermodynamic factors on the process.

7.5. Sensitivity of Solid Explosives to Shock-Wave Effects

One can single out two main directions taken by experimental investigations of shock-wave initiation of detonation. In the first of these directions, the critical parameters of the shock waves causing detonation of explosives are determined or the correlation is established between the shock compression pressure and the length of the pre-detonation portion of the wave trajectory. The second direction is based upon measurement of pressure or particle velocity wave profiles, reflecting the transition of the initiating shock wave to detonation. The former investigations are oriented mainly toward direct practical utilization of the results whereas the latter investigations yield more detailed information on the process and may, in the final analysis, have a much broader sphere of application.

The length of the pre-detonation portion of the wave trajectory is determined from measurements of the evolution of the initiating shock wave. At present, no method is available for continuous measurement of kinetic parameters of a shock discontinuity. In order to determine the velocity of an unsteady shock wave, its trajectory is measured in distance–time, x–t, coordinates. The differentiation of the x–t diagram gives the law of variation of the shock-wave velocity as it approaches the detonation mode. The measurements are performed by means of variable-resistance transducers or using the wedge-test technique.

Figure 7.14 shows the results of measurements by Dremin et al., 1971, of the evolution of shock waves in TNT having a density of 1 g/cm^3 with grain sizes of 0.1 mm and 0.37–1.0 mm. One can see that the length of the pre-detonation portion of the wave trajectory decreases with the grain size. From a comparison of the curves obtained at different initial pressures one can see the absence of a single regularity of shock-wave amplification. The results of the experiments further indicate that, at equal pressures of the initiating shock wave, the length of the pre-detonation portion of the wave trajectory increases with the explosive charge density. The corresponding data in Figure 7.15 are taken from Campbell et al., 1961.

Note that the effect of the grain size on the initiation of detonation is nonmonotonic. Investigations by Afanasenkov and Danilenko, 1975, of the initiation of detonation of degassed mixtures of RDX with liquid fillers have revealed that the minimum pressure required for initiation of detonation rises from 2.5–4.0 GPa at a particle size of 135 μm to 7.5–9.5 GPa at 800 μm. However, the reduction of particles of explosives to micron size is accompanied by an appreciable decrease of sensitivity. This is clear from Fig. 7.16 illustrating the results of measurements of the length of the pre-detonation portion of the wave trajectory for two cast monomodal compositions of RDX and 30% polyurethane (Moulard et al., 1985). The compositions having a porosity of <0.1%

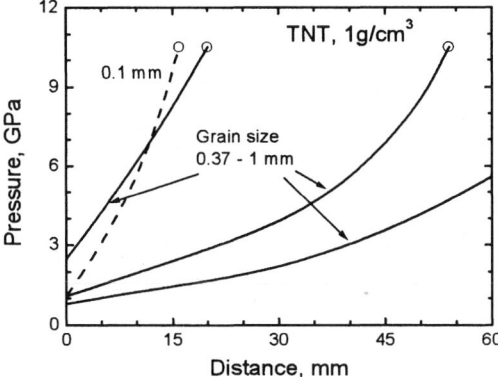

Figure 7.14. Growth of shock wave amplitude in porous TNT of different grain sizes (Dremin et al., 1971).

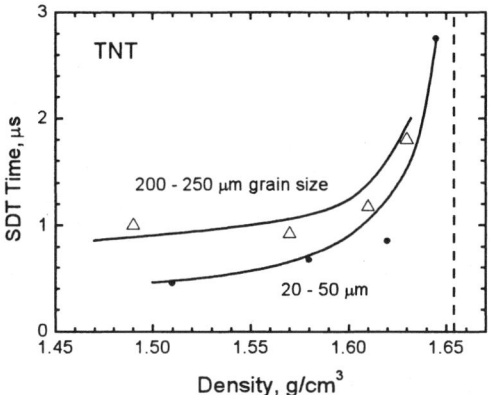

Figure 7.15. The time of the shock-to-detonation transition (SDT) in pressed TNT as a function of density and grain size (Campbell et al., 1961). The initial shock pressure is 6.0 ± 0.3 GPa.

differed by the size of particles of the explosives: 6 μm and 134 μm. The graphs demonstrate a marked increase of the propagation distance required to accomplish the shock-to-detonation transition for a fine-grain composition at shock compression pressures below 10 GPa, with the arrival at the detonation mode for this composition occurring very suddenly, as in the case of homogeneous explosives. On the other hand, the critical diameter of detonation for a composition with micron-size particles proved to be smaller. Therefore, at detonation pressures, a fine-grain composition has a shorter reaction time. The effect of intersection of dependence curves is illustrated in Fig. 7.17 presenting the results by Seitz, 1984, of measurements of the length of the shock-to-detonation transition for TATB having a porosity of about 7% and a particle size of mainly 10–25 μm and of <5 μm. The graphs also show the effect of the temperature at which the tests were conducted.

Currently, a large volume of experimental data has been accumulated on the reaction of solid explosives to shock-wave action. However, owing to differences in the actual conditions of the tests, data by different authors cannot always be compared and uniformly described. The main obstacle in the way of

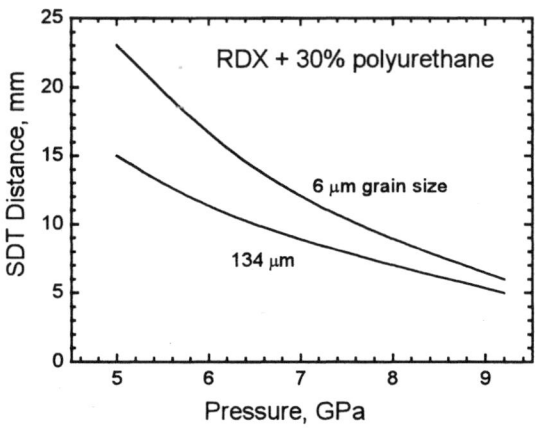

Figure 7.16. Relationships between the distance required to complete the shock-to-detonation transition and the pressure in the initiating shock wave for two full-density compositions of RDX + 30% polyurethane of different grain sizes. Data from Moulard et al., 1985.

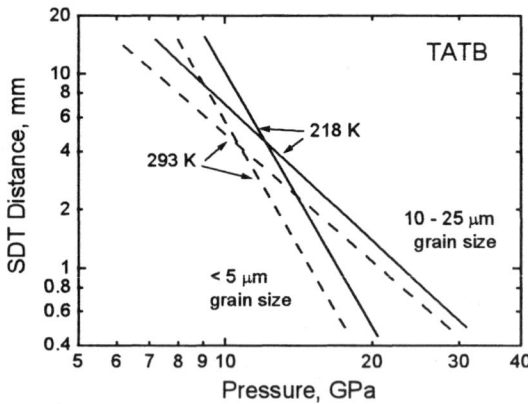

Figure 7.17. The distance of shock-to-detonation transition as a function of the pressure of the initiating shock wave, the grain size, and the initial temperature for TATB. Data from Seitz, 1984.

direct quantitative comparison of the results of measurements is, along with possible non-reproducibility of the characteristics of explosives, a difference in the conditions of shock-wave loading. In particular, the threshold conditions for initiation of detonation are determined by both the shock-wave intensity and the space–time characteristics of the entire shock-loading impulse. Various standardized schemes of shock-wave loading are proposed in order to obtain macroscopic-level characteristics of the sensitivity of explosives. One such approach is construction of the relationship between the pressure of shock-wave initiation of detonation and the duration of the initiating pulse (Walker and Wasley, 1969, Glushak et al., 1984). Pulses of uniaxial shock compression of different amplitudes and durations are generated in the samples by impact of metal flyer plates whose thickness and velocity can be varied. The initiating shock-wave pressure is calculated from the impact velocity using Hugoniots of the materials and the period of reverberation of a wave inside the impactor is taken as the duration of the loading pulse. Strictly speaking, since the dynamic impedances of metal impactors are much higher than those of explosives, full decompression of the sample is attained only after several wave reverberations in the impactor. The accepted definition of the duration of the initiating stimulus is not quite correct and is one of the reasons for the dependence of the measurement results on the impactor material.

Different kinds of variation of the shock-wave amplitude as a function of the pressure pulse duration are qualitatively shown in Fig. 7.18. Short loading pulses decay without causing a noticeable chemical reaction. For longer pulses, the energy release compensates for the pressure drop in the unloading wave and causes amplification of the shock wave until its transformation to steady detonation. A typical relationship between the threshold pressure of an initiating shock wave, p_i, and the compression pulse duration, Δt, is shown in Fig. 7.19, using cast TNT/RDX 50/50 as an example (Glushak et al., 1981). The $p_i(\Delta t)$ relationship divides the p–t plane into two regions: States in which detonation is initiated lie above the curve whereas detonation is not initiated for material in

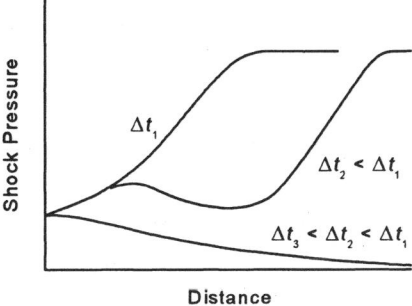

Figure 7.18. Schematics of evolution of shock waves in HE for shock pulses of different durations.

the states lying below the curve. The position of the $p_i(\Delta t)$ curve is conditional to a degree because, near the limit of initiation of detonation, its excitation is of random nature and the $p_i(\Delta t)$ relationship, generally speaking, is not unique. This can be seen in Figure 7.20, which illustrates the results of measurements for pressed TNT obtained using aluminum and steel impactors.

Several criteria have been proposed for characterizing the sensitivity of explosives to shock-wave initiation. Measurements of the relationship between the amplitude and duration of pulses capable of initiating a detonation resulted in the formulation of the concept of critical energy (Howe et al., 1976, Walker and Wasley, 1969). This criterion implies the existence of some minimum energy of the initiating pulse, per unit cross-sectional area, that must be transferred to the

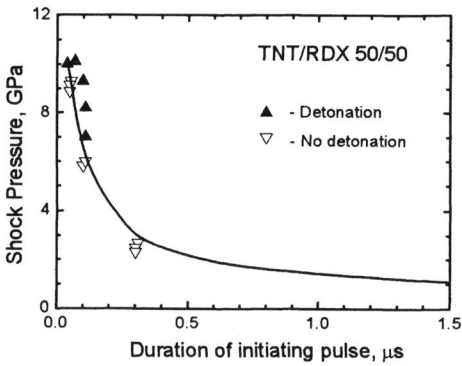

Figure 7.19. Critical parameters of initiation of TNT/RDX 50/50 composition as a function of the shock pulse duration. Data by Glushak et al., 1981.

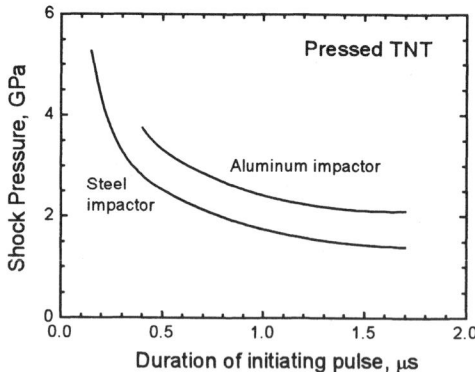

Figure 7.20. Critical pressure for initiation of detonation in pressed TNT by impact of flyer plates of different dynamic impedances as a function of the characteristic duration of the shock pulse. Data by Glushak et al., 1981.

explosive to excite detonation. The energy criterion may be presented in any of the equivalent analytical forms

$$p_i u_i \Delta t = \text{const.},$$

$$\frac{p_i^2 \Delta t}{\rho_0 D_i} = \text{const.},$$

or

$$p_i^2 \Delta t = \text{const.},$$

where p_i is the pressure in the initiating shock wave, D_i is the velocity of this wave, u_i is the particle velocity in the explosive at the given pressure, and ρ_0 is the initial density of the explosive. This criterion satisfactorily describes the experimental data in the region of low values of Δt and underestimates the threshold pressure of the initiating pulse at high values of Δt. The value of critical energy is affected by the density of explosive charge, the particle size, and other factors such as the presence of inert or chemically active additives. It also depends on the initial temperature of the explosive. Although the experimental values of the threshold pressure of initiation of detonation depend on the duration of loading, there is probably some physical limit to this quantity. The lower bound of its possible values is apparently the yield point of the given explosive. In this case, following Stresau and Kennedy, 1976, the energy criterion takes any of the equivalent forms

$$(p_i u_i - p_{\min} u_{\min}) \Delta t = \text{const.},$$

$$(p_i^2 - p_{\min}^2) \Delta t = \text{const.},$$

or

$$\left(\frac{(p_i - \sigma_{\text{HEL}})^2}{\rho_0 D_i} - p_{\min} u_{\min} \right) \Delta t = \text{const.},$$

where σ_{HEL} is the pressure required to initiate pore collapse. Fine-grain explosives are usually more difficult to initiate than coarse-grain explosives but, after the initiation of reaction, the transition to detonation is faster. Therefore, fine-grain explosives usually have higher values of $p_{\min}$ but may have lower values of critical energy.

7.6. Detonation Properties of HE Single Crystals

It follows from the previous discussion that homogeneous liquid explosives and inhomogeneous solid explosives exhibit different behavior. In homogeneous high explosives the energy release occurs by thermal explosion behind the shock front, whereas the energetic processes in solid explosives are governed by hot-spot mechanisms. The natural questions arise: how might relatively defect free single crystals behave and where is there a transition from the hot-spot mechanism to more fundamental mechanisms related to intrinsic material properties?

Whereas detonation phenomena in inhomogeneous HEs have been widely studied, only limited experimental information is available about the behavior of ultimate-density pure HE samples and single crystals. Most of the known experimental data on the behavior of HE single crystals in response to shock-wave loading have been obtained for PETN. PETN exhibits a strong orientation dependence of its initiation and detonation under shock compression. Dick (1997) and Dick et al. (1990), (1991) determined that single crystals of PETN with the <110> and <001> orientations are most sensitive. In contrast, samples of <100> or <101> orientation detonate only when initiated near the CJ conditions (Yoo et al., 2000). This was ascribed to the ability of these crystals to slip on the primary {110} slip planes. The anomalously high sensitivity of the <110> orientation correlates with light emission: Substantial light emission was recorded from a <110> crystal shocked at 4 GPa but none from adjacent crystals of <100> and <101> orientations. A very unusual double-valued behavior was observed for PETN single crystals shocked in the <110> direction: The run-distance to detonation at 4.26 GPa was the same as at 9.6 GPa but shorter than at 8.5 GPa.

Dick et al. (1994) and Dick (1997) presented time-resolved particle velocity histories for [110], [001], and [100] orientations of PETN single crystals at shock input stresses within the range ~1–7 GPa. Elastic precursor waves have been recorded for all orientations. The elastic precursor waves indicate post-yield softening and chaotic oscillations that are obviously attributable to a heterogeneous failure or faulting mechanism activated during compressive loading. At an input shock strength of 1.2 GPa the longitudinal elastic stress at the HEL is about 0.37 GPa for a [100] shock, 0.58 GPa for a [101] shock, 1.0 GPa for a [110] shock, and 1.2 for a [001] shock. The corresponding critical resolved shear stresses on the slip systems subjected to the maximum resolved shear stress are 0.11, 0.16, 0.21, and 0.22 GPa, respectively. Thus, a three-fold difference in elastic shock strength and a two-fold difference in critical resolved shear stress were observed depending on the crystal orientation. Measurements over a wide range of peak stress (see also Halleck and Wackerle, 1976) indicated that the elastic precursor amplitudes also depend on the amplitude of the input shock stress. At a 4.16 GPa input stress the elastic wave amplitude in a [110] crystal, for example, reached 2.73 GPa or more than twice that for a 1.2 GPa input stress. The recorded plastic compression waves were rather dispersive. According to the data of Halleck and Wackerle, the final states of unreacted shock-compressed PETN single crystals are in good agreement with the hydrostat, indicating a very small retained shear stress—much less than at the HEL. Such behavior is typical for brittle materials.

The two-wave structure correlates with the non-monotonic shock initiation sensitivity observed in the [110]-orientated PETN crystals: The reaction rate increases with increasing peak stress until a two-wave structure develops. At a shock stress of approximately 5 GPa, the velocity of the second, plastic wave becomes equal to the elastic precursor velocity, so the two-wave structure coalesces into a single shock. Starting from this input stress, the shock sensitivity decreases to a minimum at approximately 6 GPa followed by increasing sensitivity with increasing input stress.

Sandusky et al. (1993) studied deformation of RDX and ammonium perchlorate (AP) crystals by diamond-pyramid microindentation hardness testing and the following response of these crystals to shock-wave loading. RDX is brittle with limited slip system activity and, therefore, cracks readily to accommodate the indenter. AP is more ductile, exhibiting a larger number of slip systems, but still cracks readily because of dislocation interactions. High-speed photography of shocked crystals showed luminous crack propagation and reaction in both materials. Recovered AP crystals exhibited much more plastic deformation than RDX crystals and were often still transparent in the region opposite the face at which the shock was introduced. Recovered RDX crystals, at even the lowest shock pressure of 0.86 GPa, were uniformly white from a high density of fine cracks. Nevertheless, the reaction threshold of RDX was much higher: ~6.2 GPa versus 1.7–2.4 GPa for AP, depending on crystal orientation to the shock wave. The hardness impressions sensitized chemical decomposition of AP far from the impressions: The greatest decomposition was near the tips of cracks and along slip planes that emanated several millimeters from hardness impressions. In contrast, in the case of RDX, the extremely localized increase in defects from hardness impressions, and even a large internal crack in one crystal, did not significantly enhance chemical reaction. Much more extensive pre-shock damage is necessary for the initiation of dislocation assisted reaction at lower shock pressures in RDX.

Forbes et al. (1990) observed light emission from small RDX and HMX crystals shocked to about 10 GPa. Orthorhombic RDX crystals were shocked normal to the {210} plane. The crystals emitted very bright light, beginning at the time the shock wave first entered the crystal. The light appears to be associated with the material deformation. Bright parallel lines that could be due to twinning were seen in an HMX crystal. However, light emission was also observed in small shocked quartz crystals under the same loading, which indicates that the emission is not necessarily caused by chemical reactions in the shocked explosive crystal.

Sharma et al. (1988) investigated the nature of reaction sites in samples of TATB and TNT subjected to sub-ignition underwater shock or mechanical impact. Scanning electron microscopy revealed micron-size morphological defects in the crystal structure. In and around these microscopic defects, thin deposits of reaction products, more sensitive than the explosive, were found. The reaction products corresponded to the principal products observed in thermal decomposition.

Thus, the material microstructure influences the shock reactivity of energetic crystals. However, the reason and mechanism of this influence are not understood in detail. Dislocations are commonly considered as the reaction nucleation sites in shock-compressed crystals. According to Guang Gao et al. (1993), distortion in the electronic structure caused by defects in the lattice may be a reason for the formation of hot spots. Dick hypothesized that a "steric hindrance" to shear in a molecular crystal is the reason for high resistance to plastic deformation and simultaneously the cause of the first phase of the process of decomposition of HE molecules, so the initiation sensitivity is a consequence of the steric hindrance to inelastic shear flow. The order in measured critical resolved shear stresses is the same as that obtained in the analysis for increasing steric hindrance to shear, indicating that the relative strength of different orientations of this molecular crystal under shock conditions is governed by steric hindrance to shear. A semi-empirical quantum molecular dynamics simulation performed by Dick and Ritche, 1994 has confirmed that the shear motion is indeed responsible for the initiation along the sensitive directions.

One should say that the dislocation mechanism does not explain the non-monotonic sensitivity of PETN crystals. On the other hand, the experimental data indicate that there are probably other mechanisms of hot-spot nucleation. It seems that not only plastic deformation but also brittle fracture occurs in the initial phase of shock compression behind the elastic precursor front. During the following compression, the cracked single crystals may behave as a porous HE with the same mechanisms of hot spot formation at cracks as for other heterogeneous energetic materials.

To compare the reaction nucleation mechanisms acting at different structural levels, it is important to have quantitative macrokinetic data for explosive materials at various initial states. Meanwhile, whereas numerous such data exist for heterogeneous high explosives, kinetic information for single crystals is of a semi-quantitative character.

7.7. Evolution of Shock Waves During Initiation of Detonation of Solid Explosives

The most complete information on the process of shock-wave initiation of detonation comes from recordings of the evolution of pressure or particle velocity wave profiles inside the sample under investigation. The first measurements of this kind were performed by Dremin and Koldunov, 1967, for cast and pressed TNT using a magnetoelectric method. The non-one-dimensionality of the process and insufficient accuracy of measurements prevented bringing the results of these experiments to quantitative expression of regularities of energy release in shock-compressed explosive. Such data were obtained later by Kanel and Dremin, 1977, but even the first measurements revealed a number of important peculiarities of initiation of heterogeneous explosives. In particular, the absence of any pronounced induction period of the decomposition reaction was revealed. The energy released during the decomposition process leads to a pressure increase that advances to the shock front and amplifies it continuously until the detonation mode is established. As observed by Dremin et al., 1971, for low-density explosives, shock-wave amplification may occur even in the absence of a region of rising pressure behind the shock.

It was shown above that, if the rate of decomposition of the explosive immediately behind the shock is finite, the shock wave in the explosive may increase in amplitude even if the pressure and particle velocity behind its front decrease. If the initial reaction rate is zero, the shock-wave amplitude increases only with the pressure and particle velocity behind its front. These two types of shock-wave evolution in explosives were observed in early work involving the recording of wave profiles resulting in shock-wave initiation of detonation.

Equation (7.9) provides a way to determine the initial reaction rate by means of measurements of acceleration of the shock wave and the pressure gradient behind its front. In this way, estimates of the initial rate of energy release behind the compression shock were made by Cowperthwaite and Rosenberg, 1976.

Unfortunately, all methods of recording wave profiles at internal cross-sections of samples suffer from some inertia of transducers and electrical noise. As a result, the wave profiles in the shock region are recorded with distortions. One must carefully analyze the experimental data used to determine the initial reaction rate so that the results obtained do not contradict common sense. Lobanov, 1985, has shown that experimental errors lead to both inaccuracy of the relationship obtained and to distortion of the wave shape with the emergence of false maxima.

The development of experimental techniques stimulated the organization of quantitative investigations of the regularities of decomposition of explosives throughout the entire duration of the process. At present, two approaches to the solution of the problem have been realized. According to one of the approaches, information on the formal kinetics of conversion of the unreacted explosive to the detonation products is derived directly from analysis of the evolution of a uniaxial shock compression pulse. The second approach consists in using experimental data to check mathematical models of the process and to determine their parameters.

The direct method of obtaining kinetic data proceeds as follows. Pressure or particle velocity wave profiles reflecting the evolution of the initiating shock wave are measured at several fixed Lagrangian cross-sections of the sample. These data are used to reconstruct the ways of the change of state of isolated particles of the explosive in $p-V$ coordinates. Every point of such a "phase trajectory" corresponds to a definite moment in time. One can determine the concentration of the detonation products at every point of the phase trajectory if it is assumed that the specific volumes and energies of the unreacted explosive and the detonation products are additive. Therefore, the law of growth of the decomposition depth with time is determined for every isolated particle. The methods of construction of phase trajectories of material particles from a series of measured pressure or particle velocity profiles have been discussed by many authors. In particular, Seaman, 1974, deals with the derivatives of pressure and particle velocity along a set of "guide lines," L, or selected trajectories on the $x-t$ plane. Because the experimental data are "tied" to fixed Lagrangian coordinates, the analysis is also made in these coordinates. Consideration of the derivatives of pressure and particle velocity along the track lines, together with the equations of conservation of mass and momentum, yields the equations

$$u(h, t) = u_s - V_0 \int_{t_0}^{t} \left(\frac{dp}{dh}\bigg|_L - \frac{1}{D} \frac{\partial p}{\partial t} \right) dt$$

$$V(h, t) = V_s - V_0 \int_{t_0}^{t} \left(\frac{du}{dh}\bigg|_L - \frac{1}{D} \frac{\partial u}{\partial t} \right) dt$$

$$\mathscr{E}(h, t) = \mathscr{E}_s - V_0 \int_{t_0}^{t} p \frac{dV}{dt} dt,$$

where the subscript s designates the states behind the shock-wave front. The directions of the guide lines, L, are selected for convenience and acceptable accuracy of determination of the derivatives. In particular, they may coincide with the directions of the shock-wave trajectory and minimum gradients of flow or with the trajectories of propagation of fixed levels of the state variables being measured.

The values of the decomposition depth $\alpha(t)$ are determined from the values of p, V, and $\mathscr{E}$ found using the equations of state for the explosive and the detonation products. The question of the accuracy of the results thus obtained is important. In order to estimate the contribution introduced in the calculations by possible experimental errors, Vantine et al., 1981, disturbed the wave profiles up to ±2%. A comparison between the results of analysis of a series of disturbed particle velocity profiles with the exact analytical solution was performed for a model material. The comparison has shown that the error of determining p, V, $\mathscr{E}$, and α under the most unfavorable conditions may reach 3, 1, 20, and 60%, respectively.

Trajectories of the change of state behind the shock in cast TNT, obtained by processing the series of experimental pressure profiles given in Fig. 7.12, are shown in Fig. 7.21. As the explosive decomposes, the states of particles deviate from the Hugoniot of the unreacted explosive and approach the isentrope of the detonation products. In order to illustrate the process rate, the trajectory of the change of state has time markers on it with an interval of 0.5 μs.

Figure 7.22 illustrates concentration curves describing the variation with time of the decomposition depth of selected particles after the passage of a shock-wave front. The $\alpha(t)$ relationships are constructed from the data of Figs. 7.12 and 7.21, assuming that the mixture components are additive. Although the error of determining the decomposition depth is significant, the set of $\alpha(t)$ relationships enables one to estimate the general regularities of the process. In particular, one can see that the decomposition rate is close to zero immediately behind the compression shock. It then increases with the decomposition depth and, when the variation of pressure is not too high, passes through a maximum in the region of $\alpha = \sim 0.2 - 0.3$. The shift of the maximum reaction rate toward the beginning of the process is easy to understand assuming that the reaction nuclei are concentrated on the surface of the explosive grains and their quantity is such that the distance between them is less than the grain size. As seen from the curves $c3$ and $c5$, a sharp drop of pressure leads to a decrease of the reaction rate.

Vorthman and Wackerle, 1984, studied the effect of the initiating pulse shape (approximately square or stepped wave profile) on the rate of reaction of the HMX-based explosive, PBX-9501. It has been found that, in addition to the current state variables, the reaction rate is defined by the amplitude of the first

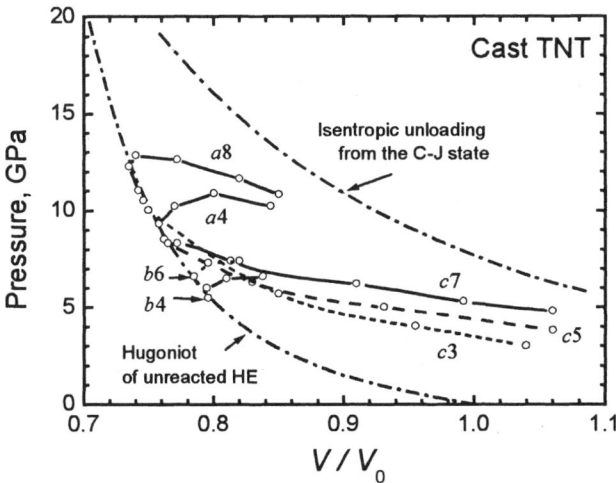

Figure 7.21. Trajectories of the change of state of TNT behind a shock front that were obtained by processing the experimental data of Fig. 7.12. The indices point to the series of experiments in accordance with Fig. 7.12 and the Lagrangian coordinate of the layer described by the given curve.

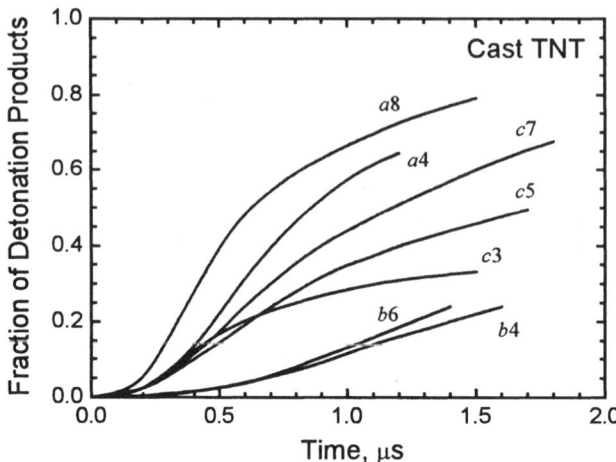

Figure 7.22. Variation of the weight fraction α of the detonation products as a function or time after shock compression. The indices correspond to those of Fig. 7.21.

shock wave. Nutt and Erickson, 1984, observed a transition from a single-stage reaction to a two-stage reaction as the detonation was developing in the composite explosive RX-26-AF, which is based on a mixture of HMX and TATB. The effect is attributed to the different rates of reaction of the components.

The approach described above requires performing numerous labor-intensive experiments under conditions of stringent requirements of measuring accuracy. A different method of obtaining macrokinetic information based on combining measurements with mathematical modeling of experimental situations is more popular. In this approach, the choice of a rational kinetic model of reaction of heterogeneous explosives is of principal importance. Unfortunately, lack of information on the properties of materials, and the size, shape, and mechanism of the formation of reaction sites, makes a detailed description of the nucleation and propagation of the reaction "from first principles" impossible at present. The absence of a rigorous, physically valid model of formation and growth of the hot spots leads to a variety of semi-empirical methods based on the most general concepts of the nature of the process. The constants of the relationships describing the decomposition of explosives (i.e., macrokinetic equations) as a function of the basic state variables are to be determined completely or partially by experiment. In order to discuss the determining factors of the reaction of explosives at hot spots, the available experimental and theoretical data on this phenomenon must be considered in more detail.

7.8. Ignition and Growth of Reaction Nuclei During Shock-Wave Initiation of Detonation

The problems of development of a theory of shock-wave initiation of inhomogeneous explosives are typical of those encountered in the mathematical description of any phenomenon based on the dynamics of various microdefects. The main difficulty lies in the lack of information on the microscopic properties of materials and on the diversity of the numerous types of defects. Nevertheless, even an estimate of possible mechanisms of the activation of reaction sites and propagation of the reaction into the material is no doubt useful for better understanding of the phenomenon, the proper choice of the basic state variables controlling the process, and their functional relationships and tendencies of their variation under conditions of varying physical structure of explosive charges.

Effects of microjetting on hard impurity particles and in pores, plastic work on the pore periphery, friction between particles, cracking of explosive grains when compacted, adiabatic compression of gas inclusions, etc., have been discussed as possible mechanisms of hot-spot formation. Note that, even under normal conditions, the thermal decomposition of solids is inhomogeneous. The process is initiated on structural nonuniformities and develops by way of growth of the reaction nuclei (Young, 1968).

The hydrodynamic mechanism of hot-spot formation during the collapse of microcavities or interaction of shock waves with hard inclusions was analyzed by Mader, 1979, by means of computer simulations. Calculations of shock-wave propagation in liquid nitromethane containing one or several non-uniformities in the form of bubbles or metal particles have demonstrated the realism of this approach. The shock-wave interaction with such inclusions leads to formation of regions of high pressure and temperature where fast reactions are possible. In the case of a spherical cavity, the size of the hot spot is close the initial diameter of the cavity. Hot spots are more effectively formed in cavities than on metal particles.

Generally speaking, the pores in solid HE usually have a distinctly nonspherical shape. In particular, in pressed charges the pores are rather star-shaped. Microcracks of different orientations are always present. When such pores collapse under the influence of shock waves, the probability of formation of cumulative jets is high. The impact of a jet on the opposite surface of the pore is accompanied by a temperature rise approximately proportional to the square of the impact velocity. The average effective jet velocity must be proportional to the particle velocity jump in the shock wave. Therefore, Stresau and Kennedy, 1976, supposed that the temperature rise T_s at the hot spot will be proportional to the square of the particle velocity behind the shock wave: $T_s \approx zu^2$. Some experimental data indicate that the critical particle velocity, u_c, behind the initiating shock wave depends very weakly on the initial charge density.

The pore-filling mechanism has a significant effect on the distribution of dissipated energy of the shock wave over the mass of the explosive. Using a model of a porous body simulated by a set of steel balls, Belyakov, 1975, has shown that the deformation of a porous sample during shock compaction is mainly localized near the surface of the particles. Because of this, considerable adiabatic heating occurs and the temperature of the surface layers may exceed the melting point. Photographs of hot spots in PBX-9404 at a shock compression pressure of 2 GPa were made by Von Holle, 1984. The temperature of these spots is estimated to be ~1500 °C.

The greatest progress in the theoretical analysis of the mechanisms of hot-spot formation was made in calculations of viscoplastic heating of material in the neighborhood of collapsing spherical pores. These studies were begun by Khasainov et al., 1980 and their results were recently summarized by Khasainov et al., 1996. These calculations demonstrated that the heating of material in shock waves at a pressure of about 1 GPa can reach 1000 K and higher, even for micron-size pores. The effective volume of the hot spots formed is sufficient for ignition of the surrounding explosive by the local thermal explosion mechanism. The ignition delay is $<10^{-7}$ s. The lower the shock-wave amplitude, the greater the size of the pores supporting the ignition of the explosive in the initiating shock wave. Some uncertainty of the model is due to melting and the associated reduction in the resistance of materials to deformation at such high temperatures.

An analysis by Taylor, 1985, helped find spatial parameters δ and λ, related to the initial size of the grains and density, which affect the characteristics of shock-wave initiation:

$$\delta = \frac{d}{W_s} \approx \frac{d\sqrt{p\rho}}{\eta}$$

$$\lambda = \frac{s}{d},$$

where W_s is the shock-wave thickness proportional to the characteristic size $x_c = \eta / \sqrt{(\rho\, p)}$, η is the viscosity coefficient, d is the average pore size, and s is the average distance between the pores. The results of the analysis have shown that, for $\delta \leq 1$, the hot-spot temperature drops and the explosive does not ignite in the hot spots.

The possibility of forming hot spots as a result of the heating of plastic deformation zones and formation of adiabatic shear bands was analyzed by Afanasiev and Bobolev, 1968, and Frey, 1981. Based on the theory of initiation of explosion in shear bands, Walker, 1985, obtained a criterion similar to the $p^2 \Delta t$ criterion.

Ever since Bowden's studies of the initiation of detonation by mechanical impact, the mechanism of initiation of detonation when explosives are ignited by adiabatically compressed gas in the pores has been discussed in the literature. The credibility of this mechanism is confirmed by experiments with RDX by Soloviev et al., 1983, in which an explosive was ignited by adiabatically compressed argon, air, or propane filling a pore having a size of the order of 1 mm. The sample with the pore was compressed by shock waves having peak pressures of 0.1–0.5 GPa and demonstrated ignition with delays of 10–100 µs. However, it has been shown by Seay and Seely, 1961, and by Koldunov et al., 1973, that the adiabatic compression of gas inclusions in stronger shock waves has no effect on the dynamics of initiation of detonation.

Therefore, there are quite a few credible mechanisms of reaction site formation due to shock-wave stimulation of explosives. The investigation by Andreev et al., 1979, and by Sharma et al., 1988, of HE samples recovered after exposure to low-amplitude shock waves revealed traces of reaction products on cracks and casting defects, in gaps near the container walls, and near the edges of crystallites; however, they gave no good grounds for choosing these effects as predominant mechanisms of nucleus formation. One should probably assume the simultaneous effect of several mechanisms. Nevertheless, the results of investigations of possible mechanisms of hot-spot formation enable one to formulate the main qualitative features of reaction of explosives in hot spots formed behind the shock-wave front.

The reaction nuclei are formed due to inelastic deformation in the vicinity of nonuniformities, mainly pores, which are almost always present in real charges of solid explosive. Therefore, there is a threshold of shock-wave initiation which exceeds the dynamic limit of elasticity of the given material.

The main cause of initiation of a fast reaction in a nucleus is local heating of the explosive. Inasmuch as competition takes place between two processes, an increase of the temperature of the nucleus due to inelastic deformation and a decrease due to heat transfer to the ambient cold explosive, the possibility of formation of a certain number of effective reaction sites increases with the compression rate. Since the maximum compression rate is reached in shock waves, the majority of reaction nuclei taking part in further development of the process are generated in this phase of the process.

The probability of ignition of the surrounding material by a nucleus increases with its size and temperature. The size of hot spots in shock-compressed matter is proportional to the size of the initial nonuniformities. Therefore, coarse-grain explosives have a lower threshold of initiation than fine-grain ones. At the same time, fine-grain explosives in their initial state contain more nonuniformities. Therefore, at a sufficiently high pressure of shock compression more reaction nuclei are formed in such explosives than in coarse-grain explosives.

The diversity of possible mechanisms of non-equilibrium local heating of materials at nonuniformities due to the action of a shock wave and the presence of pores, cracks, and other defects of structure of different types and sizes in the material in its initial state lead to the formation of hot spots of different sizes and temperatures. Accordingly, a spectrum of induction periods for initiating reaction at hot spots exists behind the shock-wave front which, as a result of averaging, is perceived in the experiments as the absence of an induction period. Nevertheless, in constructing macrokinetic models one should take into account that the fast reaction at the hot spots begins at different times. A more abrupt transition of the initiating shock wave to a detonation wave occurs in fine-grain explosives having a narrow spectrum of initiation sites.

All other things being equal, the temperature at hot spots increases with the shock-wave amplitude. Furthermore, the higher the shock compression pressure, the larger the number of mechanisms of local heating that take part in the formation of hot spots. As a result, the number of effective reaction nuclei capable of igniting the surrounding explosive increases with the shock-wave amplitude.

The major voids in the unreacted explosive are eliminated at relatively low shock compression pressures. Because of this, not all of them can form reaction nuclei during subsequent loading by a stronger shock. The homogenization of an explosive in a weak shock wave reduces its sensitivity to subsequent stronger shock-wave action. Generally speaking, this effect depends on the time interval between the preliminary and subsequent shocks. As an example, even if slow reaction in a nucleus is incapable of igniting the bulk of the explosive, it may nevertheless lead to pore growth, cracking of the material and formation of intermediate products of high sensitivity.

Apparently, the reaction propagates from the hot spots into the explosive by means of layer-by-layer combustion. From energy considerations, it follows that the fraction of initial reaction nuclei in the overall mass of explosive is small: Therefore, the process of explosive transformation behind the shock-wave front can be represented as burning over the progressively increasing surface areas of spherical nuclei. In the burning process the nuclei coalesce, after which the regressive stage of combustion commences. As a result, the macroscopic rate of reaction behind the shock-wave front at the beginning and end of the process is nearly zero whereas the average reaction rate is governed by the number of effective hot spots and the combustion velocity. The latter two parameters depend, in addition to the factors discussed above, on the initial temperature of the explosive.

7.9. Macrokinetics of Decomposition of Solid Explosives in Shock Waves

A description of the macrokinetics of reaction-site decomposition of solid explosives in shock waves can be found by constructing a coherent, consistent theory of the process with varying degrees of detail or by finding empirical dependencies.

The former approach was realized by Khasainov et al., 1980, who used the mechanism of viscoplastic formation of hot spots and the assumption that the chemical reaction evolved in the form of combustion over the surface of ignited spherical pores to construct a consistent model of shock-wave pre-detonation processes in highly dense solid explosives. The calculations have demonstrated the realism and fruitfulness of the model; however, its practical application requires determining a number of mechanical and thermophysical characteristics of shock-compressed materials which are usually known only rather vaguely. The need to include the presence of a spectrum of sites drastically increases the scope of the calculations. The excessive number of poorly defined material constants is a common drawback of physical models even in the case of a simplified physical formulation (see Khasainov et al., 1996). In any case, the models must be experimentally verified and the parameters used in these models must be determined empirically.

The difficulties associated with the construction and practical utilization of theoretical models of the process stimulates the search for empirical formal-kinetic relationships suitable for calculation of the processes of initiation and detonation in a wide range of conditions. The functional form of the empirical formal-kinetic relationships is selected on the basis of the most general and simplified concepts of the mechanism of the phenomenon. The macrokinetic characteristics included in these concepts are found empirically. The range of application of the empirical relationships is defined by the range of states investigated in the experiments.

From the consideration of hot-spot mechanisms it follows that the macroscopic reaction rate is proportional to the density, N, of effective nuclei (which depends on the initial structure of the explosive and on the shock-wave amplitude), the combustion speed $u_b(p, T)$ at current values of pressure and temperature of the explosive, and the average area Φ of combustion surface of individual nuclei varying as the explosive is burning outward:

$$\frac{\partial \alpha}{\partial t} \approx N\left[(J - J_m), \delta'\right] \cdot u_b(p,T) \cdot \Phi(\alpha) . \tag{7.16}$$

In this equation, J is a parameter characterizing the shock-wave amplitude, J_m is the threshold amplitude of the initiating shock wave, and δ' is a parameter characterizing the explosive structure (e.g., the average size of grains or pores).

Figure 7.23 shows a comparison of experimental data with calculated pressure profiles in cast TNT obtained by Utkin and Kanel, 1986, as a result of computer simulations with the simple macrokinetic relationship

$$\frac{\partial \alpha}{\partial t} = k(\mathscr{E}_s - \mathscr{E}_{st}) p \alpha^\gamma (1-\alpha)^{1-\gamma} . \tag{7.17}$$

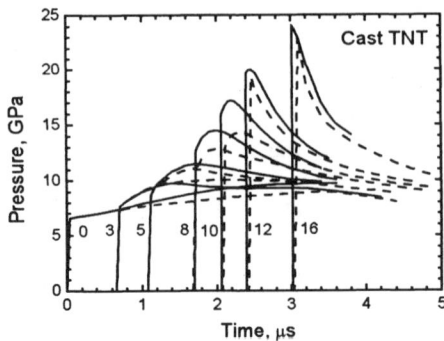

Figure 7.23. Comparison of wave profiles measured during the detonation-initiation process (dashed line) with those calculated using Eq. (7.17) (solid line). The distance from the sample surface is given in mm.

Here, the shock-wave amplitude is characterized by the jump of specific energy, $\mathscr{E}_s$. By varying the exponent γ, the concentration of detonation products at which the reaction rate is a maximum is pre-assigned. In the present example, the constants of the empirical relationship (7.17) are $k = 2.5 \times 10^{-10}$ kg·s^{-1}J^{-1}Pa^{-1} and $\gamma = 0.3$, and the initiation threshold parameter, $\mathscr{E}_{st}$, is assumed equal to zero. In view of the simplicity of this description, the agreement between the calculations and the experimental data may be regarded as quite satisfactory. The temperature is not explicitly included in Eq. (7.17). To some extent, its effect on the reaction rate is indirectly included through the pressure and energy jump at the shock wave because the process is adiabatic. The dependence of the parameters of the empirical macrokinetic equation on the initial temperature and structural parameters of the explosive is to be found experimentally.

In determining the macrokinetic equation using an approach based on comparison of the results of measurement with computer simulations of an experimental situation, there is no need to record the overall evolution of the initiating shock wave. Simpler tests that can be used for this purpose include recording of wave profiles at the boundary between the sample being tested and an inert base plate through which a shock-wave pulse with initially square pressure profile is introduced into the explosive. As an example, Fig. 7.24 presents the pressure history on the interface of the base plate and pressed TNT with a density of 1.56 g/cm^3.

The calculations were made using the empirical macrokinetic equation (7.17) with $\gamma = 0.3$ and $k = 1.25 \times 10^{-9}$ kg·s^{-1}·J^{-1}·Pa^{-1}. The reaction rate of pressed TNT is five times higher than that of cast TNT, which is in agreement with the ratio of critical diameters of detonation of these explosives (see Baum et al., 1975). The calculated values of the duration of the chemical reaction peaks are also in satisfactory agreement with the measurements.

In order to ensure a wide range of validity of empirical relationships, the discussion should include measurements made in both the initiation and detonation modes. Figures 7.25–7.27 show a comparison of the theoretical data with the results of measurements of the structure of the reaction zone in detonating charges of desensitized RDX with an initial density of 1.41–1.67 g/cm^3. In the experiments, the HE samples were covered by aluminum foil reflectors and placed in contact with water "windows" arranged immediately behind these reflectors. The velocity histories of the foil–window interfaces were recorded by VISAR or ORVIS techniques. The elastic–plastic properties of aluminum were included in the calculations. Figure 7.28 compares the computational and experimental data in the initiation mode.

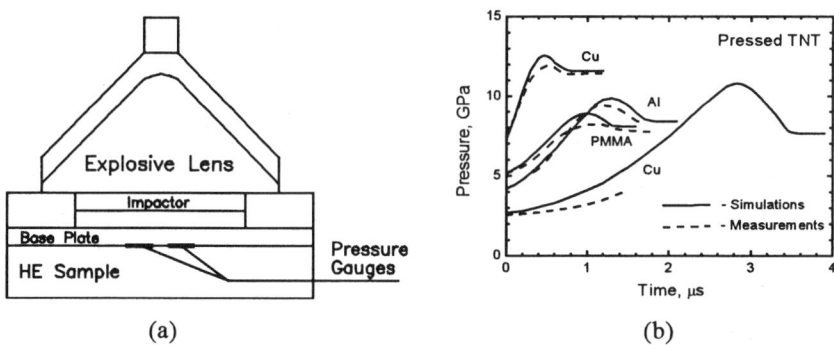

Figure 7.24. (a) Scheme of experiments for recording the pressure history during the shock to detonation transition. (b) Pressure histories recorded at the boundary of the pressed TNT explosive with the inert base plate. The base-plate material is indicated. Data by Utkin and Kanel, 1986.

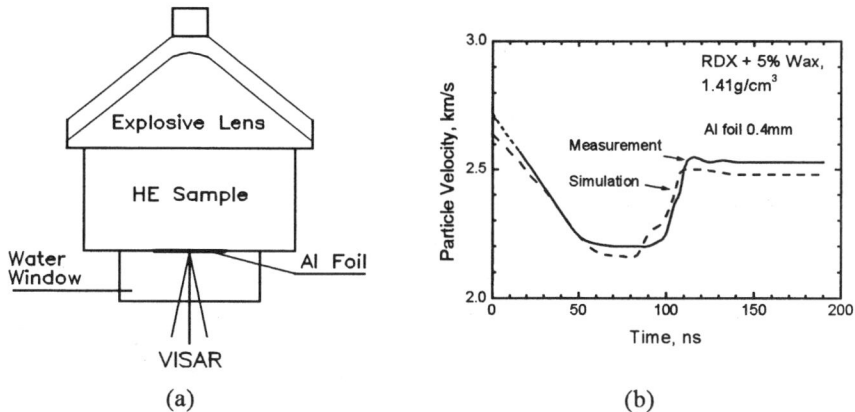

Figure 7.25. (a) Scheme of measurements of the reaction zone of a detonation wave. (b) An example of measurements in comparison with simulation data.

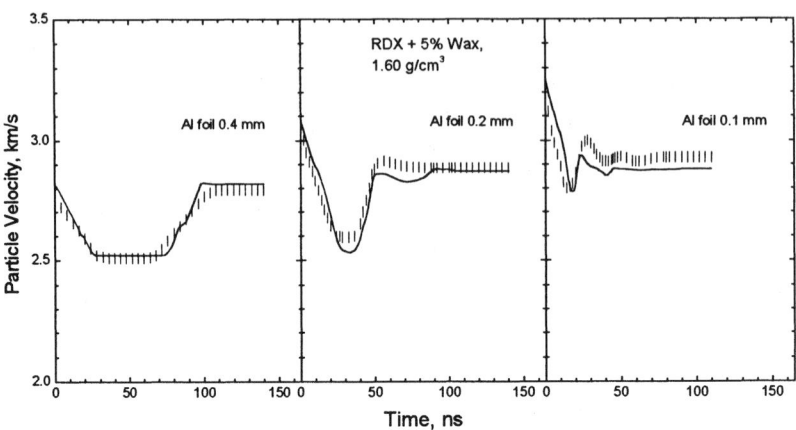

Figure 7.26. Comparison of measured (hatched areas) and simulated (solid lines) waveforms of detonating desensitized RDX of 1.60 g/cm³ density. Data by Utkin et al., 1989.

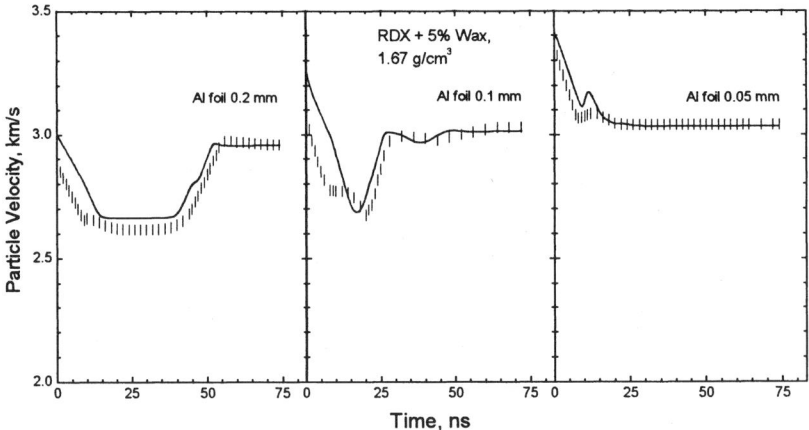

Figure 7.27. Comparison of measured (hatched areas) and simulated (solid lines) waveforms of detonating desensitized RDX of 1.67 g/cm³ density. Data by Utkin et al., 1989.

Numerous calculations have shown that the position of the pressure maximum in Fig. 7.28 depends on the k/γ ratio and corresponds approximately to the completion of the reaction at the input boundary of the explosive sample. On the other hand, integration of the macrokinetic equation (7.17) gives the estimate

$$\Delta t \approx \frac{\pi}{k \sin[\pi(\gamma - 1)] \, \overline{p} \, \mathscr{E}_s}$$

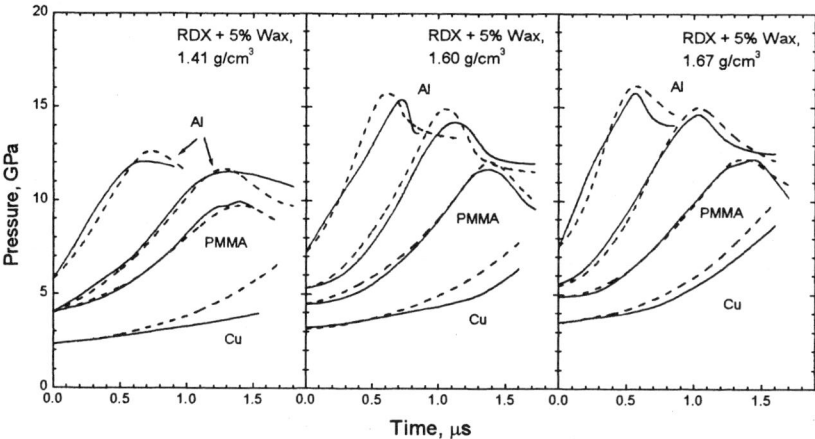

Figure 7.28. Pressure histories in desensitized RDX at the interface with a base plate. The solid lines present the results of measurements and the dashed lines show results of simulations. The base-plate materials are indicated. Data by Utkin et al., 1989.

of the reaction time in the chemical reaction zone of a steady detonation wave. In this equation, $\bar{p}$ is the mean pressure in the reaction zone. These two conditions determine the constants of the macrokinetic equation for description of the entire collection of experimental data. The values of the macrokinetic constants vary somewhat with density and are approximated in the given range by the relationships

$$\gamma = 0.13 \cdot \left(1 + \frac{0.2}{1.75 - \rho_0}\right)$$

$$k = 0.48 \cdot \left(1 + \frac{0.08}{1.75 - \rho_0}\right) \cdot 10^{-9},$$

where ρ_0 is expressed in g/cm^3 and k is expressed in SI units. In all cases, the initiation threshold pressure in the calculations was taken equal to 0.5 GPa, which is somewhat lower than the minimum pressure of 0.8 GPa cited by Soloviev et al., 1980b, for shock-wave initiation.

According to the results of computer simulations, the beginning of the regressive stage of energy release shifts to the shock-wave front as the density of the explosive decreases, which is quite explicable from the standpoint of the hot-spot mechanism of the process. In fact, the higher the porosity of the pressed material, the greater is the fraction of the surface of explosive grains neighboring the discontinuities where reaction nuclei are formed in the shock wave. In the limiting case, the reaction nuclei cover the entire surface of the grains so the combustion surface decreases throughout the process following ignition.

Figures 7.29–7.31 show results by Utkin et al., 2000a, of measurements and computer simulations of detonation phenomena in pressed pure HMX having grain sizes in the range 160–400 µm in the coarse-grain samples and 1–5 µm in the fine-grain samples. The first short pressure pulses of ~0.1 µs duration in Figs. 7.29 and 7.30 correspond to the initial shock pressure in the Teflon insulating films (~0.15 mm). Agreement between the calculated and measured profiles for both shock-wave and detonation regimes in the case of coarse-grain HMX was obtained with the macrokinetic parameters $k = 0.8 \times 10^{-9}$ Pa^{-1} J^{-1} kg·s^{-1}, $\gamma = 0.15$, and $\mathscr{E}_{st} = 4 \times 10^4$ J. To reproduce the experimental data for fine-grain HMX, we had to increase the coefficient k up to 1.2×10^{-9} Pa^{-1} J^{-1} kg·s^{-1}. It is surprising that the reaction rate for the fine-grain HMX exceeds that for the coarse-grain HMX by a factor of only 1.5, whereas the grain sizes of these materials differ by two orders of magnitude. Obviously, the HMX grains were crushed during HE compaction and that decreased the difference.

Various macrokinetic models of heterogeneous decomposition of solid explosives behind the shock-wave front are discussed in the review by Bordzilovskii et al., 1987. Here we restrict ourselves to typical examples. From the standpoint of practical validity, the empirical macrokinetic models may be divided as to the determining parameters of states used in them, as to the number of free parameters, the extrapolation possibilities, etc. In particular, one of the

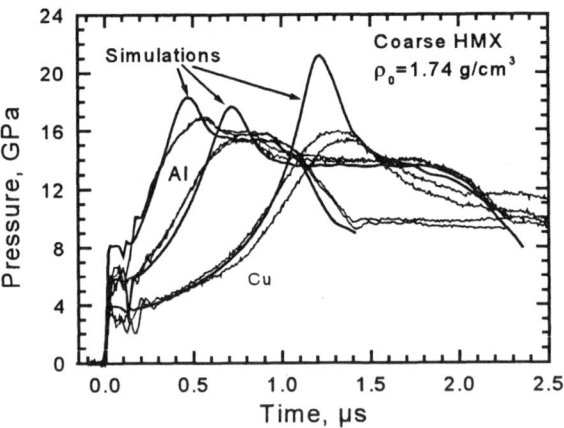

Figure 7.29. Simulated pressure profiles at the interface between the base plate and an HMX sample of 160–400 µm particle size in comparison with experimental data. Shock waves were created in the base plate by aluminum flyer plates at the impact velocity of 1.17 km/s and 1.55 km/s. The corresponding base plate material (Al or Cu) is indicated.

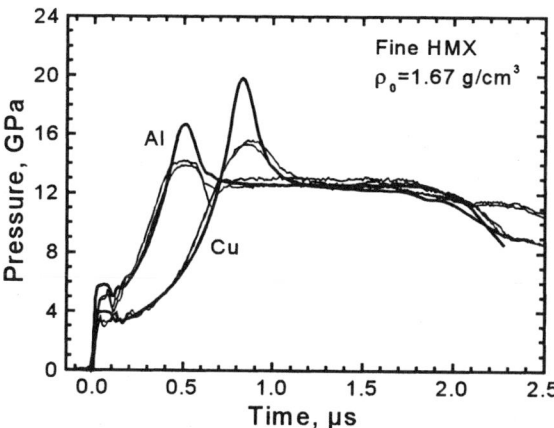

Figure 7.30. Results of simulation of the shock-wave process in fine-grain HMX in comparison with the experimental data. The HMX particles have a size varying between 1 and 5 μm.

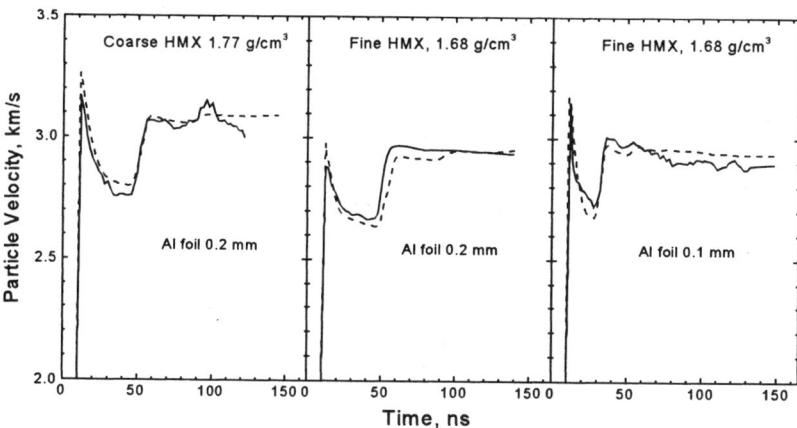

Figure 7.31. Results of measurements (solid lines) and simulations (dashed lines) of the chemical reaction zone in a steady detonation wave in coarse-grain and fine-grain HMX.

most important problems is that of the need to explicitly include the temperature of the material as the main factor determining the process rate on a macroscopic scale.

Wackerle et al., 1978, Tarver and Hallquist, 1981, and Anderson et al., 1981, constructed the macrokinetic relationship

$$\frac{\partial \alpha}{\partial t} = (1-\alpha) z_0 \, p_s^n \, G(p,t) e^{-T_*/T}$$

on the basis of Arrhenius' law with due regard for the effect of the shock-wave amplitude on the reaction rate. The constants z_0, n, and T_*, and the function $G(p, t)$ were determined empirically. Here, the temperature is regarded as some average parameter of state that correlates well with the reaction rate.

Batalova et al., 1980, have determined the concentration of active sites, N_h, as a function of the pressure, p_s, and average temperature at the shock front T, as

$$N_h = N_{0h} p_s^m e^{-T_a/T_s}$$

where the activation temperature T_a is a free parameter, and $m = 2$. Assuming that the combustion surface is spherical and the velocity of the reaction front propagating from the hot spots is a linear function of pressure, the macrokinetic equation finally takes the form

$$\frac{\partial \alpha}{\partial t} = M \, p_s^{2/3} \alpha^{2/3} \rho_{DP}^{-2/3} \rho_{HE} \, e^{-T_a/(3T_s)} . \quad (7.18)$$

Lobanov, 1980, related the concentration of active sites to the current value of thermal energy, $\mathscr{E}_T$, of the explosive. For $\mathscr{E}_T > \mathscr{E}^*$,

$$N_h = N_{0h} e^{-\mathscr{E}^*/\mathscr{E}_T} . \quad (7.19)$$

A two-stage macrokinetic equation is proposed for description of both the progressive stage of the process, namely, burning of the spherical nuclei, and the regressive after-burning of the residues when $\alpha > \hat{\alpha}$:

$$\frac{\partial \alpha}{\partial t} = 3 \frac{\alpha}{R_0} (1-\alpha) \frac{\rho_{DP}}{\rho_{DP} - \rho} \left(\frac{\rho}{\rho_0}\right)^{1/3} \left(\frac{\alpha \rho}{\rho_{DP}}\right)^{2/3} p e^{-\frac{\mathscr{E}^*}{3\mathscr{E}_T}} \quad \text{for } \alpha < \hat{\alpha}$$

$$\frac{\partial \alpha}{\partial t} = 3 \frac{\alpha}{R_0} \frac{\rho_{DP}}{\rho_0} (1-\alpha)^{1/3} \left(1 - \frac{\alpha \rho}{\rho_{DP}}\right)^{2/3} p e^{-\frac{\mathscr{E}^*}{3\mathscr{E}_T}} \quad \text{for } \alpha > \hat{\alpha}, \quad (7.20)$$

where R_0 is the average distance between the sites. The shock-front contribution to the site activation is not explicitly included in this case. The values of the constants used in the macrokinetic equations (7.18)–(7.20) depend on the equation of state used to determine the thermal component, $\mathscr{E}_T$, of the specific internal energy and the temperature. The use of another equation of state calls for correction of the constants of the macrokinetic equation. At the same time, the thermal energy and temperature of the explosive are in some way related to the current pressure and shock wave parameters because the process is adiabatic. It would be desirable to have a macrokinetic equation that would include only parameters that can be determined experimentally.

In a limited range of pressures the experimental data for cast TNT are well described by a macrokinetic equation such as that used by Agureikin et al., 1981:

$$\frac{\partial \alpha}{\partial t} = K p^{1.6} (0.08 + \alpha)(1 - \alpha)^{2.6}. \tag{7.21}$$

Calculations performed by Agureikin et al., 1981, have also shown that errors due to the disturbance introduced into the motion of the medium by transducers used to record the wave profiles at internal cross sections of the initiating explosive sample is 5%. The concentration term of Eq. (7.21) is in agreement with the results of direct analysis of the wave profile evolution but, apparently, underestimates the reaction rate at the end of the process.

At present, a model of initiation and growth of reaction sites by Lee and Tarver, 1980, has found wide application. According to this model the macrokinetic equation is sought in the form

$$\frac{\partial \alpha}{\partial t} = J(1-\alpha)^x \left(\frac{V_0}{V_s} - 1\right)^r + G(1-\alpha)^x \alpha^y p^z, \tag{7.22}$$

where the constants $x = 2/9$, $y = 2/3$ or $2/9$, $z = 1.0-2.0$, $r = 4$, and $J =$ const. The factor G is either constant or depends on the amplitude of the shock wave that has passed through the particle.

It is implied that the first term on the right-hand side of Eq. (7.22) describes the initiation of decomposition and the second term describes the reaction process. In constructing the determining relationship in the form of Eq. (7.22) with the constant factor G, the variable concentration of sites is omitted from consideration. Using the macrokinetic equation (7.22) and its modification, the dynamics of shock-wave initiation of detonation of a wide range of explosives and explosive compositions were calculated.

Various other semi-empirical macrokinetic equations describing reaction of solid explosives in shock waves are also discussed. Macrokinetic descriptions of the processes of energy release in shock waves are presently available for most of the practically important explosives. The diversity of models and determining relationships is a result of the inadequacy of our knowledge of the main regularities of this complex phenomenon. It should be borne in mind that we deal with approximate relationships. Disagreement of up to 15–20% between theory and experiment is usually regarded as quite acceptable. Apparently, the accuracy of the description may be regarded as satisfactory if one manages to reproduce in the calculations a wide range of experimental data including detonation processes and the limiting conditions of initiation by shock load pulses of different durations.

Some researchers studied regularities of the energy release in compositions containing two explosives. An appropriate analysis has been performed for the RX-26-AF composition containing 49.3% HMX, 46.6% TATB, and 4.1% binder (Nutt and Erickson, 1984, Tarver et al., 1984) and for cast TNT/RDX and TNT/HMX explosive mixtures (Bordzilovskii et al., 1983, Lobanov, 1986).

The available experience of mathematical modeling of shock-wave initiation of detonation using various empirical macrokinetic models demonstrates that introduction of the temperature or the thermal component of the specific internal energy into the macrokinetic equation in explicit form gives no great advantages in the accuracy of description of the phenomenon. Most of the models under discussion adequately describe the available experimental data on the evolution of the profiles of the initiating shock waves, the shock compression pressure dependence of the propagation distance required to complete the shock-to-detonation transition, and the relationship between the values of the amplitude and duration of the initiating pulses for shock waves of moderate intensity. Some difficulties arise when analyzing situations near the initiation threshold, with the simple introduction of a constant limit solving the problem only partially. Successful quantitative descriptions of the effect of the initial temperature and the grain size of the explosive on its macrokinetic characteristics are known (see, for example, Partom, 1988). Shock desensitization of explosives has also been modeled successfully (Johnson, 1988).

In discussing the calculations of the processes of initiation and propagation of detonation in real explosives, a question often arises: To what extent is the description of macrokinetics of explosive decomposition based on experiments with plane shock waves applicable to the analysis of three-dimensional flows of reacting matter? The final solution of the problem of validity of empirical relationships, generally speaking, can only be obtained experimentally. The corresponding calculations in two-dimensional formulations by Seitz, 1984, Tarver and Hallquist, 1981, and Vanpoperynghe et al., 1985, have demonstrated quite acceptable accuracy of determining the critical diameter of detonation and the limiting conditions of initiation by compact impactors.

The principal difference of three-dimensional flows from one-dimensional flows as regards initiation of an exothermic reaction consists in high shear deformation of the medium. High dynamic shear deformations *per se* can initiate reactions; however, because there is less energy localization, this process is much slower than those usually observed in shock waves. When an explosive charge is acted upon by a cumulative jet or compact impactor, the explosive may ignite due to surface friction with subsequent transition of combustion to detonation. This mechanism of initiation is not realized in experiments with shock waves and calls for special consideration.

7.10. Equations of State for Explosives

In calculations of shock-wave phenomena in reacting media, the macrokinetic equation supplements the system of equations of continuum mechanics including the equations of conservation of mass, momentum, and energy, and the equations of state for the unreacted explosive and the final detonation products. The principles of construction of equations of state for materials are discussed in the well-known monographs and reviews by Zel'dovich and Raizer, 1966, and McQueen et al., 1970.

The semi-empirical equations of state are constructed in a thermodynamically consistent form based on some or other material model and including material constants determined from some set of experimental data. An equation of state of the Mie–Grüneisen form

$$p(V,\mathscr{E}) = p_c(V) + \frac{\Gamma(V,\mathscr{E})}{V}[\mathscr{E} - \mathscr{E}_c(V)], \qquad (7.23)$$

is widely used. In this equation, $\mathscr{E}_c$ is the specific internal energy of elastic compression, $p_c = -d\mathscr{E}_c/dV$ is the elastic cold (at 0 K temperature) pressure component, and $\Gamma = V(\partial p/\partial \mathscr{E})_V$ is the Grüneisen coefficient that depends, in the general case, on the current values of the specific internal energy and the density. The Grüneisen coefficient is related to the coefficient of bulk thermal expansion, β, the heat capacity, c_p, and the velocity of sound, c, defined by the bulk compressibility of the material through the relation $\Gamma = \beta c^2/c_p$. The elastic component of pressure $p_c(V)$ is set in the form of potentials obtained from the analysis of some or other model of matter. The Born–Mayer potential

$$p_c = B_0 y^2 e^{B_1(1-y)} - B_2 y^4, \qquad (7.24)$$

the Birch–Murnaghan potential,

$$p_c = \frac{3}{2} B_0 (y^7 - y^5) \left[1 - \frac{3}{4}(4 - B_1)(y^2 - 1)\right], \qquad (7.25)$$

and the Morse potential

$$p_c = 3\frac{B_0}{B_1} y^2 (e^{2B_1(y-1)/y} - e^{B_2(y-1)/y}), \qquad (7.26)$$

are widely used. In these equations $y = (V_{0c}/V)^{1/3} = (\rho/\rho_{0c})^{1/3}$, ρ_{0c} is the density of cold matter at zero pressure, and B_0, B_1, and B_2 are material constants. Use is also made of expressions for the elastic pressure component in the form of polynomials or empirical relations of the type (Fortov and Dremin, 1975) of the modified TETA equation

$$p_c = \frac{B_0}{n-m}(y^n - y^m). \quad (7.27)$$

In some cases, a known arbitrary isentrope or Hugoniot for the material, rather than the cold compression curve $P_c(V)$, is selected as the reference relationship in constructing the equation of state. In particular, the equation of state of Jones, Wilkins, and Lee (JWL) (Lee and Tarver, 1980) is widely used by US authors to describe both the unreacted explosive and the detonation products. In this model, the reference curve is the isentrope of the material in the form

$$p_s = Ae^{-R_1\vartheta} + Be^{-R_2\vartheta} + C\vartheta^{-(1+\omega)}, \quad (7.28)$$

where $\vartheta = V/V_{0c}$, and A, B, C, R_1, R_2, and ω are the constants of the material.

In equations of state for explosives, the Grüneisen coefficient is assumed to be constant or to depend on the specific volume according to the equation $\Gamma = \Gamma_0 V/V_0$. In constructing more exact equations of state of condensed media use is made of the generalizing $\Gamma(V)$ relationships (Zharkov and Kalinin, 1968):

$$\Gamma(V) = \frac{\phi - 2}{3} - \frac{V}{2} \frac{\dfrac{d^2(p_c V^{2\phi/3})}{dV^2}}{\dfrac{d(p_c V^{2\phi/3})}{dV}} + d, \quad (7.29)$$

where the parameter ϕ is varied in the range 0–2 (Zubarev and Vaschenko, 1963). The normalizing constant d is found from the condition of conformity of Eq. (7.29) with the Grüneisen coefficient under normal conditions.

Accounting for the internal structure of explosives leads to considerable complication of the equation of state. The molecular crystal model by Kitaigorodskii, 1971, refines the description of thermal effects by introducing two characteristic temperatures corresponding to the low-frequency and high-frequency components of the oscillation term of the free energy. Using the Debye approximation for the low-frequency component and the Einstein approximation for the high-frequency component, Kovalev, 1984, obtained an analogue of the Grüneisen equation of state for molecular crystals.

The parameters of semi-empirical equations of state are adjusted to fit the HE Hugoniot. Measurements of shock compressibility of explosives, especially inhomogeneous and porous explosives, are complicated by the processes of shock-induced decomposition of the material. Since the shock-wave transition width is very small, it is unlikely that a noticeable fraction of the material would undergo immediate chemical reaction during shock compression. The sources of errors are associated with unsteadiness of the shock-wave due to energy release behind the shock and with the difficulty of separation of the shock-wave transition proper from the subsequent flow.

Lysne and Hardesty, 1973, derived a complete equation of state for liquid nitromethane based on measurements of its shock compressibility at different initial temperatures of the material in the range from the melting point to the boiling point. They note that the empirical equation of state is very sensitive to the errors of measurement of the thermodynamic parameters in both the initial and shock-compressed states, as well as to the choice of the relationship approximating the experimental data. In particular, this led to violation of the thermodynamic identity in some limited pressure range.

Hugoniots of some explosives are listed in Table 7.3. For approximate estimates, Afanasenkov et al., 1969, proposed the generalized Hugoniot

$$U_S = c_0 + 2u_p - \frac{1}{10c_0}u_p^2 \qquad (7.30)$$

obtained by averaging a large number of measurements of shock compressibility of organic materials.

Along with data from shock-wave measurements, the results of measurements of isothermal compressibility under hydrostatic conditions are used for developing equations of state for explosives. These measurements enable one to obtain information on the states of cold materials at high pressures and to practically eliminate the possible error due to chemical reaction of the sample in the process of compression.

Table 7.3. Hugoniots of some explosives: $U_S = a + b u_p$

Explosive	ρ_0, g/cm^3	c_0, km/s *	a, km/s	b	References
RDX	1.82		2.87	1.61	Ilyukhin and Pokhil, 1960
	1.80	2.62			Voskoboinikov et al., 1968
	1.00		0.40	2.00	Dremin et al., 1970b
TNT	1.614	2.57	2.39	2.05	Coleburn and Liddiard, 1966
	1.62		2.16	2.24	Dremin and Koldunov, 1967
	1.00		0.3	1.85	Dremin et al., 1971
PETN	1.774	2.32			Olinger and Cady, 1976
	1.75		2.26	2.32	Wackerle et al., 1976
	0.82		0.47	1.73	Dremin et al., 1970b
PBX 9404	1.84		2.69	1.72	Kennedy and Nunziato, 1976
TATB	1.937	1.43	2.9	1.68	Olinger and Cady, 1976
	1.847		2.34	2.316	Coleburn and Liddiard, 1966
TNT/RDX 40/60	1.68		2.71	1.86	Coleburn and Liddiard, 1966

*c_0 is the sound velocity at zero pressure

Figure 7.32 shows the Hugoniots of TATB and PETN explosives obtained both directly from shock-wave measurements and from measurements of static isothermal compressibility. In both cases, the compressibility inferred from the static measurements is higher. In comparing these data, one must take into account, in addition to the experimental errors, the fact that the x-ray structural analysis used by Olinger et al., 1975, to determine the density of the material under hydrostatic compression gives values of crystallographic density which are usually somewhat in excess of the density of a real polycrystalline medium with diverse structural defects. A similar method was used to determine the isotherm of solid nitromethane by Yarger and Olinger, 1970.

Returning to the question of using a shock-wave data base for the construction of equations of state for explosives, note that experiments in which the evolution of shock waves is monitored during initiation of detonation can simultaneously serve as sources of information on the compressibility of the explosive being investigated and of its detonation products. Consider, for instance, the pressure profile at the surface of an explosive sample loaded by an impact of a thick flyer plate. The typical pressure profile for this situation and the corresponding $p-u_p$ diagram are shown in Fig. 7.33. The initial pressure of shock compression of an explosive sample corresponds to the condition of equal particle velocities and equal pressures in the impactor and sample, i.e., to the point A of intersection of the Hugoniot of the decelerating impactor which had a known initial velocity u_i and the Hugoniot of the sample. Consequently, the measured first pressure jump, the known impactor velocity, and the Hugoniot of the impactor are used to determine the particle velocity behind the shock-wave front in the explosive sample. By varying the impact velocity and the impactor material,

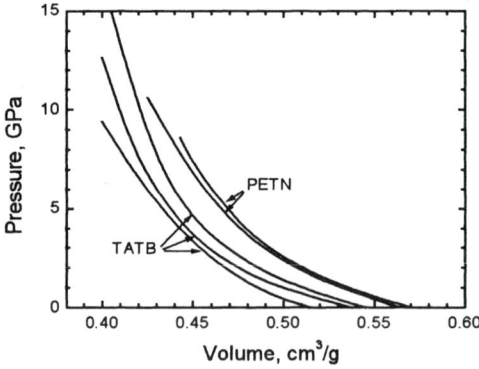

Figure 7.32. Examples of Hugoniots of high explosives. Data from papers by Wackerle et al, 1976, Vanpoperynghe et al., 1985, Olinger et al., 1975, and Coleburn and Liddiard, 1966.

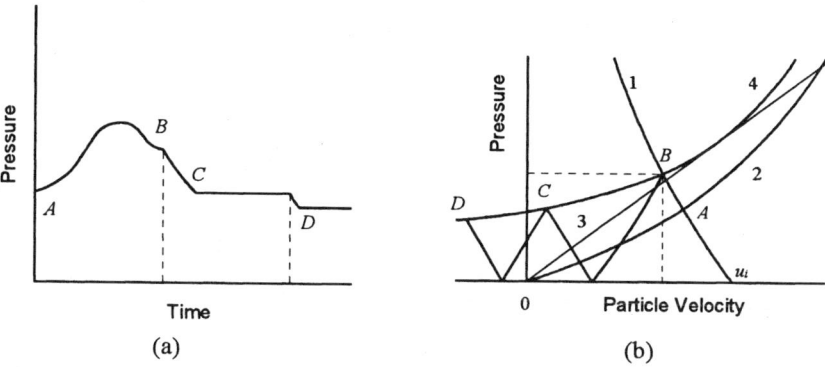

Figure 7.33. (a) A typical pressure profile in HE at the input interface with an impactor plate of finite thickness. (b) Determination of points on the HE Hugoniot and on the release isentrope of detonation products from the measured $p(t)$ history. 1 is the impactor Hugoniot, u_i is the impact velocity, 2 is the HE Hugoniot, 3 is the Rayleigh line, and 4 is the release isentrope of detonation products. The points C and D correspond to the first and second reverberations of the waves in the impactor plate.

one can get a set of points on the Hugoniot of the HE under investigation. After a steady detonation is established, the pressure on the impacted surface finally reaches a constant level, which corresponds to the intersection point of the deceleration Hugoniot of the impactor plate and the release isentrope of the detonation products. Therefore, the value of the final pressure p_B on the impact surface gives the point p_B, u_B on the release isentrope of the detonation products. If the recording time exceeds the period of wave reverberation in the impactor, a stepped release is recorded on the pressure profile. The analysis of wave interactions makes it possible to obtain additional information on the position of the release isentrope of the detonation products from the values of pressure at points C and D.

Figure 7.34 shows Hugoniots of desensitized RDX, calculated from the equations of state (7.23) and (7.25). The corresponding constants are given in Table 7.4. The experimental points at low pressures come from experiments on initiation of detonation; the top points are obtained from measurements of chemical reaction spikes of steady detonations.

The approximate equation of state for explosives in combination with reference data on heat capacity makes it possible to estimate the temperature of shock-compressed HE. In particular, in the experiments on initiation of detonation of desensitized RDX with an initial density of 1.67 g/cm^3, the shock heating in a wave with an amplitude of 5.5 GPa is approximately 110 °C. This is 100 °C below the temperature at which RDX explodes during a 5-min exposure at atmospheric pressure. The experimentally observed time of reaction in a shock wave is approximately 1 μs, which confirms the inhomogeneity of the process of energy release in a shock-compressed explosive.

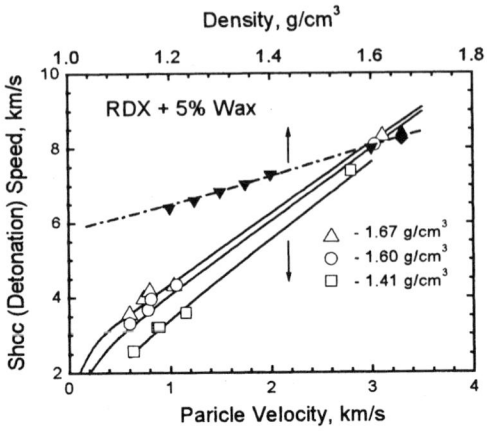

Figure 7.34. The Hugoniot of desensitized RDX (bottom) and the density dependence of detonation speed (top curve). Points on the $D(\rho_0)$ curve are from Baum et al., 1975, Voskoboinikov et al., 1974, and Al'tshuler et al., 1981.

Table 7.4. Equations of state for some explosives

HE	ρ_0, g/cm³	Reference compression curve	Grüneisen coefficient	References
RDX	1.0–1.82	Eq. (7.24) $B_0 = 4.57317$ GPa $B_1 = 10.764$ $B_2 = 6.09207$ GPa $\rho_{0k} = 1.82$ g/cm³	Eq. (7.29) $\phi = 2$ $\Gamma_0 = 2.2$	Akhmadeev, 1981
Desensitized RDX	1.41–1.67	Eq. (7.25) $B_0 = 13.2$ GPa $B_1 = 6.4$ $\rho_{0k} = 1.72$ g/cm³	$\Gamma = \Gamma_0 = 2.4$	Utkin and Kanel, 1986
TNT	1.61	Eq. (7.28) $A = 1798$ GPa $B = -93.1$ GPa $R_1 = 6.2, R_2 = 3.1$ $\omega = 0.8926$	$\Gamma = \Gamma_0 = 0.8926$	Lee and Tarver, 1980
PETN	1.75	Eq. (7.28) $A = 3746$ GPa $B = -131.3$ GPa $R_1 = 7.2, R_2 = 3.6$ $\omega = 1.173$	$\Gamma = \Gamma_0 = 1.173$	Lee and Tarver, 1980

Table 7.4. Equations of state for some explosives (cont.)

PETN	1.60	Eq. (7.28) A = 2188 GPa B = −58 GPa $R_1 = 7.8, R_2 = 3.9$ $\omega = 0.3468$	$\Gamma = \Gamma_0 = 0.3468$	Lee and Tarver, 1980
	1.00	Eq. (7.28) A = 1312 GPa B = −7.84 GPa $R_1 = 11, R_2 = 5.5$ $\omega = 0.2027$	$\Gamma = \Gamma_0 = 0.2027$	Lee and Tarver, 1980
TATB	1.90	Eq. (7.28) A = 10820 GPa B = −240.6 GPa $R_1 = 8.2, R_2 = 4.1$ $\omega = 1.251$	$\Gamma = \Gamma_0 = 1.251$	Lee and Tarver, 1980
PBX-9404	1.842	Eq. (7.28) A = 20080 GPa B = −260.8 GPa $R_1 = 9.0, R_2 = 4.5$ $\omega = 0.4541$ $E_0 = 0.556$ GPa·cm^3/cm^3	$\Gamma = \Gamma_0 = 0.4541$	Green et al., 1978

Kovalev, 1984, and Voskoboinikov et al., 1967, performed more consistent calculations of the shock compression temperature of a number of explosives using the molecular crystal model. Voskoboinikov and Voskoboinikova, 1988, approximated the results of calculations by linear equations:

$T = 407 + 58.6p$ for liquid TNT,

$T = 247 + 42.3p$ for monocrystalline TNT,

$T = 322 + 28.2p$ for RDX,

$T = 293 + 25.7p$ for HMX,

$T = 303 + 35.6p$ for PETN, and

$T = 367 + 66.7p$ for liquid nitromethane,

where T is the temperature in Kelvins and p is the pressure in GPa. The equations are valid for $p > 8$ GPa. Note that the shock compression temperature is a more sensitive parameter of the equation of state than compressibility. Delpuech et al., 1986, measured the temperature of shock-compressed explosives by recording the spectra of Raman scattering of monochromatic laser radiation. The measurements performed in experiments with samples of monocrystalline RDX yielded temperature values of 405 K at a shock compression pressure of 9.5 GPa and 485 K at 13.5 GPa. No shock-induced morphological evolution was observed on the spectra of Raman scattering for RDX, which is regarded as a proof of a constant chemical composition and structure of the explosive. The temperature measurements above 1000 K are performed in a simpler way by recording the spectra of self-radiation of the shock-compressed material. Such measurements can be made under detonation conditions.

7.11. Equations of State for Detonation Products

Determination of the equation of state for detonation products is vital both as regards the solution of the problems of initiation and of the effects of an explosion and, which is no less important, from the standpoint of thermodynamic calculation of the detonation parameters at the Chapman–Jouguet point. In the latter case, one must know the thermal equation of state $p = p(V, T)$. At present, two approaches exist for constructing equations of state for explosion products, namely, (i) a description based on summation of the contributions by individual components of real detonation products and (ii) an averaged description in which the composition of the detonation products is not taken into consideration. In the former case, the state of the detonation products is described not only by the pressure, volume, and temperature, but also by their composition. The main components are H_2O, N_2, CO_2, CO, H_2, and C. The fractions of components in the detonation products are found from the condition of chemical equilibrium.

The detonation speed, D, pressure, p, and temperature, T, at the Chapman–Jouguet point and their dependencies on the HE density are the basic experimental data with which calculated results are compared. Existing experimental techniques make it possible to determine D with an accuracy of 1% or better. The accuracy of determining the pressure is somewhat better than 5%, due partly to the uncertainty of the experimental determination of the Chapman–Jouguet point. Shvedov, 1987, discusses this problem in detail. Additional information at larger rarefaction of the detonation products is evaluated from tests in which the expansion of metal tubes produced by internal detonation of HE is monitored. Measurements of the detonation temperature are possible only for transparent homogeneous explosives and are associated with problems of opacity and emissivity, calibration, etc. As a result, it seems that temperature data are less certain and accurate than kinematic data.

The possibilities of theoretical description of gaseous states having densities comparable to those of the condensed phase are very limited. Only systems with the simplest interaction potentials between particles, in particular, the hard-sphere model, have been studied in sufficient detail. Various numerical calculations made over a wide range of parameters for this model can be best presented by the Carnagan–Starling (1969) formula

$$\frac{pV}{RT} = \frac{1+y+y^2-y^3}{(1-y)^3}, \qquad (7.31)$$

where $y = N\pi\sigma^3/6V = \sqrt{2}\pi V_s/6V$, N is the number of particles in the system, σ is the hard sphere diameter, and $N\sigma^3/\sqrt{2} = V_s$ is the volume of a hard-sphere system when most densely packed.

Despite the limiting assumptions, the hard-sphere model enables one to adequately describe the gas behavior at high densities. Feng et al., 1985, used Eq. (7.31) to calculate the parameters at the Chapman–Jouguet point and the composition of the detonation products for eleven explosives including TNT, TATB, nitromethane, RDX, HMX, and PETN. The soft-sphere model using the exponential repulsive potential is more accurate for detailed description of the properties of gases at high pressures. Heuzé et al., 1985, used this model to describe the detonation properties of gas mixtures.

Along with these equations of state, numerous empirical equations that approximate the experimental data quite well have been proposed. The Brinkley–Kistiakowsky–Wilson (BKW) equation,

$$p = \frac{\rho}{M_{av}} RT(1 + \chi e^{\beta \chi}),$$

is an empirical equation of state for the detonation products of condensed explosives that is widely used for practical calculations. In this equation M_{av} is the average molecular weight of the detonation products, $\chi = \rho K (T+\theta)^{-\zeta}$, and K, β, θ, and ζ are empirical constants.

Kondrikov and Sumin, 1987, derived equations of state for the main detonation products that are compounds of the elements C, N, H, and O by processing the available experimental data on dynamic compression of condensed matter. In order to attain the highest accuracy of description of the experimental data, on the one hand, and to use previously obtained parameters of the intermolecular interaction potential, on the other hand, the pressure and internal energy were divided into potential and thermal components. The equation of state proposed by Kondrikov and Sumin accurately describes the dependence of the detonation speed and temperature on the initial density for a number of explosives.

The Kuznetsov–Shvedov, 1966, equation for the detonation products of RDX is one of the thermal equations which ignores the molecular composition of the detonation products, but does feature a well-grounded approach to the thermal components of pressure and energy. Using this equation, Kuznetsov and Shvedov, 1969, calculated the detonation parameters and Hugoniots of the detonation products of RDX of different densities.

The JWL equations of state (7.28) in the Mie–Grüneisen form are as widely used for detonation products as they are for unreacted explosives. The Grüneisen coefficient is usually assumed constant and equal to ω in Eq. (7.28) (Finger et al., 1976). However, as shown by Green et al., 1988, the experimental $D(\rho_0)$ relationship can be satisfactorily described throughout a wide range of densities only if Γ is assumed to be a function of V.

In the JWL equations of state the $p_s(V)$ isentrope in the form of Eq. (7.28) is preassigned as the reference curve. For some explosives, the simpler volume dependence,

$$p_s(V) = Ae^{-aV} + BV^{-n}, \qquad (7.32)$$

is also acceptable (Evstigneev et al., 1976). In this equation A, B, a, and n are constants. Assumptions on the behavior of the Grüneisen coefficient are usually close to those made in constructing equations of state for explosives; in particular, Γ is regarded as a function of volume alone. However, allowance is made for the fact that, at low densities, the detonation products must be described by the ideal-gas equation of state with a characteristic value of the Grüneisen coefficient in the range 0.3–0.4.

The Grüneisen coefficient is found from the empirical relationship between the detonation speed and the initial density of the explosive. Based on the condition of tangency between the Rayleigh line and the detonation product Hugoniot at the Chapman–Jouguet point, the derivative of the detonation speed, D, with respect to the initial HE density, ρ_0, is given by Baum et al., 1975, for a polytropic isentrope $pV^k = \text{const.}$, as

$$\frac{d\ln D}{d\ln \rho_0} = \frac{k(k-1-\Gamma)}{2k-\Gamma}. \qquad (7.33)$$

The polytropic exponent, k, at the CJ point is a function of volume in accordance with the equation of state adopted. Therefore, the condition $dD/d\rho_0 = \text{const.}$ can be satisfied, generally speaking, only if Γ is also assumed to be a function of volume. In this sense, Eq. (7.33) can be regarded as a formula to be used for determining the constant Grüneisen coefficient of the detonation products. Figure 7.34 shows a comparison between the experimental data and the initial-density dependence of the detonation velocity of desensitized RDX, calculated from the equation of state for the detonation products in the form of Eq. (7.32). Good agreement is obtained at $\Gamma = 0.3$ which practically coincides with estimates based on Eq. (7.33).

In using the equations of state, one should take into account that the internal energy of the material, $\mathscr{E}$, is determined with an accuracy of the arbitrary constant $\mathscr{E}_0$. For the unreacted explosive, the constant $\mathscr{E}_0$ is usually found from the condition of $\mathscr{E} = 0$ at $p = 0$ and $V = V_0$. The detonation products result from chemical reaction of the explosive. Therefore, in order to ensure energy conservation, the internal energies of the unreacted explosive, $\mathscr{E}_{HE}$, and of the detonation products, $\mathscr{E}_{DP}$, must be brought into agreement with one another.

Akhmadeev, 1981, found the constant $\mathscr{E}_0$ in the equation of state for the detonation products from an analysis of the energy balance during the detonation of an explosive in a calorimetric bomb: $\mathscr{E}_{HE}(\rho_0, T_0) - \mathscr{E}_{DP}(\rho_0, T)$, where T_0 and T are the initial temperature and the temperature of the detonation products, respectively. By extracting the heat of detonation, $Q(\rho_0, T_0)$, from $\mathscr{E}_{HE}$ and decreasing the temperature of the detonation products from T to T_0, he obtained

$$\mathscr{E}_{HE}(\rho_0, T_0) = \mathscr{E}_{DP}(\rho_0, T) + Q(\rho_0, T_0). \tag{7.34}$$

From this relation one finds the normalizing constant, $\mathscr{E}_0$, for the equation of state for the detonation products. In Eq. (7.34) it is assumed that, in a detonation, all chemical reactions occur at constant volume. In the experimental measurement of the heat of detonation in a calorimetric bomb, the detonation products expand. According to Akhmadeev, 1981, the correction for this is 5–8% of the value of Q.

The constant $\mathscr{E}_0$ in the equation of state for the detonation products can also be found from the known detonation parameters of the explosive of initial density ρ_0. For a steady detonation wave, the law of conservation of energy is written as

$$\mathscr{E}_{DP}(p_{CJ}, V_{CJ}) - \mathscr{E}_{HE}(p=0, V_0) = (p_{CJ}/2)(V_0 - V_{CJ}),$$

where p_{CJ} and V_{CJ} are the pressure and specific volume at the Chapman–Jouguet point. At $\mathscr{E}_{HE}(p=0, V_0) = 0$,

$$\mathscr{E}_{DP}(p_{CJ}, V_{CJ}) = D^2/2(k+1)^2, \tag{7.35}$$

where D is the detonation speed, and k is the polytropic exponent ($pV^k = \text{const.}$) of the detonation products at the Chapman–Jouguet point. With the preset form of the equation of state for the detonation products, Eq. (7.35) defines the normalizing constant $\mathscr{E}_0$. In particular, for the ideal-gas equation of state,

$$\mathscr{E} = \frac{pV}{k-1} - \text{const.} \tag{7.36}$$

We find from Eq. (7.35) that

$$\text{const.} = \frac{D^2}{2(k^2-1)} \tag{7.37}$$

which, in the case of equality of the constant to the heat of explosion Q, corresponds to the known (Baum et al., 1975) expression for the detonation speed. The constant of the equation of state thus determined is, generally speaking, a function of the initial density of the explosive charge and varies somewhat with ρ_0. However, in many cases this dependence can be ignored.

Equations of state for the detonation products of some explosives are listed in Table 7.5.

Table 7.5. Equations of state of the detonation products of selected explosives

Explosive	ρ_0, g/cm^3	Reference compression curve	Grüneisen coefficient	References
RDX	1.0–1.82	Eq. (7.24) $B_0 = 5.62957$ GPa $B_1 = 9.0024$ $B_2 = 0.570347$ $\rho_{0c} = 0.82$ g/cm^3	Eq. (7.29) $\phi = 2$ $\Gamma_0 = 2.2$	Akhmadeev, 1981
Desensitized RDX	1.41–1.67	Eq. (7.32) $A = 896.16$ GPa $B = 3.98$ GPa·cm^6/g^2 $a = 10.06$ g/cm^3 $n = 2$	$\Gamma = \Gamma_0 = 0.3$	Utkin and Kanel, 1986
TNT	1.63	Eq. (7.32) $A = 521.7$ GPa $B = 1.762$ GPa·(cm^3/g)^n $a = 7.876$ g/cm^3 $n = 1.6$	$\Gamma = \Gamma_0 = 0.9$	Evstigneev et al., 1976
TNT	1.61	Eq. (7.32) $A = 371.2$ GPa $B = 3.2306$ GPa $R_1 = 4.15$ $R_2 = 0.95$ $\omega = 0.3$ $E_0 = 7.0$ GPa·cm^3/g	$\Gamma = \Gamma_0 = 0.3$	Lee and Tarver, 1980
	1.63	Eq. (7.28) $A = 373.8$ GPa $B = 3.747$ GPa $C = 0.734$ GPa $R_1 = 4.15$ $R_2 = 0.90$ $\omega = 0.35$ $E_0 = 6.0$ GPa·cm^3/g	$\Gamma = \Gamma_0 = 0.35$	Finger et al., 1976
PETN	1.75–1.77	Eq. (7.28) $A = 617$ GPa $B = 16.926$ GPa $C = 0.699$ GPa $R_1 = 4.4$ $R_2 = 1.2$ $\omega = 0.25$ $E_0 = 10.2$ GPa·cm^3/g	$\Gamma = \Gamma_0 = 0.25$	Lee and Tarver, 1980 Finger et al., 1976

Table 7.5. Equations of state of the detonation products of selected explosives (cont.)

Explosive	ρ_0, g/cm^3	Reference compression curve	Grüneisen coefficient	References
PETN	0.2–1.8	Eq. (7.28) $A = 1032.158$ GPa $B = 90.57014$ GPa $C = 3.72735$ GPa $R_1 = 6.0$ $R_2 = 2.6$ $\omega = 0.2027$ $E_0 = 10.8$ GPa·cm^3/g $\rho_{0c} = 1.763$ g/cm^3	Γ^* $A_1 = 0.145$ $A_2 = -6$, $A_3 = 0.64$, $B_1 = 0.1$ $B_2 = -10$ $B_3 = 1.22$ $C_1 = 0.47$ $C_2 = -6$ $C_3 = 2.3$ $D_1 = 0.2$	Green et al., 1988
TATB	1.9	Eq. (7.28) $A = 654.67$ GPa $B = 7.1236$ GPa $R_1 = 4.45$ $R_2 = 1.2$ $\omega = 0.35$ $E_0 = 6.9$ GPa·cm^3/g	$\Gamma = \Gamma_0 = 0.35$	Lee and Tarver, 1980
PBX-9404	1.842	Eq. (7.28) $A = 852.4$ GPa $B = 18.02$ GPa $R_1 = 4.6$ $R_2 = 1.3$ $\omega = 0.38$ $E_0 = 10.2$ GPa·cm^3/g	$\Gamma = \Gamma_0 = 0.38$	Green et al., 1978
HMX	1.891	Eq. (7.28) $A = 778.3$ GPa $B = 7.071$ GPa $C = 6.43$ GPa $R_1 = 4.2$ $R_2 = 1.0$ $\omega = 0.3$ $E_0 = 10.5$ GPa·cm^3/g	$\Gamma = \Gamma_0 = 0.3$	Finger et al., 1976
TNT/RDX 50/50	1.67	Eq. (7.32) $A = 453.9$ GPa $B = 1.94$ GPa $a = 7.281$ g/cm^3 $n = 1.6$	$\Gamma = \Gamma_0 + lVe^{-kV}$ $\Gamma_0 = 1/3$ $k = 4.95$ g/cm3 $l = 7.425$ g/cm3	Zhernokletov et al., 1969, Zubarev and Evstigneev, 1984,

$^*\,\Gamma = A_1\{1+\tanh[A_2(A_3-\rho/\rho_{0c})]\} + B_1\{1+\tanh[B_2(B_3-x)]\} + C_1\{1+\tanh[C_2(C_3-\rho/\rho_{0c})]\} + D_1$

7.12. Calculation of States of Mixtures of Explosives and Detonation Products

In the problems of detonation we deal with a two-component medium consisting of the unreacted explosive and its detonation products. For numerical solution of detonation problems, some additional hypotheses and rules are required to describe the behavior and interaction of the components. The additivity of the specific volume and internal energy,

$$V = \alpha V_{DP} + (1-\alpha)V_{HE} \tag{7.38}$$

$$\mathscr{E} = \alpha \mathscr{E}_{DP} + (1-\alpha)\mathscr{E}_{HE}, \tag{7.39}$$

is taken to be universally valid. As a rule, the phase pressures are also assumed equal to one another:

$$p_{HE}(V_{HE}, \mathscr{E}_{HE}) = p_{DP}(V_{DP}, \mathscr{E}_{DP}). \tag{7.40}$$

Finally, the fourth and last assumption is not so unique and is connected with the solution of the problem of thermal equilibrium of the components of the mixture. The condition of additivity of entropy is apparently the most consistent one (Bordzilovskii et al., 1987, Cowperthwaite and Tarver, 1976). In this case, difficulties may arise because it is necessary to have a complete equation of state for each of the components since, in addition to p, V, and $\mathscr{E}$, one must know the entropy. In some calculations, the phase temperatures are assumed equal. The resulting set of equations is convenient for iterative calculation of the pressure from known values of $\mathscr{E}$, V, and α. This approach is also accompanied by the difficulties of development and use of complete equations of state of the components. Therefore, more often, the calculations are based on the hypothesis of constant entropy of explosives after shock-wave compression:

$$d\mathscr{E}_{HE} = -p\,dV_{HE}. \tag{7.41}$$

This hypothesis is advantageous in that it does not require any additional information other than the $p-V-\mathscr{E}$ equation of state for the components.

The ambiguity in the choice of the hypothesis on thermal interaction between the components of the mixture of HE and DP does not lead to marked disagreement of the results obtained using different equations of state for the mixture. Johnson et al., 1985, studied the process of development of detonation under conditions of pre-assigned kinetics of decomposition of explosives and different assumptions on the thermal equilibrium of the components of the mixture. In one case, the temperatures were assumed to be equal to one another whereas, in the other case, the state of an explosive particle was assumed to be isentropic after shock compression. It was found that the disagreement between the calculated pressure histories obtained under different assumptions is much less than the experimental error.

The hypotheses expressed by Eqs. (7.38)–(7.41) pre-assign the energy of the mixture as a function of p, V, and α. In the general case, it cannot be reduced to one equation of state in the Mie–Grüneisen form when the coefficient $\Gamma = V(\partial p/\partial \mathscr{E})_{V,\alpha}$ is itself a function of both the specific volume and specific internal energy and of the mixture composition as well. In other words, in the mixture we have $\Gamma \neq$ const., even if the Grüneisen coefficient is assumed constant for the unreacted explosives and for the detonation products. The thermal effect of the reaction, $Q_{p,V} = -(\partial \mathscr{E}/\partial \alpha)_{p,V}$, which is often used in applications of the detonation theory, also depends on p, V, and α. Zel'dovich, 1940, expressed $Q_{p,V}$ through the heats of reaction under conditions of constant volume of the mixture, $Q_V = -(\partial \mathscr{E}/\partial \alpha)_{T,V}$, and pressure, $Q_p = -(\partial \mathscr{H}/\partial \alpha)_{T,p}$, where $\mathscr{H}$ is the enthalpy of the mixture:

$$Q_{p,V} = \frac{c_p Q_V - c_V Q_p}{c_p - c_V}$$

where c_p and c_V are the heat capacities.

Since the shock-wave amplitude is one of the main factors determining the rate of decomposition of an explosive, the shock-wave discontinuity needs to be explicitly embedded into the numerical simulations. Embedding the shock wave presents no difficulties in one-dimensional geometry. In the algorithms of hydrodynamic calculation with artificial viscosity one may use, for instance, the differential analyzer method. More exact schemes are available in which the position of the shock-wave front is regarded as one of the boundaries of the spatial region which is simulated at a given moment. Boundary conditions are set by applying the conservation laws to the shock discontinuity. Damamme, 1984, discussed an algorithm for embedding shock waves in two-dimensional flows.

7.13. Detonation Properties of High Explosives Containing Metal Particles

Metal powders, mainly aluminum, have commonly been incorporated as additives in composite high explosives to enhance the performance of the energetic mixtures. Formation of the metal oxides increases the temperature of the detonation products, alters the equilibrium product distribution, and directly influences the rate of energy delivery from the explosive. As an illustration, Table 7.6 presents the heat of detonation for some high-powered explosives with and without aluminum.

The basic peculiarities of detonation of systems in which several exothermic and endothermic processes may proceed at different rates, or in which chemical reactions are accompanied by slow relaxation processes, were analyzed by Khasainov et al., 1979, Kuznetsov and Kopotev, 1986, Lubyatinsky and Loboiko, 1996, and Khasainov and Veyssiere, 1988. These analyses have shown

Table 7.6. Heat of explosion of HE with aluminum

High Explosive	Oxygen Balance %	Heat of Detonation kcal/kg
RDX	−21.6	1320
RDX + 20% Al		1760
TNT	−74	1150
TNT + 15% Al		1320
HMX	−21.6	1290
HMX+25%Al		1830

that detonation in these systems may feature nonunique velocities, parameters significantly lower than the ideal CJ parameters, relaxation instability of the detonation front, generation of compression waves and secondary shock waves in the material behind the steady detonation zone, and steady two-wave configurations. However, the real behavior of metal additives in detonation of high-density HE has not yet been described in detail by the theory.

Figure 7.35 illustrates the case of a weak ("undercompressed") detonation that was observed, for example, in RDX + wax, HMX + wax, and PETN + wax compositions by Al'tshuler et al., 1981, 1983. In these explosive mixtures, the chemical transformation occurs in two stages: A fast exothermic decomposition of the explosive component is followed by an endothermic reaction of the wax heated by the HE decomposition products. The detonation velocity in this case is determined by the slope of the Rayleigh line OA which is tangent to the Hugoniot, S_m, of maximum energy release at point G, whereas the final pressure at the end of the reaction zone for the steady detonation complex should be determined by point F of intersection of the Rayleigh line with the Hugoniot, S_f, of the final detonation products. As the perturbation velocity between points G and F is lower than the detonation velocity, the perturbation cannot penetrate the zone of fast reaction AG to reach the detonation front.

In the case when the fast exothermic reaction is followed by a relatively slow exothermic process, two detonation velocities may be recorded depending on the length of the explosive charge (Soloviev et al., 1980a). In this case, the detonation development is governed by the dimensionless duration of the second stage, or the after-burning time, $\tau = t_a/(l_0 D_0)$, where t_a is the time of energy release in the second stage, D_0 is the initial detonation velocity, and l_0 is the HE charge length. If $\tau > 1$, the after-burning does not influence the detonation wave speed or other parameters of the detonation wave. If $0.1 < \tau \leq 1$, acceleration of the detonation wave occurs as a result of the energy release during after-burning. If $\tau < 0.1$, a steady detonation complex is formed which corresponds to a total energy release $Q_{max} = Q_0 + Q_a$, where Q_0 and Q_a are the energy release in the first and second phases, respectively. Figure 7.36 presents the pressure–volume diagram of states of the material in this case.

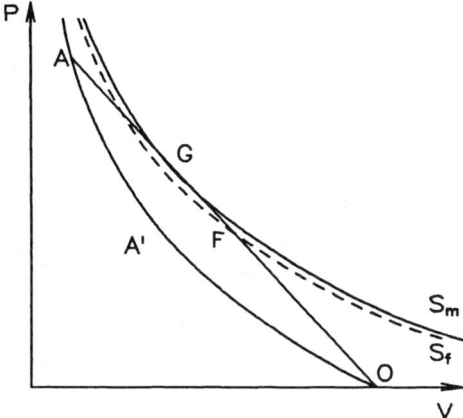

Figure 7.35. The pressure–volume diagram of detonation with energy absorption. The curve $OA'A$ is the HE Hugoniot, S_m is the detonation product Hugoniot corresponding to maximal energy release, S_f is the Hugoniot corresponding to final states of the detonation products, and G is the Chapman–Jouguet point.

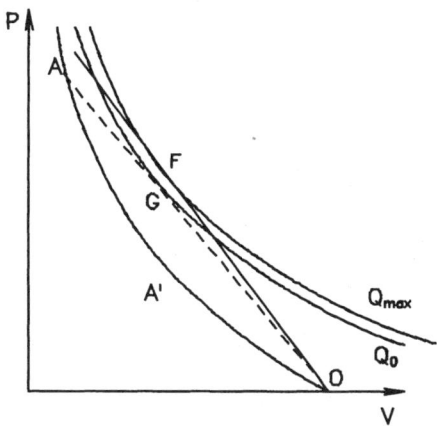

Figure 7.36. The pressure–volume diagram of detonation with after-burning. The curve Q_0 is the detonation product Hugoniot corresponding to the energy release, Q_0, in the fast first phase, Q_{max} is the Hugoniot for steady detonation corresponding to the maximum energy release, and G is the Chapman–Jouguet point at dimensionless time $\tau > 1$, and F is the Chapman–Jouguet point of the final steady detonation wave.

Metals have been used as additives enhancing the detonation characteristics of high explosives since the beginning of the twentieth century and currently find extensive application in both industrial and military explosives. It has been revealed that compositions of high explosives with aluminum, boron, or beryllium, as a rule, have an increased heat of detonation. The positive effect is provided by oxidation of metals by reduction of the CO_2 and H_2O content in the detonation products. Reactions of metals with carbon and nitrogen are not profitable and give a negative effect. The influence of metals on the detonation velocity and pressure can be different depending on the metal concentration and particle size. As a rule, the detonation parameters of molecular HEs with negative oxygen balance decrease when the explosives are compounded with any solid non-explosive component.

Detonation of composite explosives is accompanied by relaxation processes leading to mechanical and thermal equilibration behind the detonation wave front. Equilibration in velocities of condensed particles and detonation products can be described by the equation (Boyko et al., 1983):

$$F_c = m_c \frac{du_c}{dt} = \frac{\pi}{8} \rho_c d_c^2 C_D(\text{Re}, M)(u_{DP} - u_c) \cdot |u_{DP} - u_c|,$$

where m_c, u_c, d_c, and ρ_c are respectively, the mass, velocity, diameter, and density of the condensed particle, u_{DP} is the velocity of the gaseous detonation products, $C_D(\text{Re}, M)$ is the drag coefficient, $\text{Re} = \rho_c d_c (u_{DP} - u_c)/\eta$ is the Reynolds number, and η is the viscosity of the detonation products. The velocity of the condensed particles is less than the velocity of the gas in the acceleration phase and exceeds the latter in the deceleration phase. Based on measurements of acceleration of metal wires by shock waves in water by Al'tshuler et al, 1977, and Kim, 1984, we may roughly estimate the velocity equilibration time for aluminum particles of ~100 μm size to be ~100 ns or less. The specific energy lost for acceleration of condensed particles, $E_{ck} = 0.5 \beta u^2$ (β is the mass fraction of condensed particles in the HE) is usually a small fraction of the detonation energy and does not have a significant influence on the detonation parameters. However, the kinetic energy of the condensed particles may exert a significant influence on the effects of the detonation.

The temperature equilibration between the detonation products and the condensed particles is determined by the particle size and the thermal conductivity of both particles and detonation products. A rough estimate of the time of particle heating using the Smolukhovski formula is $t_h \approx d_c^2/a$, where a is the thermal diffusivity (Dremin et al., 1960). For metals $a \approx 10^{-4}$ m^2/s and that means that particles of ~100 μm size are heated to the temperature of the detonation products within ~100 μs, whereas the micron-size particles need only ~10 ns to equilibrate their temperature with that of the detonation products. Thus, small particles may absorb the detonation-product energy in the reaction zone and thereby influence the detonation parameters. More precise calculations made by Boyko et al., 1983, and Medvedev et al., 1982, account for the convective heat exchange, radiation, melting, and vaporization but now we are discussing mainly the qualitative side of the phenomenon and make only rough estimates.

If there are no chemical or thermal interactions between the detonation products and the added metal particles, the detonation parameters are estimated using the additive equation of state of the mixture. In this approach, the specific volume, V, of the mixture is calculated as the sum of specific volumes of the components:

$$V(p) = (1-f)V_{HE}(p) + fV_c(p),$$

where V_{HE} and V_c are specific volumes of high explosive (or detonation products) and metal particles in the mixture, respectively, and f is the mass fraction of the metal. For rough estimates they use the Hugoniot of the added metal as the $V_c(p)$ dependence and the Hugoniot of the HE or the isentrope of the detonation products as $V_{HE}(p)$. The detonation parameters are then determined by the condition of tangency of the Rayleigh line to the Hugoniot of the mixture of detonation products with the metal. Such estimates provide quite reasonable agreement with experimental data in the case of larger-size particles when the heat exchange time is much greater than the HE decomposition time (Afanasenkov et al., 1970).

In the case of thermal equilibration between the detonation products and the added particles, the detonation products loose the energy, q, given by the equation

$$q = \beta \int_{T_c}^{T_D} c_c(T) dT ,$$

where T_c and T_D are the temperatures of shock-compressed added metal particles and the detonation products, respectively, and c_c is the heat capacity of the added metal. The temperature of the detonation products decreases as the metal particles are heated. This results in displacement of the Hugoniot of the mixture of detonation products and metal toward a smaller volume and a decrease in the detonation velocity.

A comprehensive review of the detonation properties of HE mixtures with aluminum has been published by Aniskin, 1986. The conclusion is that, in the reaction zone of high-powered explosives, aluminum particles of > 1 μm size do not react with the detonation products and their influence on the detonation parameters is analogous to that of inert additives. The energy release associated with oxidizing aluminum particles of ~5 μm size becomes significant ~4 μs behind the detonation front and continues for at least 20–30 μs. The effect of aluminum on later stages of the detonation process was confirmed by recording the expansion rate for metal tubes filled with the HE and by measurements of the velocities at which metal plates were launched. The enhanced detonation parameters were recorded only for mixtures of aluminum with strong oxidizers for which the component grains were of micron size. These oxidizers are not individually explosives and are characterized by a relatively large (2–4 μs) reaction zone duration in detonation waves.

Thermochemical calculations of the CJ parameters by Imkhovik and Soloviev, 1995, Cherét, 1974, and Hobbs and Baer, 1993, have not shown a marked increase in the detonation velocity or pressure for mixtures of powerful explosives with aluminum. In general, calculations with completely and partially equilibrated detonation parameters show that neither the detonation velocity nor the CJ pressure determined assuming fully equilibrated products differ much from those determined for products with Al as an inert material, and only the equilibrium detonation temperature is much higher than the nonequilibrium temperature. As an illustration, Fig. 7.37 shows the results of a calculation by Imkhovik and Soloviev, 1995, and their comparison with measurements of the detonation velocity for RDX + AL compositions. The larger decrease of the detonation velocity in the case of smaller metal particles means that, in the latter case, the thermal equilibration occurs within the RDX chemical reaction zone.

Bicomponent composite explosives consisting of PETN mixed with either 5- or 18-μm-diameter spherical aluminum particles were studied by Tao et al., 1992. The immediate conversion to detonation products in PETN due to a very narrow reaction zone (less than 1 ns) allowed monitoring the interaction between the aluminum and the PETN detonation products. The recorded variations in acceleration of tantalum plates demonstrated that some or all of the Al reacts with CO_2 and/or H_2O in the detonation product gases during the microsecond time frame. Lubyatinsky and Loboiko, 1996, measured the detonation reaction zone for five RDX-based explosives containing up to 19% aluminum of particle size from 2 μm to 20 μm. Aluminum particle size was found to have no appreciable effect on the reaction zone length, which increases from 0.34 mm to 0.58 mm as the aluminum content increases from 0 to 19%.

Many fewer investigations have been conducted on mixtures of high explosives with other metals. Unlike aluminum, products of boron oxidation, B_2O_3, HBO_2, B_2O_2, and BO, are gases. Owing to that, one might hope for a significant increment in the detonation velocity and pressure for HE compositions with boron. However, the detonation product content, besides carbon and the boron oxides, includes boron nitride, BN, and boron carbide, B_4C, which are in a condensed phase and should decrease the detonation parameters. Figure 7.38 presents the data by Pepekin et al., 1972, for the heat of explosion as a function of the boron content in the RDX + B mixtures. The heat of explosion increases to 1880 kcal/kg with addition of 16.5 wt.% B and, after that, decreases with further increase of the amount of boron in the mixture. In the case of PETN, the heat of explosion reached 2050 kcal/kg at 22 wt.% boron whereas, for pure PETN, this value is 1400 kcal/kg.

Results of measurements of the detonation velocity for PETN + B mixtures have been published by Akimova et al., 1972. Experiments were done with HE of ~5 μm grain size and boron particles of 0.1 μm diameter. It has been shown that the detonation velocity decrease for the PETN + 11 wt.% B mixture amounts to ~3% as compared to the pure PETN of the same density (0.83–1.73 g/cm³). At high density, a remarkable growth in the detonation velocity was observed with an increase of the charge diameter to 3 cm.

For pure HE, information about the macrokinetics of energy release is evaluated from an analysis of the pressure or particle velocity profiles of shock and detonation waves. Utkin, Kanel et al., 2000a, studied effects of boron in the

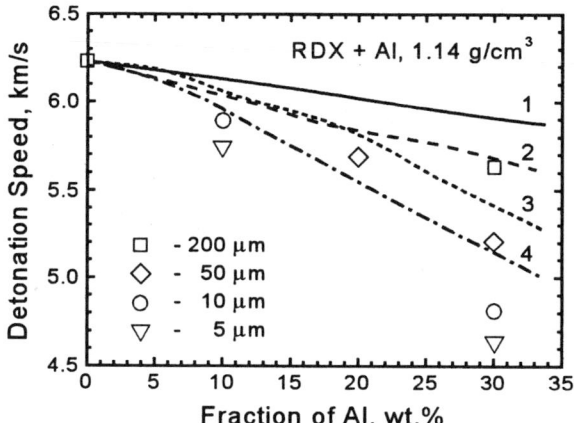

Figure 7.37. Results of calculations of the detonation velocity for compositions of RDX + AL by Imkhovik and Soloviev, 1995, in comparison with experimental data of Aniskin and Shvedov, 1979. Calculations were done supposing 1: The Al remains inert and does not participate in heat transfer, 2: The Al is heated to the equilibrium temperature of the detonation products, 3: The Al reacts but only 50% of it is completely burned, and 4: complete thermodynamic equilibrium is attained in the products.

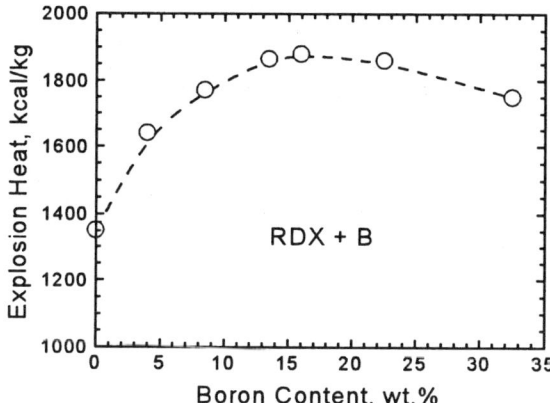

Figure 7.38. The heat of explosion of RDX + B mixtures. Data by Pepekin et al., 1972.

energy release process both in the shock-wave initiation and the detonation regimes. The HE mixture tested was composed of fine-grain HMX with the particle size varying between 1 and 5 µm and amorphous boron powder with the particle size in the range from 0.1 µm to 1 µm. The boron content of 16.4 wt. % in the HMX + B mixture corresponded to a maximum heat of explosion. Samples were compacted by pressing to density of 1.56 g/cm^3.

For analyzing the detonation and macrokinetic data, equations of state of the HMX + B mixture and its detonation products are needed. As a rough approximation, Mie–Grüneisen equations of state were constructed for both the unreacted HMX + B mixture and the detonation products. These equations of state are based on the Hugoniots of HMX and boron published by Simpson et al., 1993, and the LASL Shock Hugoniot Data published by Marsh, 1980, and on the assumption of additivity of specific volumes of components in the mixtures. Since the experiments cover only a limited pressure range, the reference isentrope was chosen in a simplest form, $p_s = B/V^k$, where B and k are constants. The Grüneisen coefficient was also assumed constant. The equations of state of the detonation products were fitted to the detonation speed $D(\rho_0)$ dependencies, and to the pressure histories in detonating HE. The results of fitting are presented in Figs. 7.39 and 7.40.

Figures 7.39 and 7.40 show that the calculated detonation parameters of pure HMX agree well with the experimental data. For the HMX + B mixture, the additive equation of state gives values of the detonation speed and pressure that are much higher than the measured values. This means that heat exchange occurs between the detonation products and the boron particles, or the boron reaction with the detonation products results in an increase of the final product density. In order to provide agreement between the calculated and measured detonation properties of the HMX + B mixture, it was necessary to reduce the value of the effective heat of detonation, Q, for the postulated detonation products.

Figure 7.40 compares the pressure profiles generated in thick Teflon witness plates by normal detonation of HMX and the HMX + B mixture. The profiles do not demonstrate any specific effects of the HMX + B mixture.

Figures 7.41 and 7.42 present examples of the pressure profiles in initiating shock waves. An attempt was made to use the additive EOS for the detonation products of the HMX + B mixture. Since the temperature of shock-compressed mixtures was low at the onset of shock-wave initiation, and since the fine-grain HMX is the main energetic component, we expected that these experiments with the HMX + B mixture could be at least partly simulated with the decomposition-kinetic parameters of pure fine-grain HMX and with the additive equation of state (without reduction in the Q value) for the detonation products. Figure 7.41 shows, however, that this approach leads to a faster pressure growth and faster shock-to-detonation transition than was observed in the experiments. This means

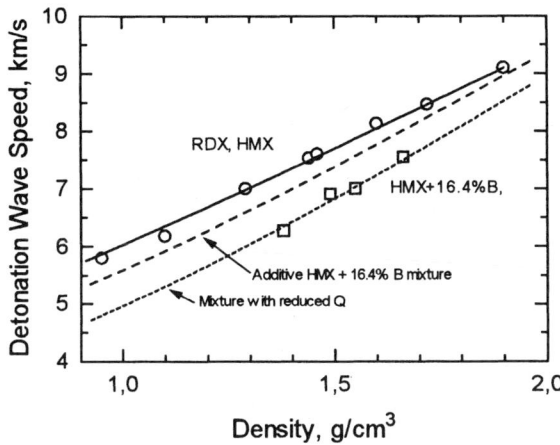

Figure 7.39. Calculated detonation speed as a function of the HE density for pure HMX and an HMX + 16.4% B mixture in comparison with experimental data of Dremin et al., 1970a, and Finger et al., 1976, for pure HMX, and our data for the mixture.

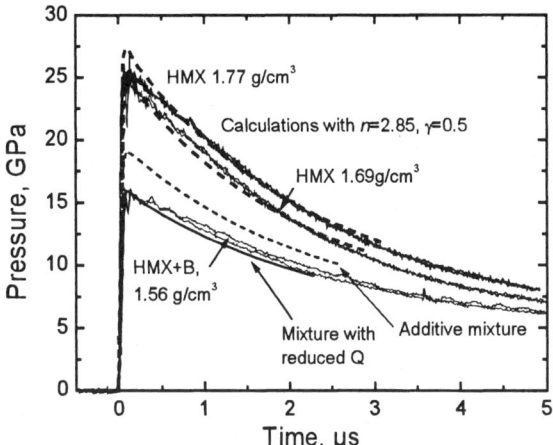

Figure 7.40. Results of a computer simulation of impact of normal detonation waves upon Teflon witness plates. The thin lines present experimental data.

that the boron particles participate in the energy release process even in the low-pressure shock waves. This is possible only for a very uniform mixture. We recall that the HMX particle size was 1–5 μm whereas the boron particle size was between 0.1 and 1 μm. Using the equation of state with reduced Q value and the same kinetic parameters, we have obtained a better agreement with the

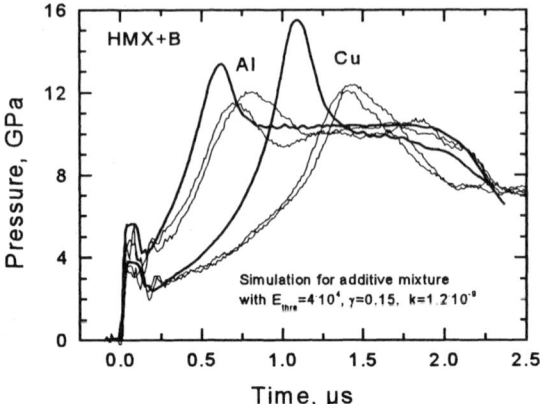

Figure 7.41. Results of simulations (thick lines) of the shock-to-detonation transition in the HMX + B mixture with additive equation of state and the same decomposition kinetics as for pure fine-grain HMX. The experimental pressure profiles are shown by thin lines. The base-plate materials are indicated.

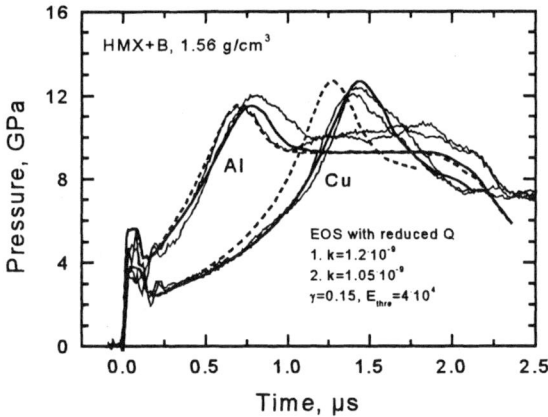

Figure 7.42. Results of simulations of the shock-to-detonation transition in an HMX + B mixture when the modified equation of state with reduced q value was used. The kinetics of fine-grain HMX lead to the pressure histories shown by the dashed lines. Solid lines present the pressure profiles calculated with $k = 1.05 \times 10^{-9}$ $Pa^{-1}J^{-1}kg \cdot s^{-1}$, $\gamma = 0.15$, and $\mathscr{E}_{th} = 4 \times 10^4$ J.

experiment (Fig. 7.42) but optimal reproduction of the experiments required a 10% reduction of the decomposition rate coefficient, k. Simulations with these equation of state and macrokinetic parameters reproduce the pressure growth and the pressure maximum position very well but lead to a lower pressure level at the plateau in the final phase of the shock-to-detonation transition. Moreover,

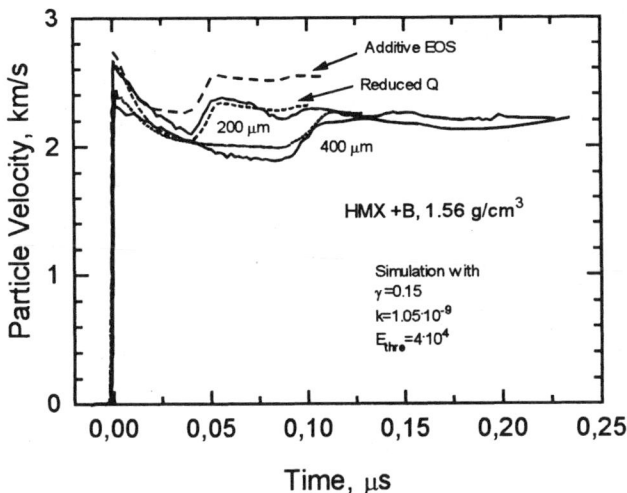

Figure 7.43. Results of simulations (dotted and dashed lines) of measurements of the chemical reaction zone in steady detonation wave in the HMX + 16.4% B mixture. The conditions correspond to those shown in Fig. 7.7 at aluminum foil reflectors of 200 μm and 400 μm thickness.

the measurements indicate an increasing pressure in this part of the process whereas, in the simulations, there is a plateau of constant pressure. We may conclude that the macrokinetic model describes only the heat exchange between the HE and its decomposition products and the boron particles but does not describe the chemical reactions between the boron and the HMX detonation products.

Figure 7.43 presents results of simulations and measurements of the reaction zone in the mixture. As could be expected, calculations with the additive equation of state and the energy release kinetics of pure fine-grain HMX lead to higher peak and final particle velocities and to shorter peak duration. Simulations with the corrected equation of state and decomposition rate give better agreement with the experimental data. However, whereas the calculated reaction zone was somewhat longer than the observed one for pure HMX, in the case of a mixture the recorded state of the material in the detonation wave continues to evolve after the final state is established in the model. This can be taken as evidence of a slower energy release as a result of the boron combustion.

Thus, the positive effect of boron appears in the residual pressure growth observed in the shock-wave experiments shown in Figs. 7.41 and 7.42. However, the boron has a negative effect on the detonation parameters. It seems that the initial decomposition rate and the detonation parameters in the mixture are controlled by heat exchange rather than by a chemical interaction between the HE decomposition products and the boron particles. The additional energy of the boron burn in the detonation products is released at later times ranging from ~0.1–0.2 μs to 1–2 μs.

7.14. Conclusion

The efforts of numerous investigators have yielded extensive information on the macrokinetic regularities of energy release and equations of state for explosives. Although many important details of the process of reaction initiation and growth to detonation behind the shock-wave front are not yet fully clear, the semi-empirical approach to the solution of the problem proved to be successful and made it possible to obtain the quantitative data required for calculations of shock-wave phenomena in explosives. In spite of the numerous explicit and implicit approximations, often intuitive, the macrokinetic equation derived on the basis of this approach, combined with approximate equations of state, provide adequately high accuracy of calculations of various situations in one- and three-dimensional geometry. Nevertheless, the study of the nature of the reaction sites and regularities of their formation does not lose its urgency. It is only on the basis of a rigorous physical theory that one can reliably predict the effect of various factors on the initiation and propagation of detonation in charges of specific configurations.

A number of aspects of this problem call for further investigation. In particular, the effect of the initial temperature of the charge is no doubt of interest. It can manifest itself both as a contribution to the temperature of hot spots behind the shock wave and the rate of propagation of combustion waves from the sites, as well as variation of the physical and chemical properties of the unreacted explosive which, in turn, may lead to a change of the mechanisms and limits of initiation of detonation. It is necessary to clarify the relationship between the properties of various additives (sensitizing or desensitizing, etc.) and macrokinetic regularities of the process. Formation of the reaction nuclei can be affected by both the size and shape of the explosive grains. In particular, the question of possible anisotropy of sensitivity of structured (e.g., cast) explosive charges to shock-wave stimulation appears to be of interest. A very interesting problem is associated with chemical reactions in detonation products on a microsecond time scale. Wide application of the techniques of computer simulations calls for equations of state that permit reduction of the scope of calculations while maintaining their accuracy. In this respect, it appears promising to search for a form of the equation of state that ensures a uniform description of the unreacted explosive, the detonation products, and mixtures of these components.

References

Afanasenkov, A.N., V.M. Bogomolov, and I.M. Voskoboinikov (1969). "Generalized shock adiabat of condensed matter," *J. Appl. Mech Tech. Phys.* **10(4)**, pp. 660–664 [trans. from *Zh. Prikl. Mekh. Tekh. Fiz.* **10(4)**, p. 137 (1969)].

Afanasenkov, A.N., Bogomolov, V.M., and Voskoboinikov, I.M. (1970). "Calculations of the detonation parameters for mixtures of high explosives with inert additives." *Comb. Expl. Shock Waves* **6(2)**, pp. 163–166 (1970) [trans. from *Fiz. Goreniya Vzryva* **6(2)**, pp. 182–186 (1970)].

Afanasenkov, A.N., and V.A. Danilenko (1975). "Initiation of shock-wave detonation of hexogen mixtures with liquid fillers," *Comb., Expl. Shock Waves* **11(6)**, pp. 778–784 [trans. from *Fiz. Goreniya Vzryva* **11(6)**, pp. 915–922 (1975)].

Afanasiev, G.T., and V.K. Bobolev (1968). *Initiation of Solid Explosives*. Nauka, Moscow. (in Russian)

Agureikin, V.A., B.P. Kryukov, and V.N. Postnov (1981). "The estimation of influence of gauges on the matter flow under shock-wave initiation of heterogeneous HE," in: *Detonation. Proceedings of the Second All-Union Workshop on Detonation*, Inst. of Chem. Phys., Chernogolovka, pp. 12–15. (in Russian).

Akhmadeev, N.Kh. (1981). "Simulation of detonation waves in solid explosives," *Comb. Expl. Shock Waves* **17(1)**, pp. 87–93. [trans. from: *Fiz. Goreniya Vzryva* **17(1)**, pp. 109–117 (1981).]

Akimova, L.N., A.Ya. Apin, and L.N. Stesik (1972). "Detonation of explosives containing boron and its organic derivatives." *Comb. Expl. Shock Waves* **8(4)**, pp. 387–390. [trans. from *Fiz. Goreniya Vzryva* **8(4)**, pp. 475–479 (1972).]

Al'tshuler, L.V., V.K. Ashayev, G.S. Doronin, A.D. Levin, et al. (1980). "Experimental investigation of states within the chemical reaction zone of detonation wave," in: *"Chemical Physics of Processes of Combustion and Explosion. Detonation"* (Proceedings of 6th All-Union Symposium on Combustion and Explosion), Inst. Chem. Phys., Chernogolovka, pp. 8–11. (in Russian).

Al'tshuler, L.V., V.K. Ashaev, V.V. Balalaev, G.S. Doronin, and V.S. Zhuchenko (1983) Parameters and Detonation Modes of Condensed HE. *Comb. Expl. Shock Waves* **19(4)**, pp. 515–520. [trans. from *Fiz. Goreniya Vzryva* **(4)**, pp. 153–159 (1983).]

Al'tshuler, L.V., V,V. Balalayev, G.S. Doronin, V.S. Zhuchenko, and A.S. Obukhov (1981). "Particularities of detonation of desensitized HE," in: *Detonation. Materials of the II All-Union Workshop on Detonation*, pp. 36–39, Inst. of Chem. Phys., Chernogolovka. (in Russian)

Al'tshuler L.V., G.I. Kanel, and B.S. Chekin (1977). "New measurements of viscosity of water behind of the shock wave fronts." *Sov. Phys.–JETP* **45(2)**, pp. 348–350. [trans. from *Zh. Eksp. Teor. Fiz.* **45(2)**, pp. 663–665 (1977).]

Anderson, A.B., M.J. Ginsberg, W.L. Seitz, and J. Wackerle (1981). "Shock initiation of porous TATB," in: *Proc. Seventh Symp. (International) on Detonation*, Report MP 82-334, Naval Surface Warfare Center, White Oak, MD, (1981).

Andreev, K.K. (1966). *Thermal Decomposition and Combustion of Explosives*. Nauka, Moscow. (in Russian)

Andreev, S.G, M.M. Boiko, I.F. Kobylkin, and V.S. Sokov'ev (1979). "Formation of sites in trotyl and tetryl with a weak shock action," *Comb. Expl. Shock Waves* **15(6)**, pp. 810–814. [trans. from *Fiz. Goreniya Vzryva* **(6)**, pp. 143–147 (1979).]

Aniskin, A.I. (1986). "Detonation of mixtures of high explosives with aluminum." in: *Detonation an Shock Waves. Proceedings of VIII Soviet Symposium on Combustion and Explosion*, Inst. Chem. Phys., Chernogolovka, pp. 26–32. (in Russian)

Aniskin, A.I., and K.K. Shvedov (1979). in: *Detonation. Critical Phenomena. Physico-Chemical Transformations in Shock Waves*, Inst. Chem. Phys., Chernogolovka, pp. 26–30. (in Russian)

Apin, A.Ya., and L.N. Stesik (1955) in: *The Physics of Explosion*, collection of papers, No. 3, USSR Academy of Sciences, p. 87. (in Russian)

Ashaev, V.K., G.S. Doronin, and A.D. Levin (1988). "Detonation front structure in condensed high explosives," *Comb. Expl. Shock Waves* **24(1)**, pp. 88–92 [trans. from *Fiz. Goreniya Vzryva* **24(1)**, pp. 95–99 (1988)].

Batalova, M.V., S.M. Bakhrakh, and V.N. Zubarev (1980). "Excitation of a detonation in heterogeneous explosives by shock waves," *Comb. Expl. Shock Waves* **16(2)**, pp. 227–231 [trans. from *Fiz. Goreniya Vzryva* **(2)**, pp. 105–109 (1980)].

Baum F.A., L.P. Orlenko, K.P. Stanyukovich, V.P. Chelyshev, and B.I. Shekhter (1975). *Physics of Explosion*, Nauka, Moscow. (in Russian)

Belyakov, G.V. (1975). "Shock deformation of granular media," *Sov. Phys.–Dokl.* **19(10)**, pp. 667–668. [trans. from: *Dokl. Akad. Nauk SSSR* **218(6)**, pp. 1280–1282 (1974).]

Belyayev, A.F., and R.Kh. Kurbangalina (1960). "Influence of the initial temperature on the failure diameter of nitroglycerine and trotil," *Russ. J. Phys. Chem.* **34(3)**, pp. 285–289. [trans. from *Zh. Fiz. Khim.* **34(3)**, p. 603 (1960).]

Bobolev, V.K. (1947). *Dokl. Akad. Nauk SSSR*, **57**, p. 789. (in Russian)

Bordzilovskii, S.A., and S.M. Karakhanov (1985). "Effects of loading rate on the predetonation length for TG 50/50," *Comb. Expl. Shock Waves* **21(6)**, pp. 752–755. [trans. from *Fiz. Goreniya Vzryva* **21(6)**, pp. 109–113 (1985).]

Bordzilovskii, S.A., S.M. Karakhanov, and V.F. Lobanov (1987). "Modeling the shock initiation of detonation of heterogeneous explosives," *Comb. Expl. Shock Waves* **23(5)**, pp. 624–638. [trans. from *Fiz. Goreniya Vzryva* **23(5)**, pp. 132–147 (1987).]

Bordzilovskii, S.A., V.F. Lobanov, and S.M. Karakhanov (1983). "The transition processes at shock initiation of trotil–hexogen and trotil–octogen alloys," *Comb. Expl. Shock Waves* **19(4)**, pp. 499–501. [trans. from *Fiz. Goreniya Vzryva* **(4)**, pp. 136–139 (1983).]

Bowden, F.P., and A.D. Yoffe (1952). *Initiation and Growth of Explosion in Liquids and Solids*, Cambridge Univ. Press, Cambridge.

Boyko V.M., V.V. Grigoryev, S.A. Zhdan, A.A. Karnaukhov, and A.N. Papyirin (1983). "Acceleration and heating of a metal particle behind a detonation wave." *Comb. Expl. Shock Waves* **19(4)**, pp. 496–499 [trans. from *Fiz. Goreniya Vzryva* **19(4)**, pp. 133–136 (1983)].

Campbell, J.R., W.C. Davis, J.B. Ramsay, and J.R. Travis (1961). "Shock initiation of solid explosives," *Phys. Fluids* **4(4)**, p. 498.

Carnagan, N., and K. Starling (1969). *J. Chem. Phys.* **51**, p. 635.

Chaiken, R. (1978). in: *Behavior of Dense Media under High Dynamic Pressures, Symp. H.D.P.*, Gordon and Breach, New York, pp. 41–54.

Chapman, D.L. (1899). "On the rate of explosion in gases." *Phil. Mag.* **47**, pp. 90–104.

Chéret, R. (1974). "Le code ARPEGE: application a l'etude d'um explosif a l'aluminium." *Acta Astronautica* **1**, pp. 893–898.

Chéret, R. (1993). *Detonation of condensed Explosives*, Springer-Verlag, New York.

Coleburn, N.L, and.T.P. Liddiard, Jr (1966). "Hugoniot equations of state of several unreacted explosives," *J. Chem. Phys.* **44(5)**, p. 1929.

Cowperthwaite, M., and J. Rosenberg (1976). "A multiple Lagrange gage study of the shock initiation process in cast TNT," in: *Proc. Sixth Symp. (International) on Detonation*, Report ACR-221, Office of Naval Research, Arlington, VA (1976), p. 786.

Cowperthwaite, M, and C.M. Tarver (1976). "On hydrodynamic effects of exothermic power in condensed explosives," *Acta Astronautica* **3**, p. 201.

Damamme, G. (1984). "A new method to simulate shocks, detonations and transitions from shock to detonation," in: *Shock Waves in Condensed Matter—1983* (eds. J.R. Asay, R.A. Graham, and G.K. Straub) North-Holland, Amsterdam, p. 575.

Dick, J.J. (1986). "Pop-plot and Arrhenius parameters for <110> pentaerythritol tetranitrate single crystals," in: *Shock Waves in Condensed Matter* (ed. Y.M. Gupta) Plenum Press, New York, p. 903.

Dick, J.J. (1997). "Anomalous shock initiation of detonation in pentaerythritol tetranitrate crystals." *J. Appl. Phys.* **81(2)**, pp. 601–612.

Dick, J.J., R.N. Mulford, W.J. Spencer, D.R. Pettit, E. Garcia, and D.C. Shaw (1991). "Shock response of pentaerythritol tetranitrate single crystals." *J. Appl. Phys.* **70(7)**, pp. 3572–3587.

Dick, J.J., D.R. Pettit, and W.J. Spencer (1990). "Crystal orientation effects in PETN explosive with 4 GPa shocks." in: *Shock Compression of Condensed Matter— 1989* (eds. S.C. Schmidt, J.N. Johnson, and L.W. Davison) North-Holland, Amsterdam, pp. 713–716.

Dick, J.J., and J.P. Ritche (1994). "Molecular mechanics modeling of shear and crystal orientation dependence of the elastic precursor shock strength in pentaerythritol tetranitrate." *J. Appl. Phys.* **76(5)**, pp. 2726–2737.

Dick, J.J., M.C. Whitehead, and A.R. Martinez (1994). "Crystal orientation dependence of elastic precursor strength in pentaerythritol tetranitrate." in: *High-Pressure Science and Technology—1993* (eds. S.C. Schmidt, J.W. Shaner, G.A. Samara, and M. Ross) American Institute of Physics, New York, pp. 1373–1376.

Dremin, A.N., and S.A. Koldunov (1967). "Initiation of detonation by shock waves in cast and pressed trotyl," *Vzryvnoe Delo*, **63/20**, p. 37. (in Russian)

Dremin, A.N., S.A. Koldunov, and K.K. Shvedov (1971). "Shock wave initiation of detonation in low density explosive charges," *Comb. Expl. Shock Waves* **7(1)**, pp. 87–92 [trans. from *Fiz. Goreniya Vzryva* **7(1)**, pp. 103–111 (1971)].

Dremin A.N., P.F. Pokhil, and M.I. Arifov (1960). "The influence of aluminum on the detonation parameters of trotyl." *Dokl.–Chem. Tech. Section*, **131(5)**, pp. 73–75. [trans from *Dokl. Akad. Nauk SSSR* **131(5)**, pp. 1140–1142 (1960).]

Dremin, A.N., S.O. Savrov, V.S. Trofimov, and K.K. Shvedov (1970a). *Detonation Waves in Condensed Media*, Nauka, Moscow. (in Russian).

Dremin, A.N., K.K. Shvedov, and O.S. Avdonin (1970b). "Compressibility and temperatures of some porous explosives under shock loading," *Comb. Expl. Shock Waves* **6(4)**, pp. 449–455. [trans. from *Fiz. Goreniya Vzryva* **6(4)**, pp. 520–529 (1970).]

Delpuech, A., A. Mentic, and B. Pouligny (1986). in: *Shock Waves in Condensed Matter*, (ed. Y.M. Gupta) Plenum Press, New York, p. 877.

Emanuel, N.M. and D.T. Knorre (1974). *A Course of Chemical Kinetics*, Vyshaya Shkola, Moscow. (in Russian)

Engelke R. and S.A. Sheffield (1998). "Initiation and propagation of detonation in condensed-phase high explosives." in: *High-Pressure Shock Compression of Solids III* (eds. L. Davison and M. Shahinpoor) Springer-Verlag, New York.

Erskine, D.J., L. Green, and C. Tarver (1990). "VISAR wave profile measurements in supra-compressed HE." in: *Shock Compression of Condensed Matter—1989* (eds. S.C. Schmidt, J.N. Johnson, and L.W. Davison) North-Holland, Amsterdam, pp. 717–720.

Evstigneev, A.A., M.V. Zhernokletov, and V.N. Zubarev (1976). "Isentropic broadening and equation of state of trotyl explosion products," *Comb. Expl. Shock Waves* **12(5)**, pp. 678–682 [trans. from *Fiz. Goreniya Vzryva* **12(5)**, pp. 758–763 (1976)].

Feng, K.K., W.K. Chung, and B.C.-Y. Lu (1985). "Calculation of detonation products by means of the hard-sphere equation of state," in: *Proc. Eighth Symp. (International) on Detonation*, Report NSWC MP 86-194, Naval Surface Weapons Center, White Oak, MD, p. 139.

Finger, M., E. Lee, F.H. Helm, B. Hayes, H. Horning, R. McGuire, M. Kahara, and M. Guidry (1976). "The effect of elemental composition on the detonation behavior of explosives." *Proc. Sixth Symp. (International) on Detonation*, Report ACR-221, Office of Naval Research, Arlington, VA, pp. 710–722.

Forbes, J.W., D.G. Tasker, R.H. Granholm, and P.K. Gustavson (1990). "Direct observation of shocked explosive crystals immersed in liquids." in: *Shock Compression of Condensed Matter—1989* (eds. S.C.Schmidt, J.N. Johnson, and L.W. Davison) North-Holland, Amsterdam, pp. 709–712.

Fortov, V.E., and A.N. Dremin (1975). "Semiempirical Equation of State of Trinitrotoluene," *Dokl.-Phys. Chem.* **222(1)**, pp. 463–466 [trans. from *Dokl. Akad. Nauk SSSR* **222(1)**, pp. 162–165 (1975)].

Frey, R.B. (1981). "Cavity collapse in energetic materials," in: *Proc. Seventh Symp. (International) on Detonation*, Report MP 82-334, Naval Surface Warfare Center, White Oak, MD.

Glushak, B.L., S.A. Novikov, A.P. Pogorelov, et al. (1981), "Initiation of solid heterogeneous explosives by shock waves," *Comb. Expl. Shock Waves* **17(6)**, pp. 660–665 [trans. from *Fiz. Goreniya Vzryva* **17(6)**, pp. 90–95 (1981)].

Glushak, B.L., S.A. Novikov, and A.P. Pogorelov (1984). "Initiation of solid heterogeneous explosives by shock waves," *Comb. Expl. Shock Waves* **20(4)**, pp. 429–436. [trans. from *Fiz. Goreniya Vzryva* **(4)**, pp. 77–85 (1984).]

Grady, D.E. (1973). "Experimental analysis of spherical wave propagation." *J. Geophys. Res.* **73**, pp. 1299–1307.

Green, L.G., E.L. Lee, D. Breithaupt, and J. Walton (1988). "The equation of state of PETN detonation products," in: *Shock Waves in Condensed Matter—1987* (eds. S.C. Schmidt and N.C. Holmes) North-Holland, Amsterdam, pp. 507–510.

Green, L., E. Nidick, E. Lee, and C. Tarver (1978). "Reactions in PBX-9404 from low amplitude shock waves," in: *Behavior of Dense Media under High Dynamic Pressures, Symp. H.D.P.*, Gordon and Breach, New York, p. 115.

Guang Gao, R. Pandey, and A.B. Kunz (1993). "*Ab inito* study of electronic structure of RDX molecular crystal." in: *Structure and properties of energetic materials* (eds. D.H. Liebenberg, R.W. Armstrong, and J.J. Gilman) *MRS Proceedings* **296**, Materials Research Society, Pittsburgh, pp. 149–154.

Halleck, P.M., and J. Wackerle (1976). "Dynamic elastic-plastic properties of single-crystal pentaerythritol tetranitrate." *J. Appl. Phys.* **47(3)**, pp. 976–982.

Hardesty, D.R. (1976). *Combust. Flame* **27**, p. 229.

Heuzé, O., P. Bauer, H.N. Presles, and C. Brochet (1985). "The equation of state of detonation products and their incorporation into the quatuor code," in: *Proc. Eighth Symp. (International) on Detonation*, Report NSWC MP 86-194, Naval Surface Weapons Center, White Oak, MD, pp. 762–769.

Hobbs, M.L., and M.R. Baer (1993). "Calibrating the BKW-EOS with a large product species data base and measured C-J properties." *Proceedings Tenth International Detonation Symposium*. Report ONR 33395-12, U.S. Office of Naval Research, Arlington, VA, pp. 409–418.

Von Holle, W.G. (1984). "Shock wave diagnostics by time-resolved infrared radiometry and non-linear Raman spectroscopy," in: *Shock Waves in Condensed Matter—1983* (eds. J.R. Asay, R.A. Graham, and G.K. Straub). North-Holland, Amsterdam, p. 283.

Howe, P., R. Frey, B. Taylor, and V. Boyle (1976). "Shock Initiation and the Critical Energy Concept," in: *Proc. Sixth Symp. (International) on Detonation*, Report ACR-221, Office of Naval Research, Arlington, VA, (1976), p. 11.

Ilyukhin, V.S., P.F. Pokhil, O.K. Rozanov, and N.S. Shvedova (1960). "Measurements of shock adiabates of cast trotyl, crystalline hexogen and nitromethane," *Sov. Phys.–Dokl.* **5(2)**, pp. 337–340. [trans. from *Dokl. Akad. Nauk SSSR* **131(4)**, pp. 793–796 (1960).]

Imkhovik, N.A., and V.S. Soloviev (1995). "Oxidation of powdered aluminium in detonation products of condensed high explosives." *Proceedings of the Twenty-First International Pyrotechnics Seminar. 11–15 Sept., 1995*. Moscow. Russia. pp. 316–331. (in Russian)

Johansson, C.H., and P.A. Persson (1970). *Detonics of High Explosives*, Academic Press, New York.

Johnson, J.N. (1988). "Hot-spot reaction in unsustained shocks," in: *Shock Waves in Condensed Matter—1987* (eds. S.C. Schmidt and N.C. Holmes) North-Holland, Amsterdam, p. 527.

Johnson, J.N., P.K. Tang, and C.A. Forest (1985). "Shock-wave initiation of heterogeneous reactive solids," *J. Appl. Phys.* **57(9)**, p. 4323.

Jouguet, E. (1905). *J. Math. Pure Appliq.* **1**, p. 347. (in French)

Kanel, G.I. (1977). "On experimental determining the kinetics of relaxation processes at shock compression of condensed matter." *Appl. Mech. Tech. Phys.* **18(5)**, pp. 685–689 [trans. from: *Zh. Prikl. Mekh. Tekh. Fiz.* **18(5)**, pp. 117–122 (1977)].

Kanel, G.I. (1978). "Kinetics of the decomposition of cast trotyl in shock waves." *Comb. Expl. Shock Waves* **14(1)**, pp. 88–91 [trans. from *Fiz. Goreniya Vzryva* **14(1)**, pp. 113–117 (1978)].

Kanel, G.l., and A.N. Dremin (1977). "Decomposition of cast trotyl in shock waves." *Comb. Expl. Shock Waves* **12(1)**, pp. 71–77 [trans. from *Fiz. Goreniya Vzryva* **13(1)**, pp. 85–92 (1977)].

Kennedy, J.E., and J.W. Nunziato (1976). "Shock-wave evolution in a chemically reacting solid," *J. Mech. Phys. Solids* **29(2/3)**, p. 107.

Khariton, Yu.B. (1947). *Voprosy Teorii Vzryvchatih Veschestv*, No.1, USSR Academy of Sciences, Moscow. (in Russian)

Khasainov, B.A., A.B. Attetkov, and A.A. Borisov (1996). "Shock-wave initiating of porous energetic materials and a viscous-plastic model of hot spots." *Chem. Phys.* **15(7)**, pp. 53–125.

Khasainov, B.A., A.A. Borisov, B.S. Ermolayev, and A.I. Korotkov (1980). "Self-consistent model of shock-wave initiation of detonation in high-density HE ," in: *The Chemical Physics of the Combustion and Explosion Processes. Detonation. (Proceedings of 6th All-Union Symposium on Combustion and Explosion)*, Institute of Chemical Physics, Chernogolovka, p. 52. (in Russian)

Khasainov, B.A., B.S. Ermolaev, A.A. Borisov, and A.I. Korotkov (1979). "Effect of exothermic reactions downstream of the C-J plane on detonation stability." *Acta Astronautica* **6**, pp. 557–568.

Khasainov, B.A., and B. Veyssiere (1988). "Steady, plane, double-front detonations in gaseous detonable mixtures containing a suspension of aluminium particles." *Dynamics of explosions. V. 114. Progress in Astronautics and Aeronautics. AIAA.* Washington. pp. 284–299.

Kim, G.Kh. (1984). Viscosity measurement for shock-compressed water. *J. Appl. Mech. Tech. Phys.* **25(5)**, pp. 692–695 [trans. from *Zh. Prikl. Mekh. Tekh. Fiz.* **25(5)**, pp. 44–48].

Kipp, M.E., J.W. Nunziato, and R.E. Setchell (1981). "Hot spot initiation of heterogeneous explosives," in: *Proc. Seventh Symp. (International) on Detonation*, Report MP 82-334, Naval Surface Warfare Center, White Oak, MD.

Kitaigorodskii, A.I. (1971). *Molecular Crystals*. Nauka, Moscow. (in Russian).

Koldunov, S.A., K.K. Shvedov, and A.N. Dremin (1973). "Decomposition of porous explosives under the effect of shock waves," *Comb. Expl. Shock Waves* **9(2)**, pp. 255–262 [trans. from *Fiz. Goreniya Vzryva* **9(2)**, pp. 295–304 (1973)].

Kondrikov, B.N., and A.I. Sumin (1987). "Equation of state of a gas at high pressure," *Comb. Expl. Shock Waves* **23(1)**, pp. 105–113 [trans. from *Fiz. Goreniya Vzryva* **23(1)**, pp. 114–122 (1987)].

Kovalev, Yu.M. (1984). "Equations of states and temperature of shock compressed crystalline explosives," *Comb. Expl. Shock Waves* **20(2)**, pp. 219–223 [trans. from *Fiz. Goreniya Vzryva* **(2)**, pp. 102–107 (1984)].

Kurbangalina, R.Kh. (1969). "A dependence of the detonation failure diameter of liquid explosives on the content of powders," *J. Appl. Mech. Tech. Phys.* **10(4)**, pp. 656–659 [trans. from: *Zh. Prikl. Mekh. Tekh. Fiz.* **10(4)**, pp. 133–136 (1969)].

Kuznetsov, N.M., and V.A. Kopotev (1986). "Detonation in a relaxing gas and relaxation instability." *Comb. Expl. Shock Waves* **22(5)**, pp. 563–573 [trans. from *Fiz. Goreniya Vzryva* **22(5)** pp. 75–86 (1986)].

Kuznetsov N.M., and V.A. Kopotev (1986). "Detonation in a relaxing gas and relaxation instability," *Comb., Expl., Shock Waves*, **22(2)**, pp. 219-230.

Kuznetsov, N.M., and K.K. Shvedov (1966). "Equation of state of the detonation products of hexogen," *Comb. Expl. Shock Waves* **2(4)**, pp. 52–58 [trans. from *Fiz. Goreniya Vzryva* **2(4)**, pp. 85–96 (1966)].

Kuznetsov, N.M., and K.K. Shvedov (1969). "Detonation and shock adiabats for hexogen products," *Comb. Expl. Shock Waves* **5(3)**, pp. 52–58 [trans. from *Fiz. Goreniya Vzryva* **5(3)**, pp. 362–369 (1969)].

Lee, E.L., R.H. Sanborn, and H.D. Stromberg (1970). "Thermal decomposition of high explosives at static pressures 10–50 kilobars," in: *Proc. Fifth Symp. (International) on Detonation*, Report ACR-184, Office of Naval Research, Arlington, VA, (1970), p. 331.

Lee, E.L., and C.M. Tarver (1980). "Phenomenological model of shock initiation in heterogeneous explosives," *Phys. Fluids.* **23(2)**, p. 2362.

Lysne, P.C, and D.R. Hardesty (1973). "Fundamental equation of state of liquid nitromethane to 100 kbar," *J. Chem. Phys.* **59(12)**, p. 6512.

Lobanov, V.F. (1980). "Simulation of detonation waves in heterogeneous condensed HE," *Fiz. Goreniya Vzryva* **16(6)**, pp. 113–116 (1980).

Lobanov, V.F. (1985). "Dynamics of the basic initiating-wave parameters for TG 50/50," *Comb. Expl. Shock Waves* **21(6)**, pp. 756–760. [trans. from *Fiz. Goreniya Vzryva* **21(6)**, pp. 113–118 (1985).]

Lobanov, V.F. (1986). "Initiating-wave parameter determination for TG 50/50," *Comb. Expl. Shock Waves* **22(5)**, pp. 589–594 [trans. from *Fiz. Goreniya Vzryva* **22(5)**, pp. 104–111 (1986)].

Lubyatinsky, S.N., and B.G. Loboiko (1996). "Reaction zone measurements in detonating aluminized explosives." in: *Shock Compression of Condensed Matter—1995* (eds. S.C.Schmidt and W.C.Tao) American Institute of Physics, New York, pp. 779–782.

Mader, C.L. (1979) *Numerical Modelling of Detonations*, University of California Press, Berkeley, CA.

Marsh, S.P., Editor (1980). *LASL Shock Hugoniot Data*. University of California Press, Berkeley, CA.

McQueen, R., S. Marsh, J.W. Taylor, J.N. Fritz, and W.J. Carter (1970). "The Equation of State of Solids from Shock Wave Studies." in: *High-Velocity Impact Phenomena* (ed. R. Kinslow) Academic Press, New York, pp. 293–417.

Medvedev A.E., A.V. Fedorov, and V.M. Fomin (1982). "Mathematical modeling of ignition of metal particles ignition in the high-temperature flow behind a shock wave." *Comb. Expl. Shock Waves* **18(3)**, pp. 261–265 [trans. from *Fiz. Goreniya Vzryva* **18(3)**, pp. 5–9 (1982)].

Merzhanov, A.G., V.V. Barzykin, and V.T. Gontkovskaya (1963). "A problem of local thermal explosion," *Dokl. Akad. Nauk SSSR* **148(2)**, p. 380.

Moulard, H., J.W. Kury, and A. Delclos (1985). "The effect of RDX particle size on the shock sensitivity of cast PBX formulation," in: *Proc. Eighth Symp. (International) on Detonation*, Report NSWC MP 86-194, Naval Surface Weapons Center, White Oak, MD, pp. 902–913.

Nigmatulin, R.N., (1987). *Dynamics of Multiphase Medium*, Nauka, Moscow. (in Russian)

Nunziato, J.W. (1973). "One-dimentional shock waves in a chemically reacting mixture of elastic materials," *J. Chem. Phys.* **58(3)**, p. 961.

Nutt, G.L., and L.M. Erickson (1984). "Reactive flow Lagrange analysis in RX-26-AF," in: *Shock Waves in Condensed Matter—1983*, (eds. J.R. Asay, R.A. Graham, and G.K. Straub) North-Holland, Amsterdam, (1984), p. 605.

Olinger, B., and H.G. Cady (1976). "The hydrostatic compression of explosives and detonation products to 10 GPa (100 kbars) and their calculated shock compression: results for PETN, TATB, CO_2 and H_2O," in: *Proc. Sixth Symp. (International) on Detonation*, Report ACR-221, Office of Naval Research, Arlington, VA, pp. 700–709

Olinger, B., P.M. Halleck, and H.G. Cady (1975). "The isothermal linear and volume compression of pentaerythritol (PETN) to 10 GPa (100 kbar) and the calculated shock compression," *J. Chem. Phys.* **62(1)**, p. 4480.

Partom, Y. (1988). "Modeling the crossover in reaction rate for micronized TATB," in: *Shock Waves in Condensed Matter—1987* (eds. S.C. Schmidt and N.C. Holmes) North-Holland, Amsterdam, (1988), p. 535.

Pepekin, V.I., M.N. Makhov, and A.Ya. Apin (1972). "The reactions of boron in the presence of an explosion." *Comb. Expl. Shock Waves* **8(1)**, pp. 109–111 [trans. from *Fiz. Goreniya Vzryva* **8(1)**, pp. 135–138 (1972)].

Sandusky, H.W., B.C. Beard, B.C. Glancy, W.L. Elban, and R.W. Armstrong (1993). "Comparison of deformation and shock reactivity for single crystals of RDX and ammonium perchlorate." in: *Structure and properties of energetic materials* (eds. D.H. Liebenberg, R.W. Armstrong, and J.J. Gilman) *MRS Proceedings* **296**, Materials Research Society, Pittsburgh, pp. 93–98.

Seaman, L. (1974). "Lagrangian analysis for multiple stress or velocity gages in attenuating waves." *J. Appl. Phys.* **45(10)**, pp. 4303–4314.

Seay, G.E., and L.B. Seely (1961). "Initiation of a low-density PETN pressing by a plane shock wave," *J. Appl. Phys.* **32(6)**, p. 1092.

Seitz, W.L. (1984). "Short-duration shock initiation of triaminotrinitrobenzene," in: *Shock Waves in Condensed Matter—1983* (eds. J.R. Asay, R.A. Graham, and G.K. Straub) North-Holland, Amsterdam, (1984), p. 531.

Setchell, R.E. (1981). "Ramp-wave initiation of granular explosives," *Combust. Flame* **43(3)**, p. 255.

Sharma, J., J.W. Forbes, C.S. Coffey, and T.P. Liddiard (1988). "The nature of reaction sites and sensitization centers in TATB and TNT." in: *Shock Waves in Condensed Matter—1987* (eds. S.C. Schmidt and N.C. Holmes) North-Holland, Amsterdam, pp. 565–568.

Sheffield, S.A., D.D. Bloomquist, and C.M. Tarver (1984). "Subnanosecond measurements of detonation fronts in solid high explosives," *J. Chem. Phys.* **80(8)**, p. 3831.

Shipitsyn, L.A. (1980). "Thermal explosion of octogen at high pressures," *Comb. Expl. Shock Waves* **16(6)**, pp. 677–679 [trans. from *Fiz. Goreniya Vzryva* **16(6)**, pp. 85–87 (1984)].

Shvedov, K.K. (1987). "Detedrmination of the Chapman-Jouguet parameters in the detonation of condensed explosives," *Comb. Expl. Shock Waves* **23(4)**, pp. 464–474 [trans. from *Fiz. Goreniya Vzryva* **23(4)**, pp. 94–104 (1987)].

Shvedov, K.K., and Koldunov, S.A. (1980) "On decomposition of the tetranitromethane in shock waves," in: *Chemical Physics of Processes of Combustion and Explosion. Detonation*" (Proceedings of 6th All-Union Symposium on Combustion and Explosion), published by Institute of Chemical Physics, Chernogolovka, p. 60. (in Russian).

Simpson, R.L., F.H. Helms, and J.W. Kury (1993). *Propellants, Explosives, and Pyrotechnics* **18**, p. 150.

Soloviev, V.S., S.G. Andreev, M.M. Boyko, and A.I. Chernov (1980a). "On Detonation of HE with After-Burning." *Proceedings of VI Soviet Symposium on Combustion and Explosion, Detonation*, Institute of Chemical Physics, Chernogolovka, pp. 21–23. (in Russian).

Soloviev, V.S., V.V. Lazarev, and S.G. Andreev (1983). "Ignition of crystalline hexogen at adiabatic compression of adjoining gas cavity," *Fiz. Goreniya Vzryva* **(4)**, p. 130 (1983).

Soloviev, V.S., I.F. Kobylkin, S.G. Andreev, et al. (1980b). "Features of decomposition of explosives in weak shock waves," in: *The Chemical Physics. of the Combustion and Explosion Processes. Detonation, (Proceedings of 6th All-Union Symposium on Combustion and Explosion)*, Chernogolovka, p. 48. (in Russian)

Sophy, J. (1966). *C. R. Acad. Sci. France* **263**C, p. 698.

Stresau, R.H., and J.E. Kennedy (1976). "Critical conditions for shock initiation of detonation in real system," in: *Proc. Sixth Symp. (International) on Detonation*, Report ACR-221, Office of Naval Research, Arlington, VA, pp. 68–75.

Tao, W.C., C.M. Tarver, and D.R. Breithaupt (1992). "Fundamental Chemical Interactions in Metal-Filled Composite Explosives." in: *Shock Compression of Condensed Matter—1991* (eds. S.C. Schmidt, R.D. Dick, J.W. Forbes, and D.G. Tasker) North-Holland, Amsterdam, pp. 655–658.

Tarver, C.M., L.M. Erickson, and N.L. Parker (1984). "Shock initiation, detonation wave propagation and metal acceleration measurements and calculations for RX-26-AF," in: *Shock Compression of Condensed Matter—1983* (eds. J.R. Asay, R.A. Graham, and G.K. Straub) North-Holland, Amsterdam, p. 609.

Tarver, C.M. and J.O. Hallquist, (1981). "Modeling two-dimensional shock initiation and detonation wave phenomena," in: *Proc. Seventh Symp. (International) on Detonation*, Report MP 82-334, Naval Surface Warfare Center, White Oak, MD, (1981) p. 488.

Taylor, P.A. (1985). "The effects of materil microstructure on the shock sensitivity of porous granular explosives," in: *Proc. Eighth Symp. (International) on Detonation*, Report NSWC MP 86-194, Naval Surface Weapons Center, White Oak, MD, pp. 26–34.

Todes, O.M. (1939). *Zh. Fiz. Khim.* **13**, p. 868. (in Russian)

Utkin, A.V., and G.I. Kanel (1986). "Investigations of the decomposition kinetics for TNT and retarded RDX in the shock and detonation waves." in: *Detonation and Shock Waves. Proceedings of VIII Soviet Symposium on Combustion and Explosion*, Institute of Chemical Physics, Chernogolovka, pp. 13–16. (in Russian)

Utkin, A.V., G.I. Kanel, A.A. Bogach, and S.V. Razorenov (2000a). "Macrokinetics of the energy release in high explosives containing nano-size boron particles." in: *Shock Compression of Condensed Matter—1999* (eds. M.D. Furnish, L.C. Chhabildas, and R.S. Hixson) American Institute of Physics, New York, pp. 869–872.

Utkin, A.V., G.I. Kanel, and V.E. Fortov (1989). "Empirical macrokinetics of the decomposition of a desensitized hexogen in shock and detonation waves," *Comb., Expl., Shock Waves*, **25**(5), pp. 625–632

Utkin, A.V., S.V. Pershin, and V.E. Fortov (2000b). "Change in structure of a detonation wave in trinitroethyl-4,4,4-trinitrobutyrate with initial density increase." *Dokl.– Phys.* **45(10)**, pp. 520–522 [trans. from *Dokl. Akad Nauk* **374(4)**, pp. 486–488 (2000)].

Vanpoperynghe, J., J. Sorel, J. Aveille, and J. Adenis (1985). in: *Proc. Eighth Symp. (International) on Detonation*, Report NSWC MP 86-194, Naval Surface Weapons Center, White Oak, MD, p. 238.

Vantine, H.C., R.B. Rainsberger, D. Curtis, R.S. Lee, M. Cowperthwaite, and J.J. Rosenberg (1981). "The accuracy of reaction rates inferred from Lagrange analysis and *in-situ* gauge measurements," in: *Proc. Seventh Symp. (International) on Detonation*, Report MP 82-334, Naval Surface Warfare Center, White Oak, MD, p. 466.

Vorthman, J., and J. Wackerle (1984). "Multiple-wave effects on explosives decomposition rates," in: *Shock Compression of Condensed Matter—1983* (eds. J.R. Asay, R.A. Graham, and G.K. Straub) North-Holland, Amsterdam, (1984), p. 613.

Voskoboinikov, I.M., A.N. Afanasenkov, and V.M. Bogomolov (1967). "Generalized Hugoniot for organic liquids," *Comb. Expl. Shock Waves* **3(4)**, pp. 359–364. [trans. from *Fiz. Goreniya Vzryva* **3(4)**, pp. 585–593 (1967).]

Voskoboinikov, I.M., V.M. Bogomolov, and A.Ya. Apin (1968). "Calculation of the initiation pressure of shock-initiated homogeneous explosives," *Comb. Expl. Shock Waves* **4(1)**, pp. 26–28 [trans. from *Fiz. Goreniya Vzryva* **4(1)**, pp. 45–49 (1968)].

Voskoboinikov, I.M. and Gogulya, M.F. (1984). "Luminescence of shock front in liquid behind detonating HE charge," *Sov. J. Chem. Phys.* **3(7)**, p. 1036 [trans. from *Khim. Fiz.* **3(7)**, pp. 1036–1041 (1984)].

Voskoboinikov, I.M., A.N. Kiryushin, A.N. Afanasenkov, and N.F. Voskoboinikova (1974). in: *Proc. of the 1st All-Union Symp. on Pulsed Pressures*, VNIIFTRI, Moscow, vol. 1.

Voskoboinikov, I.M., and N.F. Voskoboinikova (1988). "Measurements of the transformation time in detonation wave for condensed explosives," *Sov. J. Chem. Phys.* **7(3)**, p. 406.

Wackerle, J., J. Johnson, and P. Halleck (1976). "Shock initiation of high-density PETN," in: *Proc. Sixth Symp. (International) on Detonation*, Report ACR-221, Office of Naval Research, Arlington, VA, p. 20.

Wackerle, J., R.L. Rabie, M.J. Ginsburg, and A.B. Anderson (1978). "A shock initiation study of PBX-9404," in: *Behavior of Dense Media under High Dynamic Pressures, Symp. H.D.P.*, Gordon and Breach, New York, p. 127.

Walker, E.H. (1985). "Derivation of the p^2T detonation criterion," in: *Proc. Eighth Symp. (International) on Detonation*, Report NSWC MP 86-194, Naval Surface Weapons Center, White Oak, MD, pp. 1119–1125.

Walker, F.E., and R.J. Wasley (1969). *Explosivstoffe*, **17(1)**, p. 9.

Yarger, F.L., and B. Olinger (1970). "Compression of solid nitromethane to 15 GPa at 298 K," *J. Chem. Phys.*, **85(3)** p. 1534.

Young, D. (1968). *Kinetics of Decomposition of Solids*, Mir, Moscow. (in Russian)

Yoo, C.S., N.C. Holmes, P.C. Souers, C.J. Wu, F.H. Ree, and J.J. Dick (2000). "Anisotropic shock sensitivity and detonation temperature of pentaerythritol tetranitrate single crystal." *J. Appl. Phys.* **88(1)**, pp. 70–74.

Zel'dovich, Ya.B. (1940). "On the theory of propagation of detonation in gases," *Zh. Eksp. Teor. Fiz.*, **10**, p. 542. (in Russian)

Zel'dovich, Ya.B. and Kompaneetz, A.S. (1955) *Detonation Theory*, Gostekhizdat, Moscow (English translation: Academic Press, New York, 1960).

Zel'dovich, Ya.B., and Yu.P. Raizer, *Physics of Shock Waves and High-Temperature Hydrodynamic Phenomena*, Vol. I (1966) and Vol. II (1967), Academic Press, New York. Reprinted in a single volume by Dover Publications, Mineola, New York (2002).

Zharkov, V.N., and V.A. Kalinin (1971) *Equations of State for Solids at High Pressures and Temperatures*, Consultants Bureau, New York [Trans. from *High Pressure and Temperature Equations of State of Metals*, Nauka, Moscow (1968)].

Zhernokletov, M.V., V.N. Zubarev, and G.S. Telegin (1969). "Expansion isentropes of explosion products of condensed HE," *J. Appl. Mech. Tech. Phys.* **10(4)**, pp. 650–655 [trans. from *Zh. Prikl. Mekh. Tekh. Fiz.* **10(4)**, p. 127 (1969)].

Zubarev, V.N., and A.A. Evstigneev (1984). "Equations of state of explosion products of condensed HE," *Comb. Expl. Shock Waves* **20(6)**, pp. 699–710 [trans. from *Fiz. Goreniya Vzryva* **20(6)**, pp. 114–126 (1984)].

Zubarev, V.N., and A.A. Evstigneev (1984). "On possible causes of the scattering of experimental characteristics of detonation wave," *Sov. Phys.–Dokl.* **29(8)**, pp. 635–636 [trans. from *Dokl. Akad. Nauk SSSR* **277(4)**, p. 845 (1984)].

Zubarev, V.N., and V.Ya. Vaschenko (1963). *Sov. Phys.–Solid State* **5(3)**, pp. 653–655 [trans. from *Fiz. Tverd. Tela* **5(3)**, pp. 886–890 (1963)].

CHAPTER 8

Shock Waves and Extreme States of Matter

Through the use of strong shock waves, extreme states of matter can be made available for laboratory experiments. Within the narrow zone of the shock discontinuity, the kinetic energy of the flow is transformed into energy of compression and irreversible heating of the material. Shock-wave techniques provide access to practically unlimited pressures, although the time scale for measurements is very small, being typically within the range from 10^{-9}s to $\sim 10^{-6}$s. The measurements are based on application of fundamental conservation laws that enable one to reduce determination of the thermodynamic parameters of state (pressure, density, and specific energy) to measurement of only two kinematic parameters, i.e., in fact, to the measurement of time intervals during which a shock wave or flyer plate moves through a given distance.

8.1. On Wide-Range Equations of State

The relationship between the pressure, density, and the temperature or specific energy of matter is described by an equation of state (EOS). The equation of state conveys individual material properties and is necessary for any calculations of high-energy processes. This is why the problem of wide-range equations of state has been a main stimulus for development of the physics of shock waves. For many years, study of the EOS has formed the main research direction in this field.

Figure 8.1 shows a schematic projection of the EOS onto the pressure–volume plane. Results of measurements of Hugoniots are generalized by caloric equations of state of the form $\mathscr{E}(p, V)$. The internal energy, $\mathscr{E}$, is not a thermodynamic potential when expressed as a function of the variables p, and V. In order to construct a complete thermodynamic description, the temperature function $T = T(p, V)$ is also needed. Kormer et al., 1962, developed an empirical method for constructing a complete equation of state for condensed matter based on measurements of shock compressibility of samples of different initial densities. The equation of conservation of energy for a shock wave in a material with the initial specific volume V_i takes the form

$$\mathscr{E} - \mathscr{E}_i = \tfrac{1}{2}(p + p_i)(V_i - V), \tag{8.1}$$

where $\mathscr{E}_i$, and p_i are the initial values of the specific internal energy and the pressure. Hugoniots $p(V, V_i)$ centered on different initial states of p_i, V_i, and $\mathscr{E}_i$

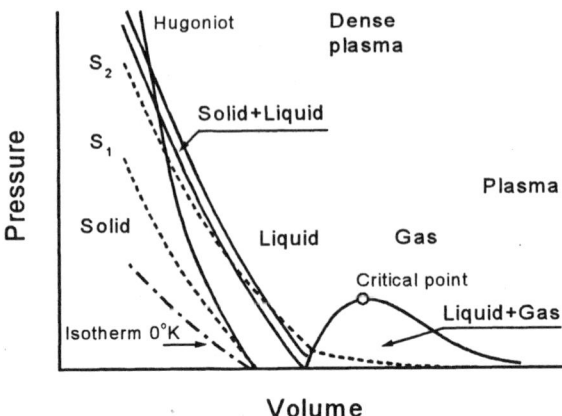

Figure 8.1. Schematic phase diagram of matter.

are found experimentally. By eliminating V_i from Eq. (8.1) using the relation $p(V, V_i)$, one obtains the empirical equation of state $\mathscr{E}(p, V)$ valid in the region covered by the measurements. A complete equation of state in the form $\mathscr{E}(S, V)$ or $\mathscr{E}(p, T)$ is determined by complementing the shock-wave measurements with equations for the specific entropy, S_i, and heat capacity in the initial state as functions of the specific volume, V_i, in this state.

Fortov and Krasnikov, 1971, have shown that thermodynamically complete equations of state of the form $\mathscr{E}(p, T)$ or $\mathscr{E}(V, S)$ can be constructed directly from dynamic measurements using only the first law of thermodynamics and the measured function $\mathscr{E}(p, V)$. This leads to the following linear inhomogeneous differential equation for $T(p, V)$:

$$\left[p + \left(\frac{\partial \mathscr{E}}{\partial V}\right)_p\right]\frac{\partial T}{\partial p} - \left(\frac{\partial \mathscr{E}}{\partial p}\right)_V \frac{\partial T}{\partial V} = T, \tag{8.2}$$

which is solved by the method of characteristics:

$$\frac{\partial p}{\partial V} = \frac{p + (\partial \mathscr{E}/\partial V)_p}{(\partial \mathscr{E}/\partial p)_V}, \qquad \frac{\partial T}{\partial V} = -\frac{T}{(\partial \mathscr{E}/\partial p)_V}. \tag{8.3}$$

Equations (8.2) and (8.3) are supplemented by boundary conditions: The temperature is specified in the low-density region in which reliable theoretical calculations can be made or experimental data are available.

It would be very attractive to build up the equation of state using only experimental data. However, the measurements are always associated with errors and scatter of the data that results in significant uncertainties of the derivatives in Eqs. (8.2) and (8.3). The experimental error in determining the Hugoniots is also a serious limitation on the validity of Kormer's approach. One must control, first, the accuracy of approximation of the experimental $p(V, V_i)$ data and, second, the validity of the fundamental thermodynamic identities and inequalities. For these reasons, the method of determining the equation of state by measuring the shock compressibility of solid and porous samples of the material can be realized only in the region of relatively high pressures where the thermal effects of shock compression significantly exceed the experimental errors. A more realistic way is the use of semi-empirical approaches in which the form of the EOS is chosen based on theoretical models. The numerical parameters appearing in the models are then adjusted to achieve the best agreement with experimental data.

Shock-wave compression of metals allows reaching very high pressures (the record pressure of 400 TPa has been obtained by Vladimirov et al., 1984) and temperatures up to 10^7 K. However, the Hugoniot is only a curve on the surface $\mathscr{E}(p, V)$ of the equation of state. Higher densities at lower temperatures are obtained by means of isentropic or quasi-isentropic compression. The latter is realized by step-like compression in a series of shock waves (see Mintsev and Fortov, 1979, and Weir et al., 1996), as illustrated in Fig. 8.2. On the other hand, the thermal effects are increased when porous samples are subjected to shock compression. This method was suggested by Zel'dovich (Zel'dovich and Raizer, 1967) and Al'tshuler, 1965. The idea is explained by the diagram in Fig. 8.3. The increment of energy $\mathscr{E} - \mathscr{E}_0$ produced by shock compression of a material initially at zero pressure is

$$\mathscr{E} - \mathscr{E}_0 = \tfrac{1}{2} p(V_0 - V).$$

At fixed pressure, p, or specific volume, V, the energy increment is greater in the case of larger initial specific volume, V_0. The value of V_0 is a function of the temperature and pressure but, for solids, it can also be varied by varying the porosity of the sample. The initial specific energy, $\mathscr{E}_0$, of a porous sample does not differ significantly from that of the pore-free matter because the surface energy of grains that is introduced is relatively small. Thus, any increase of the initial sample porosity results in an increase of the specific energy of the shock-compressed matter. This additional internal energy causes a shift of the Hugoniot toward higher pressures and larger specific volumes relative to the Hugoniot of solid matter, and this shift is larger for samples of greater porosity.

Shock-wave testing of porous samples broadens the temperature range available for measurements at high pressures. Adiabatic expansion of shock-compressed matter makes states in the high-temperature low-pressure part of the phase diagram accessible to measurement. The use of shock waves of megabar pressures gives us the chance to investigate the intermediate region between the solid and the vapor phase, including the metal–insulator transition and the satu-

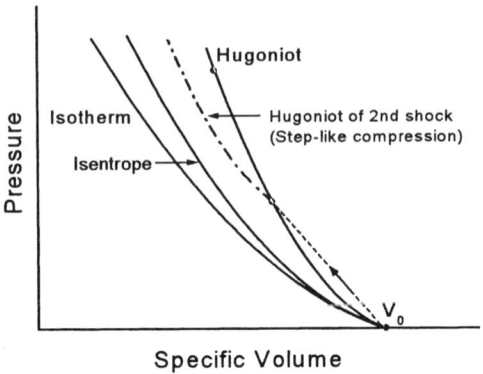

Figure 8.2. Approximation of an isentrope by a multi-shock compression process.

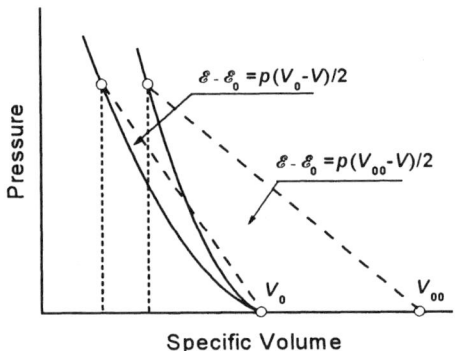

Figure 8.3. The increased thermal effect of shock compression of samples of lowered density is indicated by the area under the line connecting the initial and final states, which is proportional to the increase in specific energy produced by shock compression.

ration curve and critical point of metals. In this regard, it should be mentioned that, since the pressure and temperature at the critical points of metals are usually extremely high, common quasi-static techniques provide access to the critical parameters for only 3 alkaline metals out of ~80 metals in the periodic table. Since the pressures and temperatures of the critical points of metals are high, and close to the ionization potentials, metals evaporate directly into the non-ideal plasma state and not to the vapor state as other substances do. This can lead to exotic plasma-phase transitions, predicted by Landau and Zel'dovich, 1944, and many other theoreticians (Ebeling et al., 1991). The experimental verification of these theoretical predictions is one of the main goals of shock wave experiments with non-ideal plasmas.

8.2. Shock Waves and Non-Ideal Plasmas

Many modern tasks require simultaneous calculation of solid, liquid, and gaseous (vapor) states of the same material. A need arises to unite theoretical concepts and experimental data for different phase states. A broad high-temperature area of the phase diagram is occupied by the plasma states. The physical properties of the plasmas become simple in two asymptotic cases: (i) at extremely high pressures and (ii) at extremely high temperatures. In the first case, the inner electronic levels of atoms and ions are compressed by pressure so the Thomas–Fermi theoretical model describes the plasma properties satisfactorily. In the other extreme state of high temperature and low density, the inter-particle interaction is small and the quasi-ideal gas-like Debye–Hückel approach is valid. Strongly coupled non-ideal plasmas occupy a significant portion of the entire region of the plasma states, as illustrated in Fig. 8.4.

Low-density plasmas of relatively low temperature are partly ionized and can be considered as mixtures of ideal gases of electrons, ions, and neutral atoms. The plasma particles are moving with thermal speeds and collide with each other only occasionally. With increasing density, the distance between particles decreases and, as a result, the relative time of the inter-particle interactions increases. In other words, the average energy of inter-particle interactions is increasing with increasing density of the plasma. When the energy of inter-particle

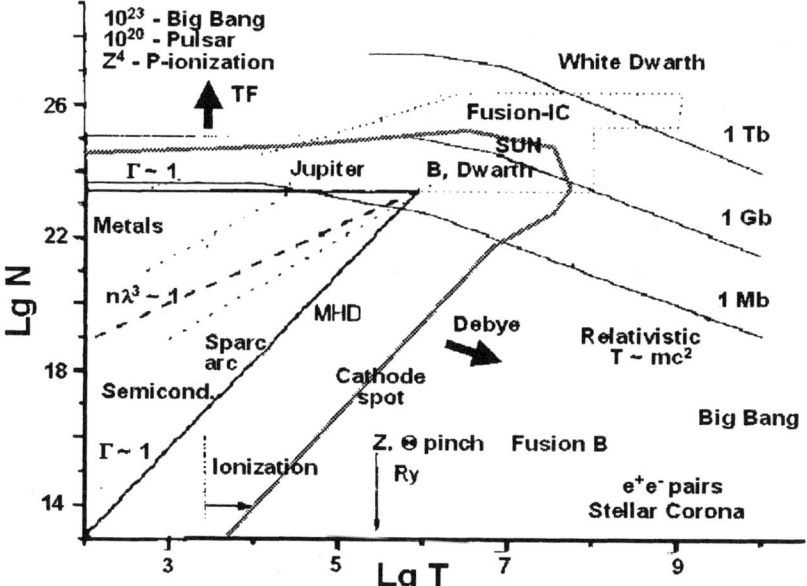

Figure 8.4. Phase diagram of the plasma states.

interactions becomes comparable to the kinetic energy of thermal motion, the plasma becomes non-ideal. The properties of such non-ideal plasmas are unusual and are not described by simple theories.

Thus, non-ideal plasmas are characterized by an essential contribution of the effects of inter-particle interactions as compared to the kinetic energy of the particles. The degree of non-ideality of plasma is characterized by the coupling parameter Γ_D, which denotes the ratio of the energy of inter-particle interactions to the kinetic energy of thermal motion of the particles:

$$\Gamma_D = \frac{2\sqrt{\pi}e^3}{(kT)^{3/2}}\sqrt{\sum_i z_i^2 n_i}, \qquad (8.4)$$

where e is the electron charge, z_i is the ion charge number, and n_i is the particle density. In the case of ideal plasmas, $\Gamma_D \ll 1$. Another widely used characteristic of plasma is the de Broglie wavelength

$$\lambda_e = \sqrt{\frac{2\pi\hbar^2}{m_e kT}}, \qquad (8.5)$$

where m_e is the electron mass. Non-ideal plasmas are a very difficult subject for theory because the strong inter-particle interactions are realized in a disordered system obeying electron statistics intermediate between the Boltzmann- and Fermi-like types.

Strong non-ideality is realized at high energy concentrations. The non-ideal plasmas occupy an exceedingly broad range of states on the phase diagram which immediately adjoins the area of the condensed phase. Obviously, information about the plasma properties is very important for calculations of hypervelocity impacts, interactions of laser and particle beams with condensed targets, and for other similar high-energy-density problems. Thus, it is necessary to be able to describe the states of matter over a wide range of the phase diagram starting from strongly compressed condensed states to the ideal Boltzmann gas, including the boiling curve and the vicinity of the critical point. In this chapter, we present only a brief overview of the problem. For more detail see, for example, Fortov and Iakubov, 1990, 1999, and Ebeling et al., 1991.

Since theoretical description of high-density plasma states is extremely sophisticated, empirical data and simplified modeling approaches are widely used. Thermophysical investigations of these states are carried out mainly for metals, because a majority of elements are metals in their normal states. Modern techniques for making quasi-static high-temperature measurements provide a capability to test materials over the range of pressures up to ~400 MPa and temperatures up to ~5000 K (Shaner and Gathers, 1979) that corresponds to the melting curve and the vicinity of the critical point for such fusible metals as Cs, Rb, and Hg. For other metals these states are not accessible when traditional experimental techniques are used. Various dynamic techniques are effective there.

8.3. Generation and Diagnosis of Dense Plasma States

Investigations of the plasma states that can now be conducted by means of shock-wave experiments include the usual EOS $\mathscr{E}(p, V)$ data, temperatures, optical, and electrophysical properties of shock-compressed matter. As a rule, the measurements are supported by theoretical analysis in a framework of various models and approaches.

The most obvious way to obtain dense plasma of a metal is to heat and compress the metal vapor, as can be done simultaneously by shock-wave compression. This method can be applied to fusible metals which may be evaporated relatively easily. The first such measurements of the equation of state and electrical conductivity were carried out with cesium using a gas shock tube (Lomakin and Fortov, 1972, 1975) and adiabatic compression facilities (Alekseev et al., 1983). In order to have a dense saturated vapor of cesium in the initial state, the experimental devices were preheated to a temperature of $\sim 900°$C.

In order to increase the energy of the shocked matter, Mintsev and Fortov, 1982, used the detonation of powerful high explosives for generation of intense shock waves in pre-compressed noble gases (see also Fortov, 1982, and Fortov and Iakubov, 1990). In this way, measurements were carried out for the EOS, electrical conductivity, and opacity of non-degenerate Boltzmann-like coupled plasmas at pressures up to 20 GPa.

Much higher shock pressures, of the order of several hundred GPa, are generated in solid targets by these explosive devices. To increase the shock pressure even further, Nabatov et al., 1979, designed a facility in which a thin sample is compressed symmetrically from both sides by simultaneous impacts of two flyer plates, each of which had been launched by its own explosive charge. Normally, the explosive facilities are able to launch metal flyer plates of reasonable thickness with ultimate velocities of up to ~6 km/s. Accumulation of the energy of explosion in thin flyer plates by means of multi-stage launchers provides a velocity increase up to 13–14 km/s (Bushman et al., 1986). The use of spherical implosions can provide impact velocities up to 20 km/s (Al'tshuler et al., 1999). The weight of similar systems is as high as 100 kg and the energy release is ~500 MJ. The maximum pressure was achieved in underground experiments with nuclear explosions (Avrorin et al., 1993). Nuclear explosives have 6 orders of magnitude higher energy density than chemical explosives. Large scale and expensive experiments with nuclear explosions gave unique information on the equation of state and optical properties of dense plasmas in the ultra-megabar pressure range.

Experimental information obtained at the highest pressures allows us to estimate the range over which extrapolation of the Thomas–Fermi model of plasma is valid. It has been shown that this model can be applied for shock pressures of approximately 10 TPa and higher. Gryaznov et al., 1982, have demonstrated that, in the case of shock compression of porous metals, the Thomas–Fermi model can be applied to description of the thermodynamics of non-ideal plasmas at lower pressures.

The first experimental information obtained on the thermodynamics of nonideal plasma (Lomakin and Fortov, 1973) was quite surprising and showed how hazardous it is to extrapolate the results and views obtained under relatively low pressures to high pressure (see also Gryaznov et al., 1980). Shock-wave experiments have shown that the pressure in strongly coupled plasma at constant temperature and/or enthalpy is much higher than the ideal-gas pressure whereas, according to standard text books, this pressure should be lower due to polarization of the plasma. It follows from analysis of the data that at least two different physical effects, screening of the charged particles (electrons and ions) and deformation of the discrete spectrum of electron energy as a result of strong inter-particle interactions, are responsible for the unusual thermodynamics.

In the following, a family of sophisticated theoretical models based on superposition of the plasma ionization approaches and the solid state cell models was developed to describe shock-wave experiments performed over a broad region of the plasma phase diagram (Bespalov et al., 1979, Gryaznov et al., 1998). In Fig. 8.5 the theoretical models are compared with Hugoniot data for aluminum over a wide pressure range. Note that, at the highest pressures of up to 400 TPa, the specific energy of the plasma is about 1 GJ/cm^3. This value is close to the energy density of a nuclear explosion, and the pressure of photon radiation is close to the kinetic pressure. This is why it is meaningless to increase the shock pressure above this limit.

On the other hand, Fig. 8.6 demonstrates that classical plasma physics provides quite reasonable agreement with the experimental Hugoniot for porous iron even in a region of high densities, which is not usual for plasma physics. It is interesting that the experimental data correspond to the metal–insulator transition region in which both temperature and pressure ionization are important for the plasma thermodynamics (see Gryaznov et al., 1998).

Fortov et al., 1990, 1992, observed rather surprising behavior of the opacity of strongly coupled plasmas. In contrast to the predictions given in classical text books, the shock-wave experiments have shown that the plasma becomes more and more transparent with compression, instead of becoming opaque. The effect has been explained in terms of the shift of discrete energy levels to the continuous spectrum as a result of strong inter-particle interaction. Theoretical models based on the "confined atom" approach or on plasma fluctuating microfields satisfactorily describe the experimental observations. The optical emission spectra of shock-compressed hydrogen (Fortov et al., 1992) and xenon (Kulish et al., 1996) plasmas show that most of the spectral lines are destroyed by strong inter-particle interactions, as illustrated in Fig. 8.7.

8. Shock Waves and Extreme States of Matter   309

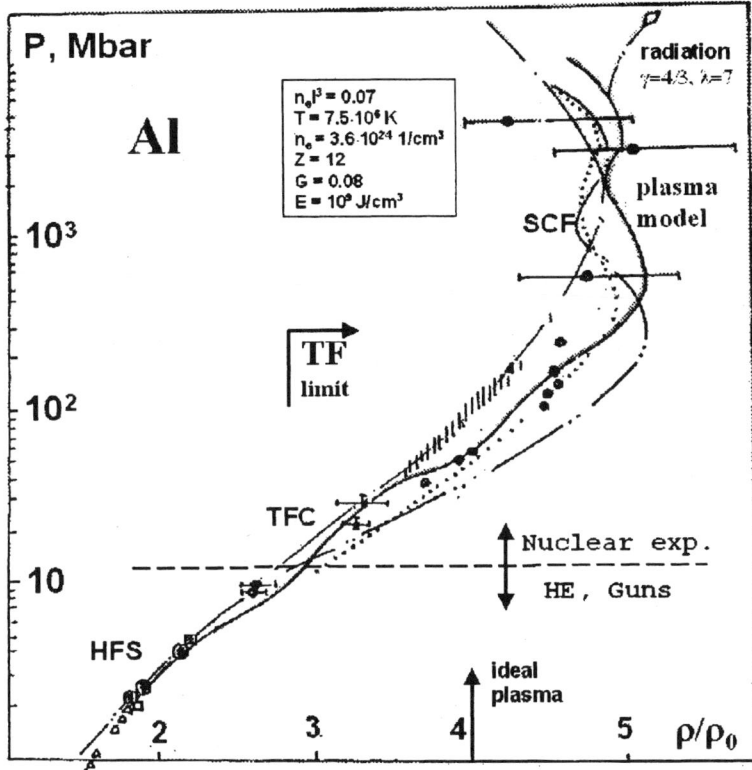

Figure 8.5. Experimental Hugoniot of aluminum in comparison with the plasma models at pressures of several Gb.

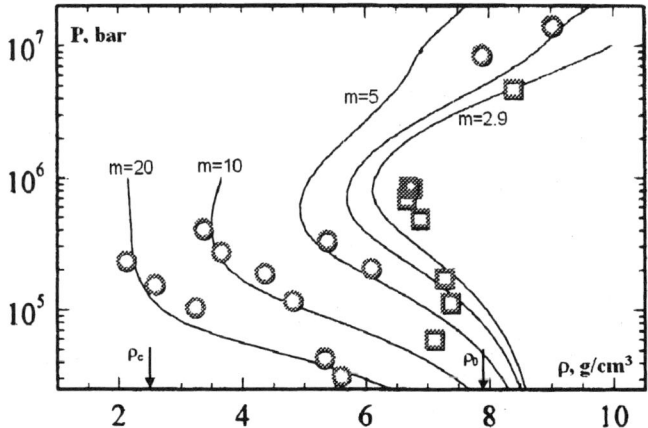

Figure 8.6. Experimental data and model predictions of the Hugoniots of porous iron. m is the initial porosity of the samples. Data by Gryaznov et al., 1980.

310 Shock-Wave Phenomena and the Properties of Condensed Matter

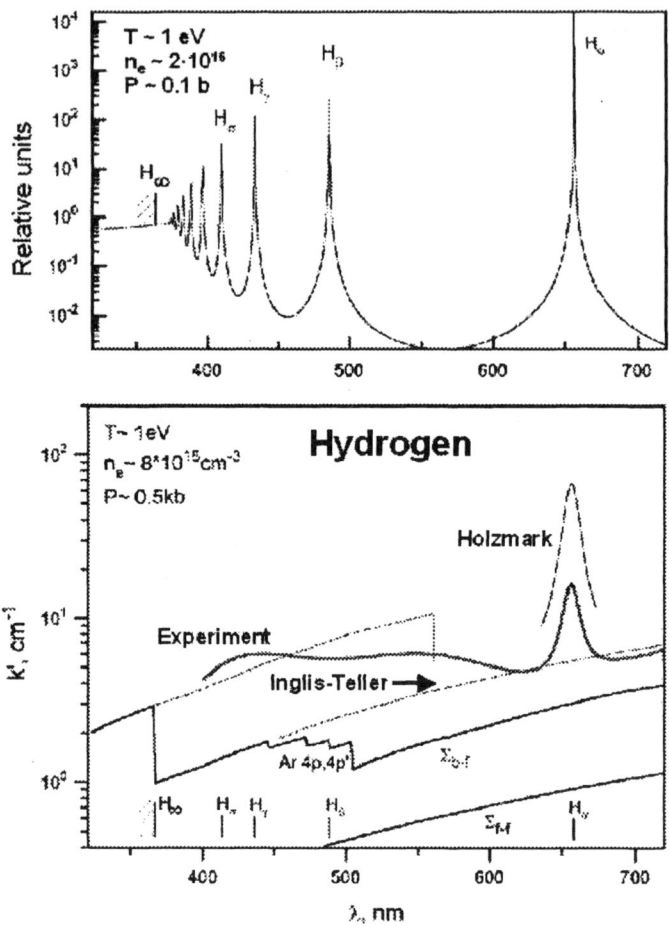

Figure 8.7. Spectral line dissolution in a high pressure hydrogen plasma. Data by Fortov et al., 1992.

Electrical conductivity of plasmas is determined by the concentration of free electrons and by electron scattering, so measurements of the conductivity of matter in shock-compressed states allow us to obtain experimental information about plasma composition. Figure 8.8 summarizes experimental data by Ivanov et al., 1973, for shock-compressed xenon. The measurements demonstrate very high conductivity of the xenon plasma which is close to the conductivity of alkaline metals. The data correspond to extremely high densities where extrapolations of common theoretical plasma models show the so-called "Coulomb collapse" predicted by Wigner, 1938, and the "Spitzer collapse" due to overestimation of Coulomb scattering.

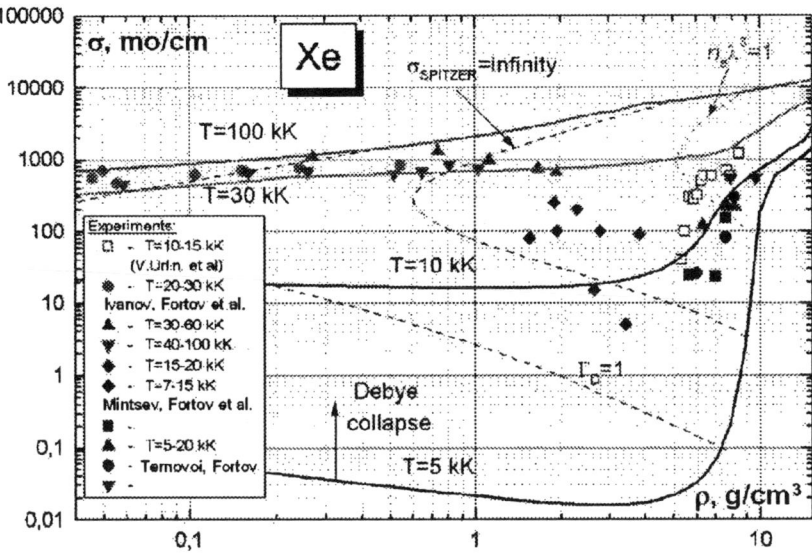

Figure 8.8. Conductivity–temperature ionization of Xe plasmas. Data by Ivanov et al., 1973, and Mintsev et al., 1979

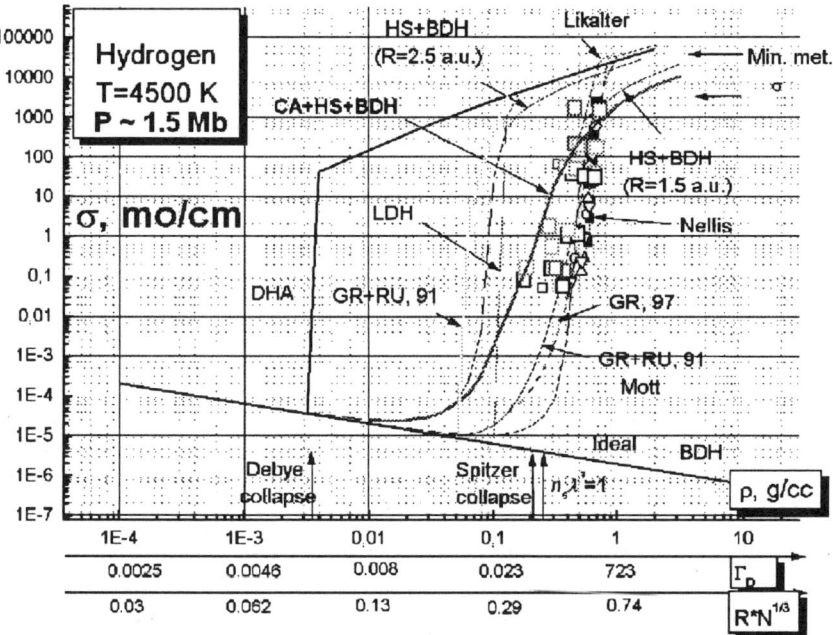

Figure 8.9. Conductivity–"pressure" ionization of hydrogen plasma at megabar pressures. Data by Weir et al., 1996, Postnov et al., 1998, and Fortov et al., 1999a.

Figure 8.9 summarizes results of measurements and theoretical predictions of the conductivity of hydrogen plasma where, as a result of step-like quasi-isentropic compression, the temperatures were much lower than that of the xenon plasma. This allowed us to investigate more definitely the pressure ionization effects. With increasing pressure, the ionization potential decreases as a result of stronger inter-particle interactions and overlapping atomic wave functions in the dense plasmas (Fortov et al., 1999a). We are currently trying to make the EOS measurements for hydrogen plasma more precise in order to understand whether the plasma phase transitions predicted by theory actually occur. At least the conductivity data shown in Figs. 8.8 and 8.9 do not contradict this hypothesis.

The last technique for studying dense plasmas to be discussed is the method of adiabatic expansion of shock-compressed metals. The irreversible energy of shock compression to megabar pressures allows us to investigate dense non-ideal plasmas located on the phase diagram in the region between solid metal and metal vapor, including the metal–insulator transition region and the high pressure saturation curve with its critical point (see Fig. 8.1). These experiments permit us to continuously connect two extreme states of matter—the high pressure condensed state and the low temperature vapor state. The experimental data for Bi plasma in Fig. 8.10 illustrate this approach. The experiments developed Zel'dovich's idea to determine the entropy and temperature of the shock-compressed condensed matter through the entropy of vapor.

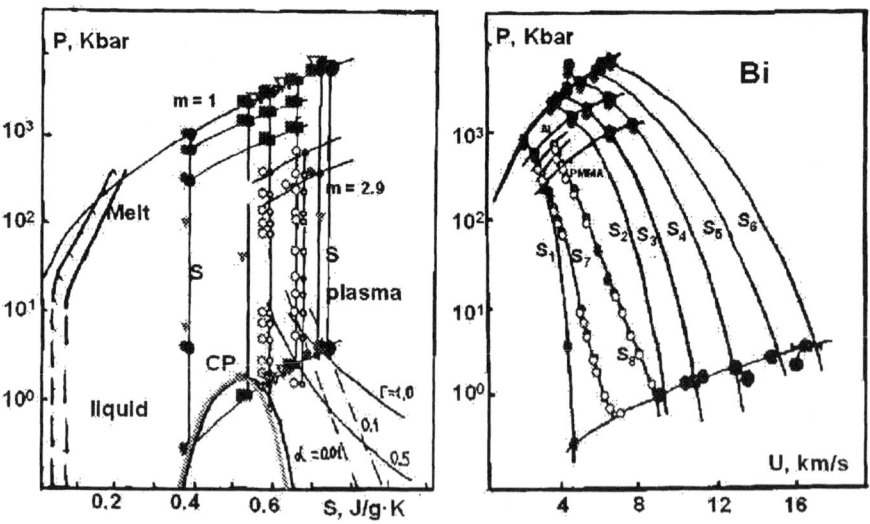

Figure 8.10. Isentropic expansion of dense bismuth plasma. Data by Glushak et al., 1989.

Time-resolved pyrometry allows us to measure the temperature and the opacity of non-ideal plasmas in states generated by unloading of shock-compressed metals. Figure 8.11 shows the temperature histories of bismuth samples measured at shock compression to 280 GPa and 360 GPa pressures. Actually, the radiation of a surface layer was recorded. As in the cases of hydrogen, argon, and xenon, the temperature data for bismuth definitely differ from the theoretical predictions as a result, obviously, of deformation of a discrete spectrum. This kind of experiment gives us an unique chance to investigate the high pressure part of the metal evaporation curve up to the critical point of metals, and to investigate the kinetics of evaporation and condensation of metals. Results of measurements in the near-critical region of metals were generalized by Fortov and Lomonosov, 1997, in a semi-empirical wide-range multiphase equation of state. Figure 8.12 demonstrates the quality of the EOS in the region of liquid + gas states of lead.

This semi-empirical equation of state was designed for quantitative description of all four (solid, liquid, vapor, and plasma) states of matter in agreement with all sets of available experimental data. At ultimate pressures and temperatures this equation of state has correct asymptotes to the theoretical Thomas–Fermi and Debye–Hückel models of highly compressed matter.

Figure 8.13 shows the general view of a semi-empirical equation of state with solid, liquid, gas, and plasma areas, phase boundaries, and asymptotes. In this form the EOS was developed for many metals and now is widely used for computer simulation of shock-wave phenomena.

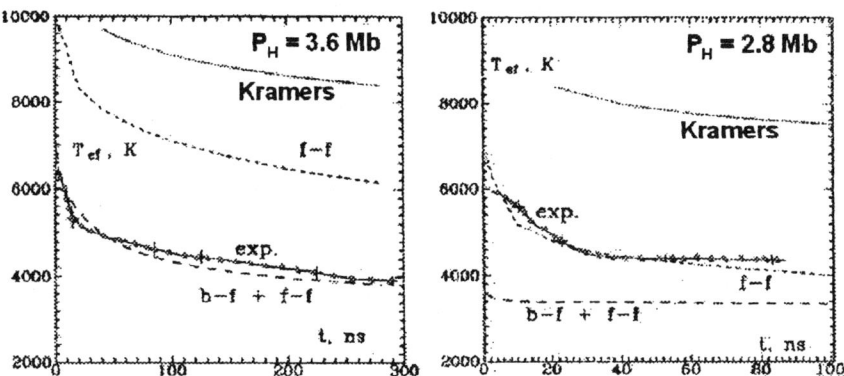

Figure 8.11. Adiabatically expanded Bi plasma brightness temperature, T_{ef}. Data by Kvitov et al., 1991.

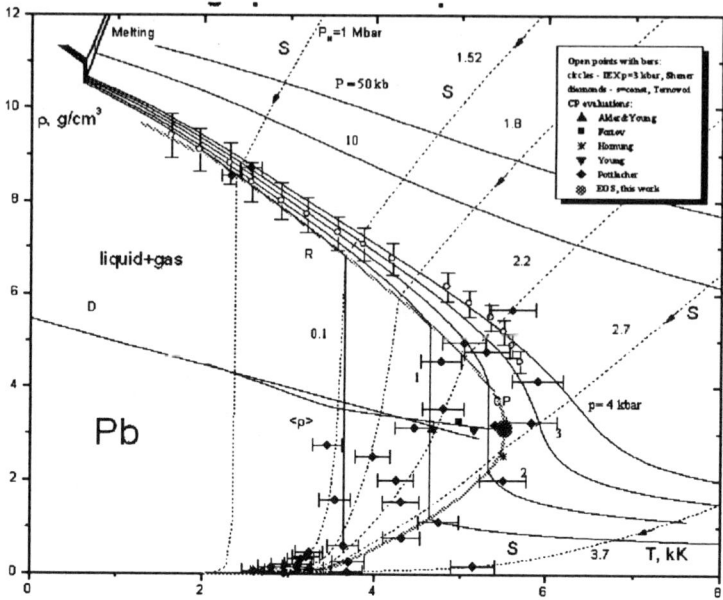

Figure 8.12. Experimental data and the equation of state of lead in the vicinity of the critical point (CP). Data by Fortov and Lomonosov, 1997.

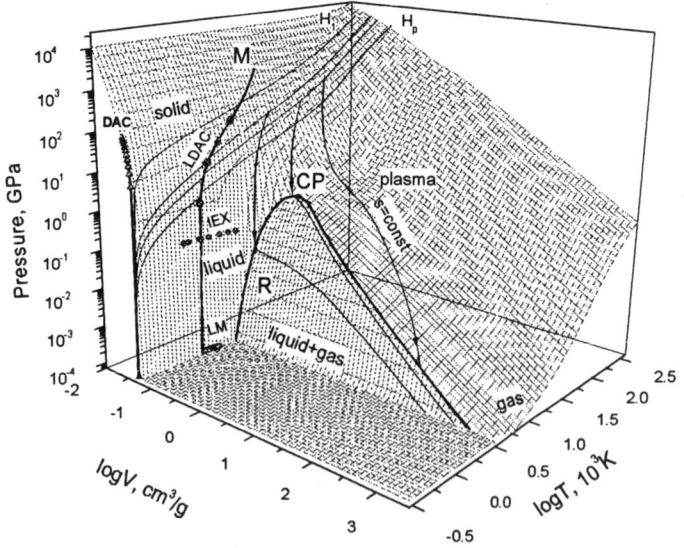

Figure 8.13. Multiphase semi-empirical equation of state by Fortov and Lomonosov, 1997. CP is the critical point, points marked IEX represent isobaric expansion data, and points marked DAC represent the data from a diamond anvil cell.

8.4. On the Metal–Insulator Transition in Shock-Compressed Lithium

It was commonly assumed that polymorphic transformations under high pressures always occur in the direction of a closer packed crystal lattice. However, Neaton and Ashcroft, 1999, have recently performed first-principle calculations of evolution of the electronic structure and crystal lattice for lithium which have demonstrated a change in character of the inter-atomic bonds at pressures of order of 100 GPa. The calculations showed that a crystal structure of lower symmetry should be more stable at high pressure and the phase transition is accompanied by a pairing of electrons that should result in a sharp decrease of electrical conductivity of lithium. After that, a series of experimental and theoretical works has been done to verify and develop this finding.

Figure 8.14 shows data by Fortov et al., 2002, who measured the resistivity history of a lithium sample during quasi-isentropic step-like compression to a pressure of 100 GPa. Simultaneously with the resistivity, the pressure history was recorded in this experiment. Similar measurements were done up to a pressure of 210 GPa and density of 2.3 g/cm^3. Figure 8.15 summarizes the experimental results in the form of dependence of the resistivity of lithium on the density under conditions of different initial temperatures and peak pressures. The temperature and density of lithium for each pressure step, measured by a manganin gauge, were calculated with a semi-empirical multi-phase equation of state.

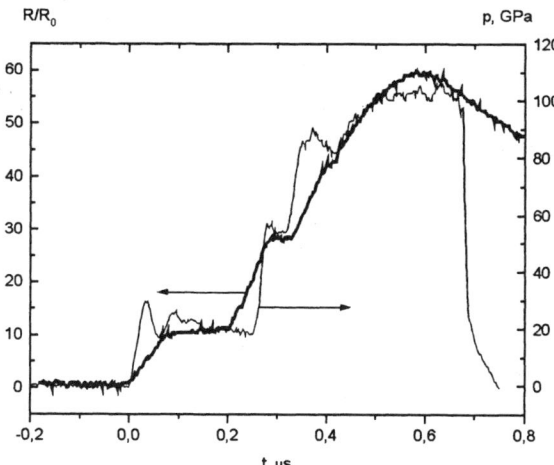

Figure 8.14. The histories of lithium resistivity and pressure during step-like compression to ~100 GPa.

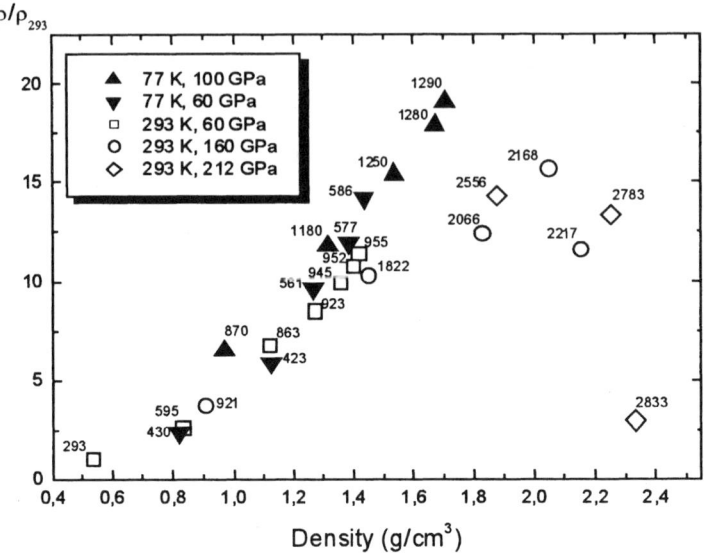

Figure 8.15. Summary of experimental data on the resistivity of quasi-isentropically compressed lithium as a function of the density and initial temperature. Data by Fortov et al. 1999b, 2002.

The results of measurements demonstrate that the lithium resistivity increases as the pressure increases up to 195 GPa (1.7 g/cm^3 density) and then it decreases as the pressure increases further to the maximum value of 212 GPa. Thus, lithium reveals anomalous behaviour under high compression. Its resistance changes weakly at pressures less then 40 GPa, rises anomalously by a factor of ten in the pressure range 40–120 GPa and decreases to metallic values as the pressure rises above 195 GPa.

8.5. Conclusion

Shock-wave physics has drastically expanded our basic knowledge of the properties of matter to a broad area of phase diagram up to pressures ten orders of magnitude higher than atmospheric and temperatures up to seven orders of magnitude higher than human body temperature. This is very impressive but one should remember the words of Voltaire, 1887: "in nature this phenomenon is perfectly natural and commonplace. The domains of some rulers in Germany and Italy, which can be circled in about a half hour, when compared with the empires of Turkey, Muscovy, or China, give only a faint idea of the remarkable contrasts that are hidden in all of nature."

References

Alekseev, V.A., V.E. Fortov, and I. Iakubov, (1983). "Physical properties of high pressure plasmas," *Sov. Phys.–Usp.* **26(2)**, pp. 99–115 [trans. from *Usp. Fiz. Nauk* **139(2)**, p. 193 (1983)].

Al'tshuler, L.V. (1965). "Application of shock waves in physics of high pressures," *Sov. Phys.–Usp.* **26(2)**, pp.52–91. [trans. from *Usp. Fiz. Nauk* **85(2)**, pp. 197–258 (1965).]

Al'tshuler, L.V., V.E. Fortov, A.I. Funtikov, R.F. Trunin, and V.D. Urlin (1999). "Development of high-pressure dynamical measurement techniques in Russia," *Usp. Fiz. Nauk* **169(3)**, pp. 323–344 (1999). (in Russian)

Avrorin, E.N., Vodolaga, B.K., Simonenko, B.A., and Fortov, V.E. (1993). "High Intensity Shock Waves and the Extreme States of Matter," *Phys.–Usp.* **36(5)**, pp. 337–364 [trans. from *Usp. Fiz. Nauk* **163(5)**, pp. 1–34 (1993)].

Bespalov, V., Gryaznov, V.K., and Fortov, V.E. (1979). "Radiation of shock-compressed argon plasma of high pressure," *Sov. Phys.–JETP* **49(1)**, pp. 71–74 [trans. from *Zh. Eksp. Teor. Fiz.* **76(1)**, pp. 140–147 (1979)].

Bushman, A.V., Glushak, B.L., V.K. Gryaznov, M.V. Zhernokletov, I.K. Krasyuk, P.P. Pashinin, A.M. Prokhorov, V.Ya. Ternovoi, A.S. Filimonov, and V.E. Fortov (1986). "Shock compression and adiabatic decompression of a dense bismuth plasma at extreme thermal-energy densities," *JETP Lett.* **44(8)**, pp. 480–483 [trans. from *Pis'ma Zh. Eksp. Teor. Fiz.* **44(8)**, pp. 375–377 (1986)].

Ebeling, V., A. Forster, V.E. Fortov, V.K. Gryaznov, and A.Ya. Polishchuk, (1991). *Thermophysical Properties of Hot Dense Plasmas,* B.G. Teubner Verlag, Leipzig.

Fortov, V.E. (1982). "Dynamic methods in plasma physics," *Sov. Phys.–Usp.* **25(11)**, pp. 781–809 [trans. from *Usp. Fiz. Nauk* **138(3)**, pp. 361–412 (1982)].

Fortov, V.E., V. Bespalov, M.I. Kulish, and S. Kuz (1990). "Experimental study of optical properties of strongly coupled plasmas," in: *Strongly Coupled Plasma Physics* (ed. S. Ichimaru), Elsevier, pp. 571–578.

Fortov, V.E., A.V. Bushman, A.C. Filimonov, S.V. Kvitov, M.I. Kulish, M.E. Lebedev, A.Ya. Polishchuk, and V.Ya. Ternovoi (1992). "Optical properties of dense plasma in shock and rarefaction waves," in: *Shock Compression of Condensed Matter— 1991* (eds. S.C. Schmidt, R.D. Dick, J.W. Forbes, and D.G. Tasker) North-Holland, Amsterdam, pp. 745–748

Fortov, V.E., and I.T. Iakubov (1990). *Physics of Nonideal Plasma.* Hemisphere, N.Y.

Fortov, V.E., and I.T. Iakubov (1999). *Physics of Nonideal Plasmas.* World. Sci. Publ., London.

Fortov, V.E., and Yu. Krasnikov (1971). "Development of thermodynamically complete equation of state for non-ideal plasma using the dynamic experiment data," *Sov. Phys.–JETP* **32**, pp. 897–902 [trans. from *Zh. Eksp. Teor. Fiz.* **59**, pp. 1645–1656 (1970)].

Fortov, V.E., and I.V. Lomonosov (1997). "Thermodynamics of extreme states of a matter," *J. Pure Appl. Chem.* **69(4)**, p. 893.

Fortov, V.E., V.I. Postnov, I.V. Lomonosov, V.V. Yakushev, K.L. Kagan, A.N. Kurjanchic, and T.N. Yakusheva, (2001). "The metal-isolator-metal transition at multiple shock loading of lithium up to 2.1 Mbar."

Fortov, V.E., V.V. Yakushev, K.L. Kagan, I.V. Lomonosov, V.I. Postnov, T.N. Yakusheva, and A.N. Kurjanchic (2002). "Abnormal electrical conductivity of lithium at high dynamic pressure," in: *Shock Compression of Condensed Matter — 2001* (eds. M.D. Furnish, N.N. Thadhani, and Y. Horie) American Institute of Physics, New York, pp. 237–240.

Fortov, V.E., V.Ya. Ternovoi, S.V. Kvitov, V.B. Mintsev, D.N. Nikolaev, A.A. Pyalling, and A.S. Filimonov (1999a). "Electrical conductivity of nonideal hydrogen plasma at megabar dynamic pressures," *JETP Lett.* **69(12)**, pp. 926–931 [trans. from *Pis'ma Zh. Eksp. Teor. Fiz.* **69(12)**, pp. 874–878 (1999)].

Fortov, V.E., V.V. Yakushev, K.L. Kagan, I.V. Lomonosov, V.I. Postnov, T.I. Yakusheva (1999b). "Anomalous electrical conductivity of lithium under quasi-isentropic compression to 60 GPa (0.6 Mbar). Transition into a molecular phase?" *JETP Lett.* **70(9)**, pp. 628–632 [trans. from *Pis'ma Zh. Eksp. Teor. Fiz.* **70(9)**, pp. 620–624 (1999)].

Glushak, B.L., Zakharov, A.P., M.V. Zhernokletov, V.Ya. Ternovi, and A.S. Filimonov (1989). "Experimental investigation of the thermodynamics of dense plasma formed from metals at high energy densities," *Sov. Phys.–JETP* **69(4)**, p. 739. [trans. from *Zh. Eksp. Teor. Fiz.* **96**, pp. 1301–1308 (1988).]

Gryaznov, V.K., I.L. Iosilevski, and V.E. Fortov (1982). "Thermodynamics of a highly compressed plasma in the megabar range," *Sov. Tech. Phys. Lett.* **8(11)**, pp. 592–593 [trans. from *Pis'ma Zh. Tekh. Fiz.* **8(11)**, pp. 1378–1381 (1982)].

Gryaznov, V.K., M.A. Zhernokletov, V.N. Zubarev, I.L. Iosilevski, and V.E. Fortov (1980). "Thermodynamic properties of nonideal plasma of argon and xenon," *Sov. Phys.–JETP* **51(2)**, pp. 288–300. [trans. from *Zh. Eksp. Teor. Fiz.* **78(2)**, pp. 573–585 (1980).]

Gryaznov, V.K., V.E. Fortov, M.A. Zhernokletov, R.F. Trunin, L.I. Trusov, and I.L. Iosilevskii, (1998). "Shock compression and thermodynamics of highly nonideal metallic plasma," *J. Exper. Theor. Phys.* **87(4)**, pp. 678–690 [trans. from *Zh. Eksp. Teor. Fiz.* **114(4)**, pp. 1242–1265 (1998)].

Ivanov, Yu.V., V.B. Mintsev, V.E. Fortov, and A.N. Dremin, (1976). "Electrical conductivity of nonideal plasma," *Sov. Phys.–JETP* **44(1)**, pp. 112–116 [trans. from *Zh. Eksp. Teor. Fiz.* **71(7)**, pp. 216–224 (1976)].

Kormer, S.B., A.I. Funtikov, V.D. Urlin, and A.N. Kolesnikova (1962). "Dynamical compression of porous metals and the equation of state with variable specific heat at high temperatures," *Sov. Phys.–JETP* **15**, pp. 477–488 (1962). [trans. from *Zh. Eksp. Teor. Fiz.* **42(3)**, pp. 686–702 (1962).]

Kvitov, S., A.V. Bushman, M.I. Kulish, I.V. Lomonosov, A.Ya. Polishchuk, A.Yu. Semenov, V.Ya. Ternovi, A.S. Filimonov, and V.E. Fortov, (1991). "Measurements of radiative properties of dense bismuth plasma in adiabatic waves," *JETP Lett.* **53(7)**, pp. 353–357. [trans. from *Pis'ma Zh. Eksp. Teor. Fiz.* **53(7)**, pp. 338–342 (1991).]

Kulish, M., V. Gryaznov, A. Mezhiba, V. Mintzev, V. Fortov, et al. (1996). "Nonideal plasma of Ar and Xe in shock waves," in: *Physics of Strongly Coupled Plasmas* (eds W.D. Kraeft and M. Schlanges), World Scientific, Singapore, pp. 337–343

Landau, L.D., and Ya.B. Zel'dovich (1944). *Zh. Eksp. Teor. Fiz.* **14**, p. 32. (in Russian)

Lomakin, B. and Fortov, V.E. (1972). "Equation of state of nonideal cesium plasma," *Zhurn. Exp. Teor. Fiz.*, **63(7)**, p. 92 (in Russian).

Lomakin, B. and Fortov, V.E. (1975). "Thermodynamics of nonideal cesium plasma," *Zhurn. Exp. Teor. Fiz.*, **69**(11), p. 1624 (in Russian)

Lomakin, B., and V.E. Fortov (1973). "Equation of state of nonideal cesium plasma," *Sov. Phys.–JETP* **36(1)**, pp. 48–53. [trans. from *Zh. Eksp. Teor. Fiz.* **63(1)**, pp. 92–103 (1972).]

Mintsev, V.B., and V.E. Fortov (1979). "Electrical conductivity of xenon under supercritical conditions," *JETP Lett.* **30(7)**, pp. 375–378. [trans. from *Pis'ma Zh. Eksp. Teor. Fiz.* **30(7)**, pp. 401–404 (1979).]

Mintsev, V.B., and V.E. Fortov (1982). "Explosive-driven shock tubes", *High Temp.* **20(3)**, pp. 623-645 [trans. from *Teplofiz. Vys. Temp.* **20(3)** pp. 745-764 (1982)].

Nabatov, S.S., A.N. Dremin, V.N. Postnov, and V.V. Yakushev (1979). "Measurements of the electrical conductivity of sulfur under super-high dynamic pressures," *JETP Lett.* **29(7)**, pp. 369–372. [trans. from *Pis'ma Zh. Eksp. Teor. Fiz.* **29(7)**, pp. 407–410 (1979).]

Neaton, J.B., and N.W. Ashcroft (1999). "Pairing in dense lithium," Nature, **400**, pp. 141–144.

Postnov, V.I., D.N. Nikolaev, V.Ja. Ternovoi, A.S. Filimonov, V.E. Fortov, and V.V. Yakushev, (1998). "The Opportunity of the use of sapphire at multiple shock-wave compression of hydrogen," in: *Shock Compression of Condensed Matter — 1997* (eds. S.C. Schmidt, D.P. Dandekar, J.W. Forbes) American institute of physics, New York, pp. 769–772.

Shaner, J.W., and G.R. Gathers (1979). in: *High Pressure Science and Technology* (eds. K.D. Timerhouse and M.S. Barber) Plenum Press, New York, p. 847.

Vladimirov, A.S., N.P. Voloshin, V.N. Nogin, A.V. Petrov, and V.A. Simonenko, (1984). "Shock compressibility of aluminum at $p > 1$ Gbar," *JETP Lett.* **39(2)**, pp. 82–84 [trans. from *Pis'ma Zh. Eksp. Teor. Fiz.* **39(2)**, pp. 69–72, (1984)].

Vladimirov, A.S., Voloshin, N.P., Simonenko, V.A. et al. (1984). "Shock compressibility of aluminum at $p > 1$ Gbar," *ZhETF Pizma*, **39**(2), p. 69 (in Russian).

Voltaire, *Micromegas, Romans*, Vol.1, Paris, 1887

Weir, S.T., A.C. Mitchell, and W.J. Nellis (1996). "Metallization of fluid molecular hydrogen at 140 GPa (1.4 Mbar)," *Phys. Rev. Lett.* **76**, p. 1860

Wigner, E. (1938). *Trans. Faraday Soc.* **34**, p. 678

Zel'dovich, Ya. B., and Yu.P. Raizer (1967). *Physics of Shock Waves and High-Temperature Hydrodynamic Phenomena*, Vol. I (1966) and Vol. II (1967), Academic Press, New York. Reprinted in a single volume by Dover Publications, Mineola, New York (2002).

Subject Index

adiabatic expansion 312

brittle materials..............................111
   axial cracking 116
      wing crack..................................116
   brittle–ductile transition...............114
   crack formation 112
   dilatancy.......................................113
   microcracking mechanisms 115
   ductile response of alumina..........167
   dynamic strength
      ceramics 145
      effect of strain rate 165
      failure criteria............................164
      single crystals............................119
   failure criteria...............................164
      failure waves in glasses.............132
   silicate glasses..............................125

computational simulation 24
conservation laws
   detonation waves..........................218
   shock waves 3
   smooth waves..................................2
coordinates
   Eulerian, see spatial
   Lagrangian 1
   spatial 1
critical point (thermodynamic) 304

detonation.......................................217
   detonation adiabat 219
   detonation product equation of state...
      270,274
   failure diameter 227
      effect of density........................228
      effect of grain size....................228
      effect of inclusions...................227
      effect of temperature 228
      failure mechanism....................227

properties of explosives containing
   metal particles 277
reaction rate 229
   kinetic constants......................232
shock initiation 229
   hot spot mechanism........... 231,249
   pressure profile evolution 234
   thermal explosion 230
state of mixtures of explosive and
   detonation products 276
detonation waves 217
   Chapman–Jouguet (CJ) theory 217
      CJ condition 218
      CJ pressure 221
   detonation wave speed............... 218
   ZND theory 218
      chemical peak.................. 218,225
      chemical reaction.................... 218
      reaction time........................... 221
      shock growth or decay....... 223,233
      wave profiles 218,222

explosives 217
   equation of state...... 263,268,270,276
   heat of explosion 217
   initiation sensitivity 236
      critical energy criterion 240
      critical shock pressure 240
      effect of density...................... 236
      effect of loading rate............... 234
      effect of particle size 236
   kinetic constants 232
   macrokinetics...................... 248,252
   selected properties 222,232,265
   shock-to-detonation transition... 233,
      236,244
   shock temperature...................... 269
   single crystals 241
elastic–plastic materials...............29,83
   Bauschinger effect 63
   bulk modulus 31

compressibility 32
constitutive equations 30,40
   Mazing model 71
dislocation dynamics 43,71,83,96
dynamic yield strength 31,33,50
   at elevated temperature 83,94,97
   at melting 88
Hugoniot elastic limit (HEL),
   see dynamic yield strength
repetitive compression 68
shear modulus 31
sound speeds 33,60,64
stepwise compression 68
strain hardening 34
strain rate effects 52
stress relaxation 39
stress state (3D) 55
stress–strain diagram 31,34
yield criterion 31
Young's modulus 31
elastic–plastic waves 29,83
   elastic precursor waves 33,35,39,85
     decay 41,50
     propagation velocity 31
     waveform 41
   free surface velocity 35,45
   relaxing materials 39
   rise time 52,54,86
   strain rate 53
   waveform structure 52
   unloading waves 47,58
equations of motion 2
equation of state 263,268,270, 274,
   276,301

fracture, see spall

Hugoniot curve 4,5,225,265,301
hydrodynamic approximation 1

instrumentation 20
   capacitor velocimeter 22
   laser velocimeter 22,179,205
   magnetoelectric particle velocity
     gauge 219
   manganan pressure gauge 23, 219
   transverse stress gauge 55
ionization 308
isentrope 4,5,9, 209,226

isentropic compression 303,312,315
isentropic flow 1
   characteristic curves 2
   dynamic impedance 3
   Riemann integrals 3
   simple wave 3
   soundspeeds 2
   unloading wave 9
isotherm 4

materials
   elastic–plastic 29,83
   polymers 209
   porous 303
   rubber 67

phase transformations 189
   melting 87,204
   metal–insulator 303,308,315
   plasma 304
   vaporization 209
plasma 304
   dense states 307
   electrical conductivity 310
   non-ideal 304,305
   opacity 308
   optical emission 308
   phase transitions 304
polymorphic transitions 189
   at negative pressure 202
   effect on shock structure 190,192
   graphite to diamond 200
   in iron and steels 191
     effect of temperature 194
     reverse transformation 193
     transformation rate 193
   in titanium 196

Rayleigh line 4,219

saturation curve 303
shock wave
   generators 20
     explosives 21,255
     nuclear explosions 307
     plate impactors 21,255,307
     proton beam 22,179
   interactions 6
   in layered plates 15,55

Index

measurements 20
   free-surface velocity 45
   gauges, see instrumentation
   interpretation 23,45
   pyrometry 313
   resistivity 315
   spatially resolved 179,205
   stress state (3D) 55
   reflection 10,17
   stability 5
spall 17,84,97
   adhesion strength 182
steady wave 5

strain
   deviatoric 31
   maximum shear 30
   strain tensor 30
   volumetric 30
stress
   deviatoric 31
   Hooke's law 31
   maximum shear 30,54,61
   pressure 30
   principal 30
   tensile 17
   tensor 20,30